Simulation and Monte Carlo

Simulation and Monte Carlo

With applications in finance and MCMC

J. S. Dagpunar

School of Mathematics
University of Edinburgh, UK

John Wiley & Sons, Ltd

Other Wiley Editorial Offices

John Wiley & Sons Inc., 111 River Street, Hoboken, NJ 07030, USA

Jossey-Bass, 989 Market Street, San Francisco, CA 94103-1741, USA

Wiley-VCH Verlag GmbH, Boschstr. 12, D-69469 Weinheim, Germany

John Wiley & Sons Australia Ltd, 42 McDougall Street, Milton, Queensland 4064, Australia

John Wiley & Sons (Asia) Pte Ltd, 2 Clementi Loop #02-01, Jin Xing Distripark, Singapore 129809

John Wiley & Sons Canada Ltd, 6045 Freemont Blvd, Mississauga, ONT, Canada L5R 4J3

Wiley also publishes its books in a variety of electronic formats. Some content that appears in print may not
be available in electronic books.

Library of Congress Cataloging in Publication Data

British Library Cataloguing in Publication Data

A catalogue record for this book is available from the British Library

ISBN-13: 978-0-470-85494-5 (HB)   978-0-470-85495-2 (PB)
ISBN-10: 0-470-85494-4 (HB)   0-470-85495-2 (PB)

Typeset in 10/12pt Times by Integra Software Services Pvt. Ltd, Pondicherry, India

This book is printed on acid-free paper responsibly manufactured from sustainable forestry
in which at least two trees are planted for each one used for paper production.

To the memory of Jim Turner, Veterinary surgeon, 1916–2006

Contents

Preface   xi

Glossary   xiii

1   Introduction to simulation and Monte Carlo   **1**
  1.1   Evaluating a definite integral   2
  1.2   Monte Carlo is integral estimation   4
  1.3   An example   5
  1.4   A simulation using Maple   7
  1.5   Problems   13

2   Uniform random numbers   **17**
  2.1   Linear congruential generators   18
     2.1.1   Mixed linear congruential generators   18
     2.1.2   Multiplicative linear congruential generators   22
  2.2   Theoretical tests for random numbers   25
     2.2.1   Problems of increasing dimension   26
  2.3   Shuffled generator   28
  2.4   Empirical tests   29
     2.4.1   Frequency test   29
     2.4.2   Serial test   30
     2.4.3   Other empirical tests   30
  2.5   Combinations of generators   31
  2.6   The seed(s) in a random number generator   32
  2.7   Problems   32

3   General methods for generating random variates   **37**
  3.1   Inversion of the cumulative distribution function   37
  3.2   Envelope rejection   40
  3.3   Ratio of uniforms method   44
  3.4   Adaptive rejection sampling   48
  3.5   Problems   52

4   Generation of variates from standard distributions   **59**
  4.1   Standard normal distribution   59
     4.1.1   Box–Müller method   59
     4.1.2   An improved envelope rejection method   61
  4.2   Lognormal distribution   62

4.3	Bivariate normal density	63
4.4	Gamma distribution	64
	4.4.1 Cheng's log-logistic method	65
4.5	Beta distribution	67
	4.5.1 Beta log-logistic method	67
4.6	Chi-squared distribution	69
4.7	Student's t distribution	69
4.8	Generalized inverse Gaussian distribution	71
4.9	Poisson distribution	73
4.10	Binomial distribution	74
4.11	Negative binomial distribution	74
4.12	Problems	75

5 Variance reduction — **79**

5.1	Antithetic variates	79
5.2	Importance sampling	82
	5.2.1 Exceedance probabilities for sums of i.i.d. random variables	86
5.3	Stratified sampling	89
	5.3.1 A stratification example	92
	5.3.2 Post stratification	96
5.4	Control variates	98
5.5	Conditional Monte Carlo	101
5.6	Problems	103

6 Simulation and finance — **107**

6.1	Brownian motion	108
6.2	Asset price movements	109
6.3	Pricing simple derivatives and options	111
	6.3.1 European call	113
	6.3.2 European put	114
	6.3.3 Continuous income	115
	6.3.4 Delta hedging	115
	6.3.5 Discrete hedging	116
6.4	Asian options	118
	6.4.1 Naive simulation	118
	6.4.2 Importance and stratified version	119
6.5	Basket options	123
6.6	Stochastic volatility	126
6.7	Problems	130

7 Discrete event simulation — **135**

7.1	Poisson process	136
7.2	Time-dependent Poisson process	140
7.3	Poisson processes in the plane	141
7.4	Markov chains	142
	7.4.1 Discrete-time Markov chains	142
	7.4.2 Continuous-time Markov chains	143

7.5   Regenerative analysis   144
7.6   Simulating a G/G/1 queueing system using the three-phase method   146
7.7   Simulating a hospital ward   149
7.8   Problems   151

8   Markov chain Monte Carlo   **157**
8.1   Bayesian statistics   157
8.2   Markov chains and the Metropolis–Hastings (MH) algorithm   159
8.3   Reliability inference using an independence sampler   163
8.4   Single component Metropolis–Hastings and Gibbs sampling   165
      8.4.1   Estimating multiple failure rates   167
      8.4.2   Capture–recapture   171
      8.4.3   Minimal repair   172
8.5   Other aspects of Gibbs sampling   176
      8.5.1   Slice sampling   176
      8.5.2   Completions   178
8.6   Problems   179

9   Solutions   **187**
9.1   Solutions 1   187
9.2   Solutions 2   187
9.3   Solutions 3   190
9.4   Solutions 4   191
9.5   Solutions 5   195
9.6   Solutions 6   196
9.7   Solutions 7   202
9.8   Solutions 8   205

Appendix 1: Solutions to problems in Chapter 1   **209**

Appendix 2: Random number generators   **227**

Appendix 3: Computations of acceptance probabilities   **229**

Appendix 4: Random variate generators (standard distributions)   **233**

Appendix 5: Variance reduction   **239**

Appendix 6: Simulation and finance   **249**

Appendix 7: Discrete event simulation   **283**

Appendix 8: Markov chain Monte Carlo   **299**

References   **325**

Index   **329**

Preface

This book provides an introduction to the theory and practice of Monte Carlo and Simulation methods. It arises from a 20 hour course given simultaneously to two groups of students. The first are final year Honours students in the School of Mathematics at the University of Edinburgh and the second are students from Heriot Watt and Edinburgh Universities taking the MSc in Financial Mathematics.

The intention is that this be a practical book that encourages readers to write and experiment with actual simulation models. The choice of programming environment, Maple, may seem strange, perhaps even perverse. It arises from the fact that at Edinburgh all mathematics students are conversant with it from year 1. I believe this is true of many other mathematics departments. The disadvantage of slow numerical processing in Maple is neutralized by the wide range of probabilistic, statistical, plotting, and list processing functions available. A large number of specially written Maple procedures are available on the website accompanying this book (www.wiley.com/go/dagpunar_simulation). They are also listed in the Appendices.[1]

The content of the book falls broadly into two halves, with Chapters 1 to 5 mostly covering the theory and probabilistic aspects, while Chapters 6 to 8 cover three application areas. Chapter 1 gives a brief overview of the breadth of simulation. All problems at the end of this chapter involve the writing of Maple procedures, and full solutions are given in Appendix 1. Chapter 2 concerns the generation and assessment of pseudo-random numbers. Chapter 3 discusses three main approaches to the sampling (generation) of random variates from distributions. These are: inversion of the distribution function, the envelope rejection method, and the ratio of uniforms method. It is recognized that many other methods are available, but these three seem to be the most frequently used, and they have the advantage of leading to easily programmed algorithms. Readers interested in the many other methods are directed to the excellent book by Devroye (1986) or an earlier book of mine (Dagpunar, 1988a). Two short Maple procedures in Appendix 3 allow readers to quickly ascertain the efficiency of rejection type algorithms. Chapter 4 deals with the generation of variates from standard distributions. The emphasis is on short, easily implemented algorithms. Where such an algorithm appears to be faster than the corresponding one in the Maple statistics package, I have given a listing in Appendix 4. Taken together, I hope that Chapters 3 and 4 enable readers to understand how the generators available in various packages work and how to write algorithms for distributions that either do not appear in such packages or appear to be slow in execution. Chapter 5 introduces variance reduction methods. Without these, many simulations are incapable of giving precise estimates within a reasonable amount of processing time. Again, the emphasis is on an empirical approach and readers can use the procedures in

---

[1] The programs are provided for information only and may not be suitable for all purposes. Neither the author nor the publisher is liable, to the fullest extent permitted by law, for any failure of the programs to meet the user's requirements or for any inaccuracies or defects in the programs.

Appendix 5 to illustrate the efficacy of the various designs, including importance and stratified sampling.

Chapters 6 and 8, on financial mathematics and Markov chain Monte Carlo methods respectively, would not have been written 10 years ago. Their inclusion is a result of the high-dimensional integrations to be found in the pricing of exotic derivatives and in Bayesian estimation. In a stroke this has caused a renaissance in simulation. In Chapter 6, I have been influenced by the work of Glasserman (2004), particularly his work combining importance and stratified sampling. I hope in Sections 6.4.2 and 6.5 that I have provided a more direct and accessible way of deriving and applying such variance reduction methods to Asian and basket options. Another example of high-dimensional integrations arises in stochastic volatility and Section 6.6 exposes the tip of this iceberg. Serious financial engineers would not use Maple for simulations. Nevertheless, even with Maple, it is apparent from the numerical examples in Chapter 6 that accurate results can be obtained in a reasonable amount of time when effective variance reduction designs are employed. I also hope that Maple can be seen as an effective way of experimenting with various models, prior to the final construction of an efficient program in C++ or Java, say. The Maple facility to generate code in, say, C++ or Fortran is useful in this respect.

Chapter 7 introduces discrete event simulation, which is perhaps best known to operational researchers. It starts with methods of simulating various Markov processes, both in discrete and continuous time. It includes a discussion of the regenerative method of analysing autocorrelated simulation output. The simulation needs of the operational researcher, the financial engineer, and the Bayesian statistician overlap to a certain extent, but it is probably true to say that no single computing environment is ideal for all application fields. An operational researcher might progress from Chapter 7 to make use of one of the powerful purpose-built discrete event simulation languages such as Simscript II.5 or Witness. If so, I hope that the book provides a good grounding in the principles of simulation.

Chapter 8 deals with the other burgeoning area of simulation, namely Markov chain Monte Carlo and its use in Bayesian statistics. Here, I have been influenced by the works of Robert and Casella (2004) and Gilks *et al.* (1996). I have also included several examples from the reliability area since the repair and maintenance of systems is another area that interests me. Maple has been quite adequate for the examples discussed in this chapter. For larger hierarchical systems a purpose-built package such as BUGS is the answer.

There are problems at the end of each chapter and solutions are given to selected ones. A few harder problems have been designated accordingly. In the text and problems, numerical answers are frequently given to more significant figures than the data would warrant. This is done so that independent calculations may be compared with the ones appearing here.

I am indebted to Professor Alastair Gillespie, head of the School of Mathematics, Edinburgh University, for granting me sabbatical leave for the first semester of the 2005–2006 session. I should also like to acknowledge the several cohorts of simulation students that provided an incentive to write this book. Finally, my thanks to Angie for her encouragement and support, and for her forbearance when I was not there.

Glossary

beta $(\alpha, \beta)^1$	A random variable that is beta distributed with p.d.f. $f(x) = \Gamma(\alpha+\beta) x^{\alpha-1} (1-x)^{\beta-1} / [\Gamma(\alpha) \Gamma(\beta)]$, $1 \geq x \geq 0$, where $\alpha > 0$, $\beta > 0$.
binomial (n, p)	A binomially distributed random variable.
$\{B(t), t \geq 0\}$	Standard Brownian motion.
c.d.f.	Cumulative distribution function.
$\mathbf{C}'$	The transpose of a matrix $\mathbf{C}$.
$\text{Cov}_f (X, Y)$	The covariance of X and Y where $f(x, y)$ is the joint p.d.f./p.m.f. of X and Y (the subscript is often dropped).
e.s.e.	Estimated standard error.
$\text{Exp}(\lambda)$	A r.v. that has the p.d.f. $f(x) = \lambda e^{-\lambda x}$, $x \geq 0$, where $\lambda > 0$.
$E_f(X)$	The expectation of a random variable X that has the p.d.f. or p.m.f. f (the subscript is often dropped).
$f_X(x)$	The p.d.f. or p.m.f. of a random variable X (the subscript is often dropped).
$F_X(x)$	The c.d.f. of a random variable X.
$\overline{F}_X(x)$	Complementary cumulative distribution function $[= 1 - F_X(x)]$.
gamma (α, λ)	A gamma distributed r.v. with the p.d.f. $f(x) = \lambda^\alpha x^{\alpha-1} e^{-\lambda x} / \Gamma(\alpha)$, $x \geq 0$, where $\alpha > 0$, $\lambda > 0$.
gig (λ, ψ, χ)	A r.v. distributed as a generalized inverse Gaussian distribution with the p.d.f. $f(x) \propto x^{\lambda-1} \exp\left(-\frac{1}{2}[\psi x + \chi/x]\right)$, $x \geq 0$.
i.i.d.	Identically and independently distributed.
negbinom (k, p)	A negative binomial r.v. with the p.d.f. $f(x) \propto p^x (1-p)^k$, $x = 0, 1, 2, \ldots$, where $0 < p < 1$.
$N(\mu, \sigma^2)$	A normal r.v. with expectation μ and variance σ^2, or the density itself.
$N(\boldsymbol{\mu}, \boldsymbol{\Sigma})$	A vector r.v. $\mathbf{X}$ distributed as multivariate normal with mean $\boldsymbol{\mu}$ and covariance matrix $\boldsymbol{\Sigma}$.
p.d.f.	Probability density function.
p.m.f.	Probability mass function.
Poisson (μ)	A r.v. distributed as a Poisson with the p.m.f. $f(x) = \mu^x e^{-\mu x}/x!$, $x = 0, 1, \ldots$, where $\mu > 0$.
$P(X < x)$	Probability that the random variable X is less than x.
$P(X = x)$	Probability that a (discrete) random variable equals x.
r.v.	Random variable.
s.e.	Standard error.

---

[1] This can also refer to the distribution itself. This applies to all corresponding random variable names in this list.

support (f)	$\{x : x \in \mathbb{R}, f(x) \neq 0\}$.
$U(0, 1)$	A continuous r.v. that is uniformly distributed in the interval $(0, 1)$.
$\text{Var}_f(X)$	The variance of a random variable X that has the p.d.f. or p.m.f f (the subscript is often dropped).
v.r.r.	Variance reduction ratio.
Weibull (α, λ)	A Weibull distributed random variable with the p.d.f. $f(x) = \alpha\lambda^\alpha x^{\alpha-1} e^{-(\lambda x)^\alpha} / \Gamma(\alpha)$, $x \geq 0$, where $\alpha > 0$, $\lambda > 0$.
x^+	$\max(x, 0)$.
$\sigma_f(X)$	The standard deviation of a random variable X that has the p.d.f. or p.m.f. f (the subscript is often dropped).
$\phi(z)$	The p.d.f. for the standard normal.
$\Phi(z)$	The c.d.f. for the standard normal.
χ_n^2	An r.v. distributed as chi-squared with n degrees of freedom, $n = 1, 2, \ldots$. Therefore, $\chi_n^2 = \text{gamma}(n/2, 1/2)$.
1_P	Equals 1 if P is true, otherwise equals 0.
$\sim$	'Is distributed as'. For example, $X \sim \text{Poisson}(\mu)$ indicates that X has a Poisson distribution.
$:= y$	In Maple or in pseudo-code this means 'becomes equal to'. The value of the expression to the right of $:=$ is assigned to the variable or parameter to the left of $:=$.

1

Introduction to simulation and Monte Carlo

A *simulation* is an experiment, usually conducted on a computer, involving the use of *random numbers*. A random number stream is a sequence of statistically independent random variables uniformly distributed in the interval [0,1). Examples of situations where simulation has proved useful include:

  (i) modelling the flow of patients through a hospital;

 (ii) modelling the evolution of an epidemic over space and time;

(iii) testing a statistical hypothesis;

(iv) pricing an option (derivative) on a financial asset.

A feature common to all these examples is that it is difficult to use purely analytical methods to either model the real-life situation [examples (i) and (ii)] or to solve the underlying mathematical problems [examples (iii) and (iv)]. In examples (i) and (ii) the systems are stochastic, there may be complex interaction between resources and certain parts of the system, and the difficulty may be compounded by the requirement to find an 'optimal' strategy. In example (iii), having obtained data from a statistical investigation, the numerical value of some test statistic is calculated, but the distribution of such a statistic under a null hypothesis may be impossibly difficult to derive. In example (iv), it transpires that the problem often reduces to evaluating a multiple integral that is impossible to solve by analytical or conventional numerical methods. However, such integrals can be estimated by *Monte Carlo* methods. Dating from the 1940s, these methods were used to evaluate definite multiple integrals in mathematical physics. There is now a resurgence of interest in such methods, particularly in finance and statistical inference.

In general, simulation may be appropriate when there is a problem that is too difficult to solve analytically. In a simulation a controlled *sampling experiment* is conducted on a computer using random numbers. *Statistics* arising from the sampling experiments (examples are sample mean, sample proportion) are used to *estimate* some parameters of interest in the original problem, system, or population.

Simulation and Monte Carlo: With applications in finance and MCMC   J. S. Dagpunar
© 2007 John Wiley & Sons, Ltd

Since simulations provide an *estimate* of a parameter of interest, there is always some error, and so a quantification of the *precision* is essential, and forms an important part of the design and analysis of the experiment.

1.1   Evaluating a definite integral

Suppose we wish to evaluate the integral

$$I_\alpha = \int_0^\infty x^{\alpha-1} e^{-x} dx \qquad (1.1)$$

for a specific value of $\alpha > 0$. Consider a random variable X having the probability density function (p.d.f.) f on support $[0, \infty)$ where

$$f(x) = e^{-x}.$$

Then from the definition of the expectation of a function of a random variable Equation (1.1) leads to

$$I_\alpha = E_f\left(X^{\alpha-1}\right).$$

It follows that a (statistical) estimate of I_α may be obtained by conducting the following controlled sampling experiment. Draw a random sample of observations $X_1, X_2, \ldots, X_n$ from the probability density function f. Construct the statistic

$$\widehat{I}_\alpha = \frac{1}{n}\sum_{i=1}^n X_i^{\alpha-1}. \qquad (1.2)$$

Then $\widehat{I}_\alpha$ is an unbiased estimator of I_α and assuming the $\{X_i\}$ are independent, the variance of $\widehat{I}_\alpha$ is given by

$$\mathrm{Var}_f\left(\widehat{I}_\alpha\right) = \frac{\mathrm{Var}_f\left(X^{\alpha-1}\right)}{n}.$$

Thus, the standard deviation of the sampling distribution of the statistic $\widehat{I}_\alpha$ is

$$\sigma_f\left(\widehat{I}_\alpha\right) = \frac{\sigma_f\left(X^{\alpha-1}\right)}{\sqrt{n}}.$$

This is the *standard error* (s.e.) of the statistic and varies as $1/\sqrt{n}$. Therefore, to change the standard error by a factor of K, say, requires the sample size to change by a factor of $1/K^2$. Thus, extra precision comes at a disproportionate extra cost.

By way of a numerical example let us estimate the value of the definite integral

$$I_{1.9} = \int_0^\infty x^{0.9} e^{-x} dx.$$

Firstly, we need to know how to generate values from the probability density function $f(x) = e^{-x}$. It will be seen in Chapter 3 that this is done by setting

$$X_i = -\ln R_i \qquad (1.3)$$

where $\{R_i : i = 0, 1, \ldots\}$ is a *random number stream* with $R_i \sim U(0, 1)$. From a built-in calculator function the following random numbers were obtained:

$$R_1 = 0.0078, \quad R_2 = 0.9352, \quad R_3 = 0.1080, \quad R_4 = 0.0063.$$

Using these in Equations (1.3) and (1.2) gives $\widehat{I}_{1.9} = 2.649$. In fact, the true answer to five significant figures (from tables of the gamma function) is $I_{1.9} = \Gamma(1.9) = 0.96177$. Therefore, the estimate is an awful one. This is not surprising since only four values were sampled. How large should the sample be in order to give a standard error of 0.001, say? To answer this we need to know the standard error of $\widehat{I}_{1.9}$ when $n = 4$. This is unknown as $\sigma_f(X^{0.9})$ is unknown. However, the sample standard deviation of $X^{0.9}$ is s where

$$s = \sqrt{\frac{1}{4-1}\left(\sum_{i=1}^{4}[x_i^{0.9}]^2 - \left\{\sum_{i=1}^{4}x_i^{0.9}\right\}^2 / 4\right)} = 1.992 \qquad (1.4)$$

and $\{x_i : i = 1, \ldots, 4\}$ is the set of four values sampled from f. Therefore, the estimated standard error (e.s.e.) is $s/\sqrt{4} = 0.9959$. In order to reduce the standard error to 0.001, an approximately 996-fold reduction in the standard error would be needed, or a sample of approximate size $4 \times 996^2 \approx 3.97 \times 10^6$. We learn from this that an uncritical design and analysis of a simulation will often lead to a vast consumption of computer time.

Is it possible to design the experiment in a more efficient way? The answer is 'yes'. Rewrite the integral as

$$I_{1.9} = \int_0^\infty xe^{-x}\left(\frac{1}{x^{0.1}}\right)dx. \qquad (1.5)$$

There is a convenient method of generating variates $\{x\}$ from the probability density function

$$g(x) = xe^{-x} \qquad (1.6)$$

with support $[0, \infty)$. It is to take two random numbers $R_1, R_2, \sim U(0, 1)$ and set

$$X = -\ln R_1 R_2.$$

Given this, Equation (1.5) can be written as

$$I_{1.9} = E_g\left(\frac{1}{X^{0.1}}\right)$$

where X has the density of Equation (1.6). Given the same four random numbers, two random values (variates) can be generated from Equation (1.6). They are $x_1 = -\ln R_1 R_2 = 4.9206$ and $x_2 = -\ln R_3 R_4 = 7.2928$. Therefore,

$$\widehat{I}_{1.9} = \frac{1}{2}(4.9206^{-0.1} + 7.2928^{-0.1}) = 0.8363. \qquad (1.7)$$

This is a great improvement on the previous estimate. A theoretical analysis shows that $\sigma_g\left(\widehat{I}_{1.9}\left(X_1,\ldots,X_n\right)\right)$ is much smaller than $\sigma_f\left(\widehat{I}_{1.9}\left(X_1,\ldots,X_n\right)\right)$, the reason being that $\sigma_g\left(1/X^{0.1}\right) << \sigma_f\left(X^{0.9}\right)$.

Now try to estimate $I_{1.99}$ using both methods with the same random numbers as before. It is found that when averaging $1/X^{0.01}$ sampled from g, $\widehat{I}_{1.99} = 0.98226$, which is very close to the true value, $I_{1.99} = 0.99581$. This is not the case when averaging $X^{0.99}$ sampled from f.

This simple change in the details of the sampling experiment is an example of a *variance reduction* technique. In this case it is known as *importance sampling*, which is explored in Chapter 5.

1.2   Monte Carlo is integral estimation

How is it that the Monte Carlo method evolved from its rather specialized use in integral evaluation to its current status as a modelling aid is used to understand the behaviour of complex stochastic systems? Let us take the example of some type of queue, for example one encountered in a production process. We might be interested in the expectation of the total amount of time spent waiting by the first n 'customers', given some initial state S_0. In this case we wish to find

$$E_f\{W_1 + \cdots + W_n | S_0\}$$

where f is the multivariate probability density of $W_1, \ldots, W_n$ given S_0. For most systems of any practical interest it will be difficult, probably impossible, to write down the density f. At first sight this might appear to rule out the idea of generating several, say m, realizations $\left\{\left\{w_1^{(i)},\ldots,w_n^{(i)}\right\} : i = 1,\ldots,m\right\}$. However, it should not be too difficult to generate a realization of the waiting time W_1 of the first customer. Given this, the known structure of the system is then used to generate a realization of W_2 given a knowledge of the state of the system at all times up to the departure of customer 1. Note that it is often much easier to generate a *realization* of W_2 (given W_1 and the state of the system up to the departure of customer 1) than it is to write down the conditional *distribution* of W_2 given W_1. This is because the value assumed by W_2 can be obtained by breaking down the evolution of the system between the departures of customers 1 and 2 into easy stages. In this way it is possible to obtain realizations of values of $W_1,\ldots,W_n$.

The power of Monte Carlo lies in the ability to estimate the value of any definite integral, no matter how complex the integrand. For example, it is just as easy to estimate $E_f\{\text{Max}\left(W_1,\ldots,W_n\right)|S_0\}$ as it is to estimate $E_f\{W_1 + \cdots + W_n | S_0\}$. Here is another example. Suppose we wish to estimate the expectation of the time average of queue length in $[0, T]$. Monte Carlo is used to estimate $E_f\left\{(1/T)\int_0^T Q(t)\,dt\right\}$, where $\{Q(t), t \geq 0\}$ is a stochastic process giving the queue length and f is the probability density for the paths $\{Q(t), t \geq 0\}$. Again, a realization of $\{Q(t), T \geq t \geq 0\}$ is obtained by breaking the 'calculation' down into easy stages. In practice it may be necessary to discretize the time interval $[0, T]$ into a large number of small subintervals. A further example is given. Suppose there is a directed acyclic graph in which the arc lengths represent random costs that are statistically dependent. We wish to find the probability that the shortest path through the network has a length (cost) that does not exceed x, say. Write this probability as $P(X < x)$. It can be seen that

$$P(X < x) = \int_{-\infty}^{\infty} f(t)(1 - 1_{x>t}) \, dt$$

where f is the probability density of the length of the shortest path and $1_{x>t} = 1$ if $x > t$; else $1_{x>t} = 0$. Since the probability can be expressed as an integral and since realizations of $1_{x>t}$ can be simulated by breaking down the calculation into easier parts using the structure of the network, the probability can be estimated using Monte Carlo. In fact, if 'minimum' is replaced by 'maximum' we have a familiar problem in project planning. This is the determination of the probability that the duration of a project does not exceed some specified value, when the individual activity durations are random and perhaps statistically dependent. Note that in all these examples the integration is over many variables and would be impossible by conventional numerical methods, even when the integrand can be written down.

The words 'Monte Carlo' and 'simulation' tend to be used interchangeably in the literature. Here a *simulation* is defined as a controlled experiment, usually carried out on a computer, that uses $U(0, 1)$ random numbers that are statistically independent. A *Monte Carlo* method is a method of estimating the value of an integral (or a sum) using the realized values from a simulation. It exploits the connection between an integral (or a sum) and the expectation of a function of a(some) random variable(s).

1.3 An example

Let us now examine how a Monte Carlo approach can be used in the following problem. A company owns K skips that can be hired out. During the nth day $(n = 1, 2, \ldots)$ Y_n people approach the company each wishing to rent a single skip. $Y_1, Y_2, \ldots$ are independent random variables having a Poisson distribution with mean λ. If skips are available they are let as 'new hires'; otherwise an individual takes his or her custom elsewhere. An individual hire may last for several days. In fact, the probability that a skip currently on hire to an individual is returned the next day is p. Skips are always returned at the beginning of a day. Let $X_n(K)$ denote the total number of skips on hire at the end of day n and let $H_n(K)$ be the number of new hires during day n. To simplify notation the dependence upon K will be dropped for the moment.

The problem is to find the optimal value of K. It is known that each new hire generates a fixed revenue of £c_f per skip and a variable revenue of £c_v per skip per day or part-day. The K skips have to be bought at the outset and have to be maintained irrespective of how many are on hire on a particular day. This cost is equivalent to £c_0 per skip per day.

Firstly, the stochastic process $\{X_n, n = 0, 1, \ldots\}$ is considered. Assuming skips are returned at the beginning of a day before hiring out,

$$Y_n = \text{Poisson}(\lambda) \quad (n = 1, 2, \ldots),$$

$$X_{n+1} = \min\{K, \text{binomial}(X_n, 1 - p) + Y_{n+1}\} \quad (n = 0, 1, 2, \ldots),$$

$$H_{n+1} = X_{n+1} - \text{binomial}(X_n, 1 - p) \quad (n = 0, 1, 2, \ldots).$$

Since X_{n+1} depends on X_n and Y_{n+1} only, and since Y_{n+1} is independent of $X_{n-1}, X_{n-2}, \ldots,$ it follows that $\{X_n, n = 0, 1, \ldots\}$ is a discrete-state, discrete-time, homogeneous Markov chain. The probability transition matrix is $\mathbf{P} = \{p_{ij} : i, j = 0, \ldots, K\}$ where $p_{ij} = P(X_{n+1} = j | X_n = i)$ for all n. Given that i skips are on hire at the end of day n the probability that r of them remain on hire at the beginning of day $n+1$ is the binomial probability $\binom{i}{r}(1-p)^r p^{i-r}$. The probability that there are $j - r$ people requesting new hires is the Poisson probability $\lambda^{j-r} e^{-\lambda} / (j-r)!$. Therefore, for $0 \le i \le K$ and $0 \le j \le K - 1$,

$$p_{ij} = \sum_{r=0}^{\min(i,j)} \binom{i}{r}(1-p)^r p^{i-r} \frac{\lambda^{j-r} e^{-\lambda}}{(j-r)!}. \tag{1.8}$$

For the case $j = K$,

$$p_{iK} = 1 - \sum_{j=0}^{K-1} p_{ij}. \tag{1.9}$$

Since $p_{ij} > 0$ for all i and j, the Markov chain is ergodic. Therefore, $P(X_n = j) \to P(X = j) = \pi_j$, say, as $n \to \infty$. Similarly, $P(H_n = j) \to P(H = j)$ as $n \to \infty$. Let $\boldsymbol{\pi}$ denote the row vector $(\pi_0, \ldots, \pi_K)$. $\boldsymbol{\pi}$ is the *stationary distribution* of the chain $\{X_n, n = 0, 1, \ldots\}$. Suppose we wish to maximize the long-run average profit per day. Then a K is found that maximizes $Z(K)$ where

$$Z(K) = c_f E(H[K]) + c_v E(X[K]) - c_0 K \tag{1.10}$$

and where the dependence upon K has been reintroduced. Now

$$E(H[K]) = \lim_{n \to \infty} \{E(H_n[K])\}$$

$$= \lim_{n \to \infty} \{E(X_n[K]) - E(\text{binomial}(X_{n-1}[K], 1 - p))\}$$

$$= E(X[K]) - E(X[K])(1 - p)$$

$$= pE(X[K])$$

This last equation expresses the obvious fact that in the long run the average number of new hires per day must equal the average returns per day. Substituting back into Equation (1.10) gives

$$Z(K) = E(X[K])(pc_f + c_v) - c_0 K$$

The first difference of $Z(K)$ is

$$D(K) = Z(K+1) - Z(K)$$

$$= \{E(X[K+1]) - E(X[K])\}[pc_f + c_v] - c_0$$

for $K = 0, 1, \ldots$. It is obvious that increasing K by one will increase the expected number on hire, and it is reasonable to assume that $E(X[K+1]) - E(X[K])$ is decreasing in K. In that case $Z(K)$ will have a unique maximum.

To determine the optimal K we require $E(X[K])$ for successive integers K. If K is large this will involve considerable computation. Firstly, the probability transition matrix would have to be calculated using Equations (1.8) and (1.9). Then it is necessary to invert a $(K+1) \times (K+1)$ matrix to find the stationary distribution π for each K, and finally we must compute $E(X[K])$ for each K.

On the other hand, the following piece of *pseudo-code* will *simulate* $X_1(K), \ldots, X_n(K)$:

```
Input K, X_0(K), λ, p, n
x := X_0(K)
For i = 1, ..., n
   y := Poisson(λ)
   r := binomial(x, 1 - p)
   x := min(r + y, K)
   Output x
Next i
```

Now $E(X[K]) = \sum_{x=0}^{K} x\pi_x$. Therefore, if $\{X_1(K), \ldots, X_n(K)\}$ is a sample from π, an unbiased estimator of $E(X[K])$ is $(1/n)\sum_{i=1}^{n} X_i(K)$. There is a minor inconvenience in that unless $X_0(K)$ is selected randomly from π, the stationary distribution, then $\{X_1(K), \ldots, X_n(K)\}$ will not be precisely from π, and therefore the estimate will be slightly *biased*. However, this may be rectified by a *burn-in* period of b observations, say, followed by a further $n - b$ observations. We then estimate $E(X[K])$ using $[1/(n-b)]\sum_{i=b+1}^{n} X_i(K)$.

In terms of programming effort it is probably much easier to use this last Monte Carlo approach. However, it must be remembered that it gives an *estimate* while the method involving matrix inversion gives the *exact* value, subject to the usual numerical roundoff errors. Further, we may have to simulate for a considerable period of time. It is not necessarily the burn-in time that will be particularly lengthy. The fact that the members of $\{X_i(K), i = 0, 1, 2, \ldots\}$ are *not independent* (they are *auto-correlated*) will necessitate long sample runs if a precise estimate is required.

Finding the optimal K involves determining $E(X[K+1]) - E(X[K])$ with some precision. We can reduce the sampling variation in our estimate of this by inducing positive correlation between our estimates of $E(X[K+1])$ and $E(X[K])$. Therefore, we might consider making $Y_n[K] = Y_n[K+1]$ and making binomial $(X_n[K], 1 - p)$ positively correlated with binomial $(X_n[K+1], 1 - p)$. Such issues require careful experimental planning. The aim is that of variance reduction, as seen in Section 1.1.

1.4 A simulation using Maple

This section contains an almost exact reproduction of a Maple worksheet, 'skipexample.mws'. It explores the problem considered in Section 1.3. It can also be downloaded from the website accompanying the book.

It will now be shown how the algorithm for simulation of skip hires can be programmed as a Maple *procedure*. Before considering the procedure we will start with a fresh worksheet by typing 'restart' and also load the statistics package by typing 'with(stats)'. These two lines of input plus the Maple generated response, '[anova, ..., transform]' form an *execution group*. In the downloadable file this is delineated by an elongated left bracket, but this is not displayed here.

> restart;
  with (stats); [*anova, describe, fit, importdata, random, statevalf,*
                    statplots, transform]

A Maple procedure is simply a function constructed by the programmer. In this case the name of the procedure is *skip* and the arguments of the function are λ, p, K, $x0$ and n, the number of days to be simulated. Subsequent calling of the procedure *skip* with selected values for these parameters will create a *sequence; hire*[1], ..., *hire*[n] where *hire*[i] = [i, x]. In Maple terminology *hire*[i] is itself a *list* of two items: i the day number and x the total number of skips on hire at the end of that day. Note how a list is enclosed by square brackets, while a sequence is not.

Each Maple procedure starts with the Maple prompt '>'. The procedure is written within an *execution group*. Each line of code is terminated by a semicolon. However, anything appearing after the '#' symbol is not executed. This allows programmer comments to be added. Use the 'shift' and 'return' keys to obtain a fresh line within the procedure. The procedure terminates with a semicolon and successful entry of the procedure results in the code being 'echoed' in blue type. The structure of this echoed code is highlighted by appropriate automatic indentation.

```
> skip: =proc (lambda, p, K, x0, n) local x, i, y, r, hire;
  randomize (5691443); # An arbitrary integer sets the 'seed'
  for the U(0, 1) random number generator.
  x:=x0; # x is the total number on hire, initially set to x0
  for i from 1 to n do;
    y:=stats[random, poisson[lambda]] (1, 'default',
     'inverse'); # Generates a random
  Poisson variate; 1 is the number of variates generated,
  'default' is MAPLE's default uniform generator, 'inverse'
  means that the variate is generated by inverting the
  cumulative distribution function [see Chapter 3]
     if x=0 then r:=0 else
  r:=stats[random, binomiald[x,1−p]] (1, 'default',
   'inverse') end if; # Generates a random binomial variate
    x:=min (r+y, K); # Updates the total number on hire
    hire[i]: =[i, x];
  end do;
  seq(hire[i], i=1 . . n); #Assigns the sequence hire[1], . . . ,
  hire[n] to procedure 'SKIP';
  end proc;
```

skip := **proc**(λ, p, K, $x0$, n)
local *x, i, y, r, hire*;
    randomize(5691443);

```
x := x0;
for i to n do
    y := stats[random, poisson[λ]](1, 'default', 'inverse');
    if x = 0 then r := 0
    else r :=stats[random, binomiald[x, 1 − p]](1, 'default', 'inverse')
    end if;
    x := min(r + y, K);
    hire[i] := [i, x]
end do;
seq(hire[i], i = 1..n)
end proc
```

The 'echoed' code is now examined. *Local* variables are those whose values cannot be transmitted to and from the procedure. The 'randomize' statement uses an arbitrarily chosen integer to set a *seed* for a $U(0, 1)$ random number generator within Maple. More will be said about such generators in Chapter 2. Initially, the total number of skips $x0$ on hire in the current day (day 0) is assigned to the variable x. This is followed by a 'for i to...do...end do' loop. The statements within this loop are executed for $i = 1, \ldots, n$. Maple contains a 'stats' package and 'random' is a subpackage of this. Note the use of the function 'poisson' within this subpackage. The generated Poisson variate is assigned to the variable y. Following this is an example of a conditional 'if...then...else...endif' statement. If the total number of skips on hire on the previous day is zero, then the number remaining on hire today (r) must be zero; otherwise r equals a random binomial variate with parameters x and $1 − p$. Following this, the value of the variable x is updated. Then the list $[i, x]$ is assigned to the variable $hire[i]$. The last executable statement forms a sequence $hire[1], \ldots, hire[n]$. Maple procedures use the convention that the result of the last statement that is executed is assigned to the name of the procedure. So when *skip* is subsequently called it will output a random realization of this sequence.

Some results will now be obtained when $\lambda = 5$ per day, $p = 0.2, K = 30, x0 = 0$, and $n = 100$ days. Calling 'skip' and assigning the results to a variable *res* gives the sequence *res*.

```
> res :=skip (5, 0.2, 30, 0, 100);
```

res := [1, 6], [2, 6], [3, 13], [4, 17], [5, 16], [6, 19], [7, 25], [8, 25], [9, 24], [10, 27], [11, 29], [12, 30], [13, 30], [14, 28], [15, 25], [16, 30], [17, 26], [18, 30], [19, 26], [20, 23], [21, 25], [22, 25], [23, 28], [24, 30], [25, 24], [26, 24], [27, 28], [28, 24], [29, 21], [30, 22], [31, 21], [32, 20], [33, 22], [34, 25], [35, 23], [36, 29], [37, 29], [38, 29], [39, 26], [40, 25], [41, 27], [42, 27], [43, 30], [44, 30], [45, 27], [46, 30], [47, 25], [48, 23], [49, 21], [50, 16], [51, 20], [52, 20], [53, 18], [54, 22], [55, 23], [56, 26], [57, 25], [58, 29], [59, 27], [60, 26], [61, 30], [62, 27], [63, 29], [64, 27], [65, 25], [66, 26], [67, 30], [68, 30], [69, 27], [70, 28], [71, 23], [72, 25], [73, 21], [74, 24], [75, 22], [76, 23], [77, 22], [78, 25], [79, 22], [80, 22], [81, 25], [82, 25], [83, 23], [84, 19], [85, 19], [86, 19], [87, 21], [88, 23], [89, 25], [90, 24], [91, 26], [92, 24], [93, 27], [94, 24], [95, 27], [96, 29], [97, 29], [98, 29], [99, 30], [100, 30]

Then the following can be plotted.

```
> PLOT (CURVES ([res]), TITLE (''Plot of total number of
  skips on hire against day''), AXESLABELS (''i=day'',
  ''x=skips''), AXESSTYLE(BOX));
```

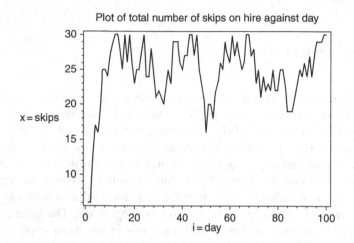

Note how the argument of 'CURVES' is a *list* (*res* has been enclosed in square brackets) of 100 points, each point expressed as a list of two values, its Cartesian coordinates. Now observe in the plot that the starting state $x = 0$ is hardly representative of a state chosen at random from the stationary distribution. However, by around day 10 the sequence $\{x_n\}$ appears to reflect near-stationary behaviour. Therefore, the remainder of the sequence will be used to estimate $E(X)$. First a sequence $x_{11}, \ldots, x_{100}$ needs to be constructed.

```
> data := seq (op (2, res[i]), i=11 .. 100);
```

data := 29, 30, 30, 28, 25, 30, 26, 30, 26, 23, 25, 25, 28, 30, 24, 24, 28, 24, 21, 22, 21, 20, 22, 25, 23, 29, 29, 29, 26, 25, 27, 27, 30, 30, 27, 30, 25, 23, 21, 16, 20, 20, 18, 22, 23, 26, 25, 29, 27, 26, 30, 27, 29, 27, 25, 26, 30, 30, 27, 28, 23, 25, 21, 24, 22, 23, 22, 25, 22, 22, 25, 25, 23, 19, 19, 19, 21, 23, 25, 24, 26, 24, 27, 24, 27, 29, 29, 29, 30, 30

Note how 'op' extracts the second element of *res[i]*, and this is repeated for items 11 through 100 of the original sequence. Next the 'describe' function will be used within the statistics package of Maple to compute the mean of the *list*, [data].

```
> describe[mean] ([data]);
```

$$\frac{455}{18}$$

The result is expressed as a rational number. To obtain the result in decimal form the function 'evalf' is called. The '%' argument is the last computed value.

```
> evalf (%);
```

$$25.27777778$$

Now the number of skips available will be changed from 30 to 28. The next execution group will calculate the mean. Notice that there are colons at the end of the first two statements. These suppress the respective outputs and only the result for the mean will be seen. Also, note that in the execution group below there are no MAPLE prompts after the first line. This is achieved by using 'shift-return' rather than 'return'.

```
> res1 :=skip (5, 0.2, 28, 0, 100) :
  data1 :=seq (op (2, res1[i]), i=11 .. 100):
  evalf (describe [mean] ([data1]));
```

$$24.34444444$$

Suppose that the number of available skips is reduced further, this time from 28 to 26.

```
> res2 : =skip (5, 0.2, 26, 0, 100) :
  data2 : =seq (op (2, res2 [i]), i=11 .. 100) :
  evalf (described [mean] ([data2])) ;
```

$$23.44444444$$

It is pleasing that the *estimates* of the expected number of skips on hire, under stationarity, decreases as the availability is decreased from 30 to 28 to 26. This must be the case for the *actual* (as opposed to the estimated) expectations. However, it is not guaranteed for the estimates. Had *different seeds* been used for the three experiments there would be less likelihood of preserving the decreasing expectations in the estimates. At least by using the *same seed* in each case we have (inadvertently?) ensured a reasonably good experimental design that attempts to reduce the variance of the difference between any two of these estimates.

Now the experiment will be repeated for various K in order to find the optimal value for given cost parameters. Each realization is based upon a larger sample of 500 days, the burn-in time being selected as 10 days. The procedure '*skipprofit*' below will deliver a sequence $[K, pf[K]], K = 20, \ldots, 30$, where $pf[K]$ is the estimate of expected daily profit when the availability is K skips.

```
> skipprofit : =proc (c0, cf, cv, p) local K, res, datab,
                 avprofit;
    for K from 20 to 30 do:
      res: =skip (5, 0.2, K, 0, 500):
      datab: =seq (op (2, res[i]), i=10 .. 500):
      avprofit [K]: = [K, (p*cf+cv) *evalf(describe[mean]
                   ([datab])) −c0*K]:
    end do:
    seq (avprofit [K], K=20 .. 30);
    end proc;
```

```
skipprofit : = proc(c0, cf, cv, p)
local K, res, datab, avprofit;
    for K from 20 to 30 do
        res := skip(5, 0.2, K, 0, 500);
        datab := seq (op(2, res[i]), i = 10 .. 500);
        avprofit[K] :=[K, (p*cf +cv)*evalf(describe[mean]([datab]))
                     −c0* K]
    end do;
    seq(avprofit[K], K = 20 .. 30)
end proc
```

This will be called with a fixed cost of £40 per skip per day, a fixed rental price of £100 per hire, and a variable rental price of £50 per skip per day, together with a 0.2 probability of return each day.

> profit :=skipprofit (40, 100, 50, 0.2);

profit := [20, 513.747454], [21, 523.217922], [22, 531.975560], [23, 536.741344], [24, 542.647658], [25, 548.696538], [26, 545.193483], [27, 535.417516], [28, 525.926680], [29, 513.869654], [30, 499.246436]

Now these results will be plotted.

> PLOT (CURVES ([profit]), TITLE (''Plot of Average
  profit per day against number of skips''), AXESLABELS
  (''skips'', ''profit per day''), AXESSTYLE(BOX));

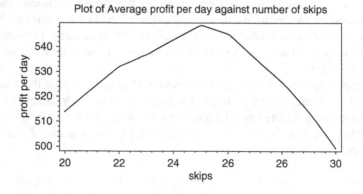

The estimated optimal strategy is to have 25 skips and the estimated long-run profit per day is approximately £549 per day. What sort of belief do we have in the proposed optimal 25 skips? This set of experiments should really be repeated with a different seed. The table below shows the seed, the optimal K, and the optimal profit for five sets of experiments in all. The optimal is $K = 25$ in three cases and $K = 24$ in two cases. Therefore, the optimal strategy is to select (with the aid of a single random number) 25 with probability 0.6 and 24 with probability 0.4. In the table below we should not be too concerned that the optimal profit shows considerable variations between experiments. In

Seed	Optimal K	Profit
5691443	25	548.7000
5691444	25	549.9800
21349109	24	536.6600
27111351	25	526.4600
3691254	24	503.8700

a decision problem such as this it is the *change* in the (unknown) expected profit resulting from a move away from the optimal decision that is important. A typical profit versus K plot (for example, the one shown above) reveals that there is a small reduction in profit and therefore a small cost associated with incorrectly choosing $K+1$ or $K-1$ rather than the optimal K.

1.5 Problems (see Appendix 1)

1. Use a Monte Carlo method, based upon 1000 random standard normal deviates, to find a 95 % confidence interval for $\int_{-\infty}^{\infty} \exp(-x^2) |\cos x| \, dx$. Use the Maple 'with(stats)' command to load the stats package. The function 'stats[random,normald](1)' will generate a single random standard normal deviate.

2. Use a Monte Carlo method to find a 95% confidence interval for

$$\int_{-\infty}^{\infty} \int_{-\infty}^{\infty} \exp\left\{ -\frac{1}{2} \left[x^2 + (y-1)^2 - \frac{x(y-1)}{10} \right] \right\} dx \, dy.$$

3. A machine tool is to be scrapped 4 years from now. The machine contains a part that has just been replaced. It has a life distribution with a time-to-failure density $f(x) = xe^{-x}$ on support $(0, \infty)$. Management must decide upon one of two maintenance strategies. The first is to replace the part whenever it fails until the scrapping time. The second is to replace failures during the first two years and then to make a preventive replacement two years from now. Following this preventive replacement, the part is replaced on failures occurring during the second half of the 4 year span. Assume that replacements are instantaneous and cost $£c_f$ on failure and $£c_p$ on a preventive basis. Simulate 5000 realizations of 4 years for each policy and find a condition on c_p/c_f for preventive replacement to be the preferred option.

4. Two points A and B are selected randomly in the unit square $[0, 1]^2$. Let D denote the distance between them. Using Monte Carlo:

   (a) Estimate $E(D)$ and $Var(D)$.

   (b) Plot an empirical distribution function for D.

   (c) Suggest a more efficient method for estimating $P(D > 1.4)$, bearing in mind that this probability is very small.

5. An intoxicated beetle moves over a cardboard unit circle $x^2 + y^2 < 1$. The (x, y) plane is horizontal and the cardboard is suspended above a wide open jar of treacle. In the time interval $[t, t + \delta t)$ it moves by amounts $\delta x = Z_1 \sigma_1 \sqrt{\delta t}$ and $\delta y = Z_2 \sigma_2 \sqrt{\delta t}$ along the x and y axes where Z_1 and Z_2 are independent standard normal random variables and σ_1 and σ_2 are specified positive constants. The aim is to investigate the distribution of time until the beetle arrives in the treacle pot starting from the point (x_0, y_0) on the cardboard.

   (a) Write a procedure that simulates n independent times between starting at the point (x_0, y_0) and landing in the treacle. The function 'stats[random,normald](1)' creates a random standard normal deviate.

   (b) Plot a histogram showing the distribution of 200 such times when $\sigma_1 = \sigma_2 = 1$, $\delta t = 0.01$, and $x_0 = y_0 = 0$. To create a histogram, load the subpackage 'statplots' using 'with(statplots)' and use the function 'histogram(a)' where a is a *list* of the 200 times.

6. The following binomial model is frequently used to mimic share price movements. Let S_i denote the price at time ih where $i = 0, 1, 2, \ldots$ and h is a positive time increment. Let μ and σ denote the growth rate and volatility respectively. Let

$$u = \frac{1}{2}\left(e^{-\mu h} + e^{(\mu + \sigma^2)h}\right) + \frac{1}{2}\sqrt{\left(e^{-\mu h} + e^{(\mu + \sigma^2)h}\right)^2 - 4},$$

$$v = u^{-1},$$

$$p = \frac{e^{\mu h} - v}{u - v}.$$

Then

$$S_i = X_i S_{i-1}$$

where $X_i, i = 0, 1, \ldots,$ are independent Bernoulli random variables with distribution $P(X_i = u) = p$, $\text{Prob}(X_i = v) = 1 - p$ for all i.

   (a) Simulate the price at the end of each week during the next year when $S_0 = 100$ pence, $\mu = 0.2$ per annum, $\sigma = 0.3$ per annum, and $h = 1/52$ years.

   (b) Now suppose there are 252 trading days in a year. Put $h = 1/252$. For any realization let $S_{\max} = \max(S_j : j = 0, \ldots, 756)$. Let $\text{loss} = S_{\max} - S_{756}$. Loss denotes the difference between selling the share at its peak value during the next 3 years and selling it after 3 years. Simulate 200 realizations of loss and construct an empirical distribution function for it. You will need to sort the 200 values. Do this by loading the 'stats' package and using the function 'transform[statsort](x)' where x is a list of the data to be sorted. Note that if the order statistics for loss are $x_{(1)}, \ldots, x_{(n)}$ then an unbiased estimate of $P(X < x_{(i)})$ is $i/(n+1)$.

7. Consider a single server queue. Let a_i denote the interarrival time between customer $i - 1$ and customer i, s_i the service time of customer i, and w_i the waiting time in the queue (i.e. the time between arrival and start of service) for customer i.

(a) Show that $w_i = \max(0, w_{i-1} - a_i + s_{i-1})$.

(b) Now consider an M/M/1 queue (that is one in which arrivals follow a Poisson process and service durations are exponentially distributed) in which the arrival rate is λ and the service rate is μ. Write a procedure that simulates $w_1, \ldots, w_n$ given w_0, λ, and μ.

(c) Experiment with different values for the traffic intensity λ/μ plotting w_i against i to demonstrate queues that achieve stationary behaviour (i) quickly, (ii) slowly, and (iii) never. In cases (i) and (ii) provide point estimates of the expectation of w_i in the steady state.

2

Uniform random numbers

All simulations work on a 'raw material' of *random numbers*. A sequence $R_1, R_2, \ldots$ is said to be a random number sequence if $R_i \sim U(0, 1)$ for all i and R_i is independent of R_j for all $i \neq j$. Some authors use the phrase 'random numbers' to include variates sampled from any specified probability distribution. However, its use here will be reserved solely for $U(0, 1)$ variates.

How can such a sequence be generated? One approach is to use a physical randomizing device such as a machine that picks lottery numbers in the UK, a roulette wheel, or an electronic circuit that delivers 'random noise'. There are two disadvantages to this. Firstly, such devices are slow and do not interface naturally with a computer. Secondly, and paradoxically, there is often a need to *reproduce* the random number stream (making it nonrandom!). This need arises, for example, when we wish to control the input to a simulation for the purpose of verifying the correctness of the programming code. It is also required when we wish to compare the effect of two or more policies on a simulation model. By using the same random number stream(s) for each of the experiments we hope to reduce the variance of estimators of the *difference* in response between any two policies.

One way to make a random number stream reproducible is to copy it to a peripheral and to read the numbers as required. A peripheral could be the hard disc of a computer, a CD ROM, or simply a book. In fact, the RAND corporation published 'A million random digits with 100 000 random normal deviates' (Rand Corporation, 1955), which, perhaps, was not that year's best seller. Accessing a peripheral several thousands or perhaps millions of times can slow down a simulation. Therefore, the preferred approach is to generate *pseudo-random numbers* at run-time, using a specified deterministic recurrence equation on integers. This allows fast generation, eliminates the storage problem, and gives a reproducible sequence. However, great care is needed in selecting an appropriate recurrence, to make the sequence *appear* random.

Simulation and Monte Carlo: With applications in finance and MCMC   J. S. Dagpunar
© 2007 John Wiley & Sons, Ltd

2.1   Linear congruential generators

These deliver a sequence of non-negative integers $\{X_i, i = 1, 2, \ldots\}$ where

$$X_i = (aX_{i-1} + c) \bmod m \quad (i = 1, 2, \ldots).$$

The recurrence uses four integer parameters set by the user. They are: $a(> 0)$ a *multiplier*, $X_0(\geq 0)$ a *seed*, $c(\geq 0)$ an *increment*, and $m(>0)$ a *modulus*. The first three parameter values lie in the interval $[0, m-1]$. The modulo m process returns the remainder after dividing $aX_{i-1} + c$ by m. Therefore, $X_i \in [0, m-1]$ for all i. The pseudo-random number is delivered as $R_i = X_i/m$ and so $R_i \in [0, 1)$. The idea is that if m is large enough, the discrete values $\frac{0}{m}, \frac{1}{m}, \frac{2}{m}, \ldots, \frac{m-1}{m}$ are so close together that R_i can be treated as a continuously distributed random variable. In practice we do not use $X_i = 0\,(R_i = 0)$ in order to avoid problems of division by zero and taking the logarithm of zero, etc. As an example consider the generator

$$X_i = (9X_{i-1} + 3) \bmod 2^4 \quad (i = 1, 2, \ldots). \tag{2.1}$$

Choose $X_0 \in [0, 15]$, say $X_0 = 3$. Then $X_1 = 30 \bmod 2^4 = 14$, $X_2 = 129 \bmod 2^4 = 1, \ldots$. The following sequence for $\{R_i\}$ is obtained:

$$\frac{3}{16}, \frac{14}{16}, \frac{1}{16}, \frac{12}{16}, \frac{15}{16}, \frac{10}{16}, \frac{13}{16}, \frac{8}{16}, \frac{11}{16}, \frac{6}{16}, \frac{9}{16}, \frac{4}{16}, \frac{7}{16}, \frac{2}{16}, \frac{5}{16}, \frac{0}{16}, \frac{3}{16}, \ldots. \tag{2.2}$$

The *period* of a generator is the smallest integer λ such that $X_\lambda = X_0$. Here the sequence repeats itself on the seventeenth number and so $\lambda = 16$. Clearly, we wish to make the period as large as possible to avoid the possibility of reusing random numbers. Since the period cannot exceed m, the modulus is often chosen to be close to the largest representable integer on the computer.

There are some number theory results (Hull and Dobell, 1962) that assist in the choice of a, c, and m. A *full period* $(=m)$ is obtained if and only if

c and m *are relatively prime* (*greatest common divisor of* c *and* m *is* 1);   (2.3)

$a - 1$ is a multiple of q for every prime factor q of m;   (2.4)

$a - 1$ is a multiple of 4 if m is.   (2.5)

Linear congruential generators can be classified into *mixed* $(c > 0)$ and *multiplicative* $(c = 0)$ types.

2.1.1   Mixed linear congruential generators

In this case $c > 0$. A good choice is $m = 2^b$ where b is the maximum number of bits used to represent positive integers with a particular computer/language combination. For example, many computers have a 32 bit word for integers. One bit may be reserved for the sign, leaving $b = 31$. Such a computer can store integers in the interval $\left[-2^{31}, 2^{31} - 1\right]$.

By choosing $m = 2^b$ the generator will have a full period ($\lambda = m$) when c is chosen to be odd-valued, satisfying condition (2.3), and $a - 1$ to be a multiple of 4, satisfying conditions (2.4) and (2.5). This explains why the generator (2.1) is a full period one with $\lambda = 16$.

In Maple there is no danger that $aX_{i-1} + c$ exceeds N, the largest positive integer that can be stored on the computer. This is because Maple performs arithmetic in which the number of digits in N is essentially limited only by the memory available to Maple. The Maple procedure 'r1' below uses parameter values $m = 2^{31}$, $a = 906185749$, $c = 1$. Note the importance of declaring 'seed' as a global variable, so that the value of this variable can be transmitted to and from the procedure. The period of the generator is $m = 2147483648$, which is adequate for most simulations. This generator has been shown to have good statistical properties (Borosh and Niederreiter, 1983). More will be stated about statistical properties in Sections 2.2 and 2.4.

```
> r1 := proc() global seed;
    seed := (906185749*seed + 1) mod 2^31;
    evalf(seed/2^31);
    end proc:
```

The code below invokes the procedure five times starting with seed = 3456:

```
> seed := 3456;
    for j from 1 to 5 do;
    r1();
    end do;
```

```
seed := 3456
.3477510815
.2143113120
.7410933147
.4770359378
.6231261701
```

The generator took about 12 microseconds per random number using a Pentium M 730 processor. The generator 'r2' below (L'Ecuyer, 1999) has a far more impressive period. The parameter values are $m = 2^{64}$, $a = 2,862,933,555,777,941,757$, $c = 1$. The period is 18, 446, 744, 073, 709, 551, 616 and the execution speed is not much slower at 16 microseconds per random number.

```
> r2 := proc() global seed;
    seed := (2862933555777941757*seed + 1) mod 2^64;
    evalf(seed/2^64);
    end proc:
```

In most scientific languages (e.g. FORTRAN90, C++) the finite value of N is an issue. For example, N may be $2^{31} - 1$ when the word size is 32 bits. Explicit calculation of $aX_{i-1} + c$ may cause a fatal error through *overflow* when $aX_{i-1} + c \notin [-2^{31}, 2^{31} - 1]$. However, in some implementations, the integer following N is stored as $-(N+1)$

followed by $-N$, and so overflow never occurs. In that case, it is necessary only to take the last 31 bits. In such cases the modulo m process is particularly simple to implement. If this feature is not present, overflow can be avoided by working in *double precision*. For example, with a 32 bit word size, double precision uses 64 bits. However, the random number generator will be somewhat slower than with single precision.

Another way of dealing with potential overflow will now be described. It is based on a method due to Schrage (1979). Let

$$u = \left\lfloor \frac{m}{a} \right\rfloor, \quad w = m \bmod a.$$

Then

$$m = ua + w.$$

Now,

$$X_i = \left\lfloor \frac{X_i}{u} \right\rfloor u + X_i \bmod u$$

and so

$$aX_i + c = \left\lfloor \frac{X_i}{u} \right\rfloor au + a(X_i \bmod u) + c$$

$$= \left\lfloor \frac{X_i}{u} \right\rfloor (m - w) + a(X_i \bmod u) + c.$$

Therefore,

$$(aX_i + c) \bmod m = \left\{ -\left\lfloor \frac{X_i}{u} \right\rfloor w + a(X_i \bmod u) + c \right\} \bmod m. \tag{2.6}$$

Now

$$0 \le \left\lfloor \frac{X_i}{u} \right\rfloor w \le \frac{X_i w}{u} \le X_i \le m - 1$$

providing that a is chosen such that

$$w \le u. \tag{2.7}$$

Similarly,

$$c \le a(X_i \bmod u) + c \le a(u - 1) + c$$

$$= m - w - a + c$$

$$\le m - 1$$

providing a and c are chosen such that

$$m - a \ge w \ge c + 1 - a. \tag{2.8}$$

Therefore, subject to conditions (2.7) and (2.8)

$$-m+1+c \le -\left\lfloor \frac{X_i}{u} \right\rfloor w + a\,(X_i \bmod u) + c \le m-1.$$

To perform the mod m process in Equation (2.6) simply set

$$Z_{i+1} = -\left\lfloor \frac{X_i}{u} \right\rfloor w + [a\,(X_i \bmod u)] + c.$$

Then

$$X_{i+1} = \begin{cases} Z_{i+1} & (Z_i \ge 0), \\ Z_{i+1}+m & (Z_i < 0). \end{cases}$$

The Maple procedure below, named 'schrage', implements this for the full period generator with $m = 2^{32}$, $a = 69069$, and $c = 1$. It is left as an exercise (see Problem 2.3) to verify the correctness of the algorithm, and in particular that conditions (2.7) and (2.8) are satisfied. It is easily recoded in any scientific language. In practice, it would not be used in a Maple environment, since the algorithm is of most benefit when the maximum allowable size of a positive integer is 2^{32}. The original generator with these parameter values, but without the Schrage innovation, is a famous one, part of the 'SUPER-DUPER' random number suite (Marsaglia, 1972; Marsaglia et al., 1972). Its statistical properties are quite good and have been investigated by Anderson (1990) and Marsaglia and Zaman (1993).

```
> schrage:=proc() local s,r; global seed;
  s:=seed mod 62183;r:= (seed-s)/62183;
  seed:=-49669*r+69069*s+1;
  if seed < 0 then seed:=seed+2^32 end if;
  evalf(seed/2^32);
  end proc;
```

Many random number generators are proprietary ones that have been coded in a lower level language where the individual bits can be manipulated. In this case there is a definite advantage in using a modulus of $m = 2^b$. The evaluation of $(aX_{i-1}+c) \bmod 2^b$ is particularly efficient since X_i is returned as the last b bits of $aX_{i-1}+c$. For example, in the generator (2.1)

$$X_7 = (9 \times 13 + 3)(\bmod 16).$$

In binary arithmetic, $X_7 = [(1001.) \times (1101.) + (11.)] \bmod 10000.$. Now $(1001.) \times (1101.) + (11.) =$

$$\begin{array}{ll} 0\,1\,1\,0\,1\,0\,0\,0. & \\ 0\,0\,0\,0\,1\,1\,0\,1. \;\; + & \qquad (2.9) \\ \underline{0\,0\,0\,0\,0\,0\,1\,1.} & \\[4pt] \underline{0\,1\,1\,1\,1\,0\,0\,0.} & \qquad (2.10) \end{array}$$

Note that the first row of (2.9) gives (1000.) × (1101.) by shifting the binary (as opposed to the decimal) point of (1101.) 3 bits to the right. The second row gives (0001.) × (1101.) and the third row is (11.). The sum of the three rows is shown in (2.10). Then X_7 is the *final* 4 bits in this row, that is 1000., or $X_7 = 8$ in the decimal system. In fact, it is unnecessary to perform any calculations beyond the fourth bit. With this convention, and omitting bits other than the last four, $X_8 = \{(1001.) \times (1000.) + (11.)\}$ mod (10000.) =

$$
\begin{array}{r}
0\,0\,0\,0. \\
1\,0\,0\,0. \;\; + \\
\underline{0\,0\,1\,1.} \\
1\,0\,1\,1.
\end{array}
$$

or $X_8 = 1011.$ (binary), or $X_8 = 11$ (decimal). To obtain R_8 we divide by 16. In binary, this is done by moving the binary point four bits to the left, giving (0.1011) or $2^{-1} + 2^{-3} + 2^{-4} = 11/16$. By manipulating the bits in this manner the issue of overflow does not arise and the generator will be faster than one programmed in a high-level language. This is of benefit if millions of numbers are to be generated.

2.1.2   Multiplicative linear congruential generators

In this case $c = 0$. This gives

$$X_i = (aX_{i-1}) \bmod m.$$

We can never allow $X_i = 0$, otherwise the subsequent sequence will be ..., 0,0,.... Therefore the period cannot exceed $m - 1$. Similarly, the case $a = 1$ can be excluded. It turns out that a maximum period of $m - 1$ is achievable if and only if

$$m \text{ is prime and} \tag{2.11}$$

$$a \text{ is a primitive root of } m \tag{2.12}$$

A multiplicative generator satisfying these two conditions is called a *maximum period prime modulus generator*. Requirement (2.12) means that

$$m \nmid a \text{ and } m \nmid a^{(m-1)/q} - 1 \text{ for every prime factor } q \text{ of } m - 1. \tag{2.13}$$

Since the multiplier a is always chosen such that $a < m$, the first part of this condition can be ignored. The procedure 'r3' shown below is a good (see Section 2.2) maximum period prime modulus generator with multiplier $a = 630360016$. It takes approximately 11 microseconds to deliver one random number using a Pentium M 730 processor:

```
> r3:= proc() global seed;
   seed:= (seed*630360016)mod(2^31 − 1);
   evalf(seed/(2^31 − 1));
   end proc;
```

At this point we will describe the in-built Maple random number generator, 'rand()' (Karian and Goyal, 1994). It is a maximum period prime modulus generator with $m = 10^{12} - 11$, $a = 427419669081$ (Entacher, 2000). To return a number in the interval $[0,1)$, we divide by $10^{12} - 11$, although it is excusable to simply divide by 10^{12}. It is slightly slower than 'r1' (12 microseconds) and 'r2' (16 microseconds), taking approximately 17 microseconds per random number. The seed is set using the command 'randomize(integer)' before invoking 'rand()'. Maple also provides another $U(0, 1)$ generator. This is 'stats[random,uniform](1)'. This is based upon 'rand' and so it is surprising that its speed is approximately 1/17th of the speed of 'rand()/10^12'. It is not advised to use this.

For any prime modulus generator, $m \neq 2^b$, so we cannot simply deliver the last b bits of aX_{i-1} expressed in binary. Suppose $m = 2^b - \gamma$ where γ is the smallest integer that makes m prime for given b. The following method (Fishman, 1978, pp. 357–358) *emulates* the bit shifting process, previously described for the case $m = 2^b$. The generator is

$$X_{i+1} = (aX_i) \bmod (2^b - \gamma).$$

Let

$$Y_{i+1} = (aX_i) \bmod 2^b, \quad K_{i+1} = \lfloor aX_i/2^b \rfloor. \tag{2.14}$$

Then

$$aX_i = K_{i+1}2^b + Y_{i+1}.$$

Therefore,

$$
\begin{aligned}
X_{i+1} &= \left(K_{i+1}2^b + Y_{i+1} \right) \bmod (2^b - \gamma) \\
&= \left\{ K_{i+1}(2^b - \gamma) + Y_{i+1} + \gamma K_{i+1} \right\} \bmod (2^b - \gamma) \\
&= \left\{ Y_{i+1} + \gamma K_{i+1} \right\} \bmod (2^b - \gamma)
\end{aligned}
$$

From (2.14), $0 \leq Y_{i+1} \leq 2^b - 1$ and $0 \leq K_{i+1} \leq \lfloor a(2^b - \gamma - 1)/2^b \rfloor \leq a - 1$. Therefore, $0 \leq Y_{i+1} + \gamma K_{i+1} \leq 2^b - 1 + a\gamma - \gamma$. We would like $Y_{i+1} + \gamma K_{i+1}$ to be less than $2^b - \gamma$ so that it may be assigned to X_{i+1} without performing the troublesome $\bmod(2^b - \gamma)$. Failing that, it would be convenient if it was less than $2(2^b - \gamma)$, so that $X_{i+1} = \{Y_{i+1} + \gamma K_{i+1}\} - (2^b - \gamma)$, again avoiding the $\bmod(2^b - \gamma)$ process. This will be the case if $2^b - 1 + a\gamma - \gamma \leq 2(2^b - \gamma) - 1$, that is if

$$a \leq \frac{2^b}{\gamma} - 1. \tag{2.15}$$

In that case, set $Z_{i+1} = Y_{i+1} + \gamma K_{i+1}$. Then

$$
X_{i+1} = \begin{cases}
Z_{i+1} & \left(Z_{i+1} < 2^b - \gamma \right), \\
Z_{i+1} - (2^b - \gamma) & \left(Z_{i+1} \geq 2^b + \gamma \right).
\end{cases}
$$

The case $\gamma = 1$ is of practical importance. The condition (2.15) reduces to $a \leq 2^b - 1$. Since $m = 2^b - \gamma = 2^b - 1$, the largest possible value that could be chosen for a is $2^b - 2$. Therefore, when $\gamma = 1$ the condition (2.15) is satisfied for *all* multipliers a. Prime numbers of the form $2^k - 1$ are called *Mersenne primes*, the low-order ones being $k = 2, 3, 5, 7, 13, 17, 19, 31, 61, 89, 107, \ldots$

How do we find primitive roots of a prime, m? If a is a primitive root it turns out that the others are

$$\{a^j \bmod m : j < m - 1, j \text{ and } m - 1 \text{ are relatively prime}\}. \qquad (2.16)$$

As an example we will construct all maximum period prime modulus generators of the form

$$X_{i+1} = (aX_i) \bmod 7.$$

We require all primitive roots of 7 and refer to the second part of condition (2.13). The prime factors of $m - 1 = 6$ are 2 and 3. If $a = 2$ then $7 = m \mid a^{6/2} - 1 = 2^{6/2} - 1$. Therefore, 2 is not a primitive root of 7. If $a = 3, 7 = m \nmid a^{6/2} - 1 = 3^{6/2} - 1$ and $7 = m \nmid a^{6/3} - 1 = 3^{6/3} - 1$. Thus, $a = 3$ is a primitive root of 7. The only j ($< m - 1$) which is relatively prime to $m - 1 = 6$ is $j = 5$. Therefore, by (2.16), the remaining primitive root is $a = 3^5 \bmod m = 9 \times 9 \times 3 \bmod 7 = 2 \times 2 \times 3 \bmod 7 = 5$. The corresponding sequences are shown below. Each one is a reversed version of the other:

$$a = 3 : \ldots, \frac{1}{7}, \frac{3}{7}, \frac{2}{7}, \frac{6}{7}, \frac{4}{7}, \frac{5}{7}, \frac{1}{7}, \ldots$$

$$a = 5 : \ldots, \frac{1}{7}, \frac{5}{7}, \frac{4}{7}, \frac{6}{7}, \frac{2}{7}, \frac{3}{7}, \frac{1}{7}, \ldots$$

For larger moduli, finding primitive roots by hand is not very easy. However, it is easier with Maple. Suppose we wish to construct a maximum period prime modulus generator using the Mersenne prime $m = 2^{31} - 1$ and would like the multiplier $a \approx m/2 = 1073741824.5$. All primitive roots can be found within, say, 10 of this number using the code below:

```
> with(numtheory):
  a:=1073741814;
  do;
  a:=primroot(a, 2^31-1);
  if a > 1073741834 then break end if;
  end do;
```

$a := 1073741814$
$a := 1073741815$
$a := 1073741816$
$a := 1073741817$
$a := 1073741827$
$a := 1073741829$
$a := 1073741839$

We have concentrated mainly on the maximum period prime modulus generator, because of its almost ideal period. Another choice will briefly be mentioned where the modulus is $m = 2^b$ and b is the usable word length of the computer. In this case the maximum period achievable is $\lambda = m/4$. This occurs when $a = 3$ mod 8 or 5 mod 8, and X_0 is odd. In each case the sequence consists of $m/4$ odd numbers, which does not communicate with the other sequence comprising the remaining $m/4$ odd numbers. For example, $X_i = (3X_{i-1})$ mod 2^4 gives either

$$\ldots, \frac{1}{15}, \frac{3}{15}, \frac{9}{15}, \frac{11}{15}, \frac{1}{15}, \ldots$$

or

$$\ldots, \frac{5}{15}, \frac{15}{15}, \frac{13}{15}, \frac{7}{15}, \frac{5}{15}, \ldots$$

depending upon the choice of seed.

All Maple procedures in this book use 'rand' described previously. Appendix 2 contains the other generators described in this section.

2.2 Theoretical tests for random numbers

Most linear congruential generators are one of the following three types, where λ is the period:

Type A: full period multiplicative, $m = 2^b$, $a = 1$ mod 4, c odd-valued, $\lambda = m$;

Type B: maximum period multiplicative prime modulus, m a prime number, $a = a$ primitive root of m, $c = 0$, $\lambda = m - 1$;

Type C: maximum period multiplicative, $m = 2^b$, $a = 5$ mod 8, $\lambda = m/4$.

The output from type C generators is *identical* (apart from the subtraction of a specified constant) to that of a corresponding type A generator, as the following theorem shows.

Theorem 2.1 *Let* $m = 2^b$, $a = 5$ mod 8, X_0 *be odd-valued,* $X_{i+1} = (aX_i)$ mod m, $R_i = X_i/m$. *Then* $R_i = R_i^* + (X_0 \bmod 4)/m$ *where* $R_i^* = X_i^*/(m/4)$ *and* $X_{i+1}^* = (aX_i^* + [X_0 \bmod 4]\{(a-1)/4\})$ mod $(m/4)$.

Proof. First we show that $X_i - X_0$ mod 4 is a multiple of 4. Assume that this is true for $i = k$. Then $X_{k+1} - X_0$ mod $4 = [a(X_k - X_0 \bmod 4) + (a-1)\{X_0 \bmod 4\}]$ mod m. Now, $4 \mid a-1$ so $4 \mid X_{k+1} - X_0$ mod 4. For the base case $i = 0$, $X_i - X_0$ mod $4 = 0$, and so by the principle of induction $4 \mid X_i - X_0$ mod 4 $\forall i \geq 0$. Now put $X_i^* = (X_i - X_0 \bmod 4)/4$. Then $X_i = 4X_i^* + X_0$ mod 4 and $X_{i+1} = 4X_{i+1}^* + X_0$ mod 4. Dividing the former equation through by m gives $R_i = R_i^* + (X_0 \bmod 4)/m$ where $X_{i+1} - aX_i = 4(X_{i+1}^* - aX_i^*) - (X_0 \bmod 4)(a-1) = 0$ mod m. It follows that $X_{i+1}^* - aX_i^* - [X_0 \bmod 4][(a-1)/4] = 0$ mod $(m/4)$ since $4 \mid m$. This completes the proof.

This result allows the investigation to be confined to the theoretical properties of type A and B generators only. Theoretical tests use the values a, c, and m to assess the quality of the output of the generator over the *entire* period. It is easy to show (see Problem 5) for both type A and B generators that for all but small values of the period λ, the mean and variance of $\{R_i, i = 0, \ldots, \lambda - 1\}$ are close to $\frac{1}{2}$ and $\frac{1}{12}$, as must be the case for a true $U(0, 1)$ random variable.

Investigation of the *lattice* (Ripley, 1983a) of a generator affords a deeper insight into the quality. Let $\{R_i, i = 0, \ldots, \lambda - 1\}$ be the entire sequence of the generator. In theory it would be possible to plot the λ *overlapping pairs* $(R_0, R_1), \ldots, (R_{\lambda-1}, R_0)$. A necessary condition that the sequence consists of *independent* $U(0, 1)$ random variables is that R_1 is independent of R_0, R_2 is independent of R_1, and so on. Therefore, the pairs should be uniformly distributed over $[0, 1)^2$. Figures 2.1 and 2.2 show plots of 256 such points for the full period generators $X_{i+1} = (5X_i + 3) \bmod 256$ and $X_{i+1} = (13X_i + 3) \bmod 256$ respectively. Firstly, a disturbing feature of both plots is observed; all points can be covered by a set of parallel lines. This detracts from the uniformity over $[0, 1)^2$. However, it is unavoidable (for all linear congruential generators) given the linearity (mod m) of these recurrences. Secondly, Figure 2.2 is preferred in respect of uniformity over $[0, 1)^2$. The minimum number of lines required to cover all points is 13 in Figure 2.2 but only 5 in Figure 2.1, leading to a markedly nonuniform density of points in the latter case. The separation between adjacent lines is wider in Figure 2.1 than it is in Figure 2.2. Finally, each lattice can be constructed from a reduced basis consisting of vectors $\mathbf{e}_1$ and $\mathbf{e}_2$ which define the smallest lattice cell. In Figure 2.1 this is long and thin, while in the more favourable case of Figure 2.2 the sides have similar lengths. Let $l_1 = |\mathbf{e}_1|$ and $l_2 = |\mathbf{e}_2|$ be the lengths of the smaller and longer sides respectively. The larger $r_2 = l_2/l_1$ is, the poorer the uniformity of pairs and the poorer the generator.

This idea can be extended to find the degree of uniformity of the set of overlapping k-tuples $\{(R_i, \ldots, R_{i+k-1 \bmod m}), i = 0, \ldots, \lambda - 1\}$ through the hypercube $[0, 1)^k$. Let $l_1, \ldots, l_k$ be the lengths of the vectors in the reduced basis with $l_1 \leq \cdots \leq l_k$. Alternatively, these are the side lengths of the smallest lattice cell. Then, generators for which $r_k = l_k/l_1$ is large, at least for small values of k are to be regarded with suspicion. Given values for a, c, and m, it is possible to devise an algorithm that will calculate either r_k or an upper bound for r_k (Ripley, 1983a). It transpires that changing the value of c in a type A generator only translates the lattice as a whole; the relative positions of the lattice points remain unchanged. As a result the choice of c is immaterial to the quality of a type A generator and the crucial decision is the choice of a.

Table 2.1 gives some random number generators that are thought to perform well. The first four are from Ripley (1983b) and give good values for the lattice in low dimensions. The last five are recommended by (Fishman and Moore 1986) from a search over all multipliers for prime modulus generators with modulus $2^{31} - 1$. There are references to many more random number generators with given parameter values together with the results of theoretical tests in Entacher (2000).

2.2.1 Problems of increasing dimension

Consider a maximum period multiplicative generator with modulus $m \approx 2^b$ and period $m - 1$. The random number sequence is a permutation of $1/m, \ldots, (m-1)/m$. The distance between neighbouring values is constant and equals $1/m$. For sufficiently large m this is small enough to ignore the 'graininess' of the sequence. Consequently, we are happy to use this discrete uniform as an approximation to a continuous

The generator X(i + 1) = 5X(i) + 3 mod256

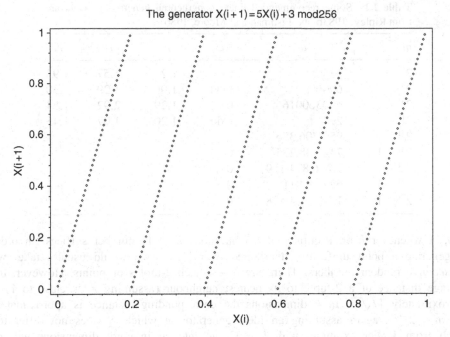

Figure 2.1 Plot of $\{(X_i, X_{i+1}), i = 0, \ldots, 255\}$ for $X_{i+1} = (5X_i + 3) \bmod 256$

The generator X(i + 1) = 13X(i) + 3 mod256

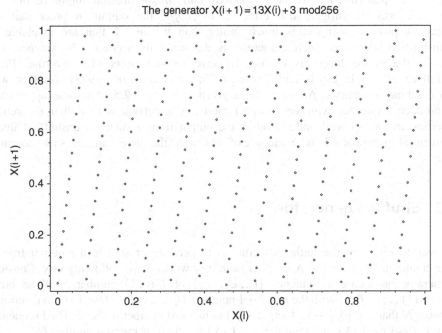

Figure 2.2 Plot of $\{(X_i, X_{i+1}), i = 0, \ldots, 255\}$ for $X_{i+1} = (13X_i + 3) \bmod 256$

Table 2.1 Some recommended linear congruential generators. (Data are from Ripley, 1983b and Fishman and Moore, 1986)

m	a	c	r_2	r_3	r_4
2^{59}	13^{13}	0	1.23	1.57	1.93
2^{32}	69069	Odd	1.06	1.29	1.30
$2^{31}-1$	630360016	0	1.29	2.92	1.64
2^{16}	293	Odd	1.20	1.07	1.45
$2^{31}-1$	950,706,376	0			
$2^{31}-1$	742,938,285	0			
$2^{31}-1$	1,226,874,159	0			
$2^{31}-1$	62,089,911	0			
$2^{31}-1$	1,343,714,438	0			

$U(0, 1)$ whenever b-bit accuracy of a continuous $U[0, 1)$ number suffices. In order to generate a point uniformly distributed over $[0, 1)^2$ we would usually take two *consecutive* random numbers. There are $m - 1$ such 2-tuples or points. However, the average distance of a 2-tuple to its nearest neighbour (assuming r_2 is close to 1) is approximately $1/\sqrt{m}$. In k dimensions the corresponding distance is approximately $1/\sqrt[k]{m} = 2^{-b/k}$, again assuming an ideal generator in which r_k does not differ too much from 1. For example, with $b = 32$, an integral in eight dimensions will be approximated by the expectation of a function of a random vector having a discrete (rather than the desired continuous) distribution in which the average distance to a nearest neighbour is of the order of $2^{-4} = \frac{1}{16}$. In that case the graininess of the discrete approximation to the continuous uniform distribution might become an issue. One way to mitigate this effect is to *shuffle* the output in order that the number of possible k-tuples is much geater than the $m - 1$ that are available in an unshuffled sequence. Such a method is described in Section 2.3. Another way to make the period larger is to use Tauseworthe generators (Tauseworthe, 1965; Toothill *et al.*, 1971; Lewis and Payne, 1973; Toothill *et al.*, 1973). Another way is to combine generators. An example is given in Section 2.5. All these approaches can produce sequences with very large periods. A drawback is that their theoretical properties are not so well understood as the output from a standard unshuffled linear congruential generator. This perhaps explains why the latter are in such common usage.

2.3 Shuffled generator

One way to break up the lattice structure is to permute or shuffle the output from a linear congruential generator. A *shuffled generator* works in the following way. Consider a generator producing a sequence $\{U_1, U_2, \dots\}$ of $U(0, 1)$ numbers. Fill an array $T(0), T(1), \dots, T(k)$ with the first $k+1$ numbers $U_1, \dots, U_{k+1}$. Use $T(k)$ to determine a number N that is $U[0, k-1]$. Output $T(k)$ as the next number in the shuffled sequence. Replace $T(k)$ by $T(N)$ and then replace $T(N)$ by the next random number U_{k+2} in the un-shuffled sequence. Repeat as necessary. An algorithm for this is:

$N := \lfloor kT(k) \rfloor$
Output $T(k)$ (becomes the next number in the shuffled sequence)
$T(k) := T(N)$
Input U (the next number from the unshuffled sequence)
$T(N) := U$

Note that $\lfloor x \rfloor$ denotes the floor of x. Since x is non-negative, it is the integer part of x. An advantage of a shuffled generator is that the period is increased.

2.4 Empirical tests

Empirical tests take a segment of the output and subject it to statistical tests to determine whether there are specific departures from randomness.

2.4.1 Frequency test

Here we test the hypothesis that $R_i, i = 1, 2, \ldots$, are uniformly distributed in $(0,1)$. The test assumes that $R_i, i = 1, 2, \ldots$, are *independently* distributed. We take n consecutive numbers, $R_1, \ldots, R_n$, from the generator. Now divide the interval $(0, 1)$ into k subintervals $(0, h), [h, 2h), \ldots, [\{k-1\} h, kh)$ where $kh = 1$. Let f_i denote the observed frequency of observations in the ith subinterval. We test the null hypothesis that the sample is from the $U(0, 1)$ distribution against the alternative that it is not. Let $e_i = n/k$, which is the expected frequency assuming the null hypothesis is true. Under the null hypothesis the test statistic

$$X^2 = \sum_{i=1}^{k} \frac{(f_i - e_i)^2}{e_i} = \sum_{i=1}^{k} \frac{f_i^2}{e_i} - n$$

follows a chi-squared distribution with $k - 1$ degrees of freedom. Large values of X^2 suggest nonuniformity. Therefore, the null hypothesis is rejected at the $100\alpha\%$ significance level if $X^2 > \chi^2_{k-1,\alpha}$ where $\alpha = P\left(\chi^2_{k-1} > \chi^2_{k-1,\alpha}\right)$.

As an example of this, 1000 random numbers were sampled using the Maple random number generator, 'rand()'. Table 2.2 gives the observed and expected frequencies based upon $k = 10$ subintervals of width $h = 0.1$. This gives

$$X^2 = \frac{100676}{100} - 1000 = 6.76.$$

From tables of the percentage points of the chi-squared distribution it is found that $\chi^2_{9,0.05} = 16.92$, indicating that the result is not significant. Therefore, there is insufficient evidence to dispute the uniformity of the population, assuming that the observations are independent.

The null chi-squared distribution is an asymptotic result, so the test can be applied only when n is suitably large. A rule of thumb is that $e_i \gtrsim 5$ for every interval. For a really large sample we can afford to make k large. In that case tables for chi-squared are

Table 2.2 Observed and expected frequencies

Interval	$[0,h)$	$[h,2h)$	$[2h,3h)$	$[3h,4h)$	$[4h,5h)$	$[5h,6h)$	$[6h,7h)$	$[7h,8h)$	$[8h,9h)$	$[9h,1)$
f_i	99	94	95	108	108	88	111	92	111	94
e_i	100	100	100	100	100	100	100	100	100	100

not available, but the asymptotic normality of the distribution can be used: as $m \to \infty$, $\sqrt{2\chi_m^2} \to N\left(\sqrt{2m-1}, 1\right)$.

A disadvantage of the chi-squared test is that it is first necessary to test for the *independence* of the random variables, and, secondly, by dividing the domain into intervals, we are essentially testing against a discrete rather than a continuous uniform distribution. The test is adequate for a crude indication of uniformity. The reader is referred to the Kolmogorov–Smirnov test for a more powerful test.

2.4.2 Serial test

Consider a sequence of random numbers $R_1, R_2, \ldots$. Assuming uniformity and independence, the *nonoverlapping* pairs $(R_1, R_2), (R_3, R_4), \ldots$ should be uniformly and independently distributed over $(0,1)^2$. If there is serial dependence between consecutive numbers, this will be manifested as a clustering of points and the uniformity will be lost. Therefore, to investigate the possibility of serial dependence the null hypothesis that the pairs $(R_{2i-1}, R_{2i}), i = 1, 2, \ldots$, are uniformly distributed over $(0,1)^2$ can be tested against the alternative hypothesis that they are not. The chi-squared test can be applied, this time with k^2 subsquares each of area $1/k^2$. Nonoverlapping pairs are prescribed, as the chi-squared test demands independence. Let f_i denote the observed frequency in the ith subsquare, where $i = 1, 2, \ldots, k^2$, with $\sum_{i=1}^{k^2} f_i = n$, that is $2n$ random numbers in the sample. Under the null hypothesis, $e_i = n/k^2$ and the null distribution is $\chi_{k^2-1}^2$.

The assumption of independence between the n points in the sample is problematic, just as the assumption of independence between random numbers was in the frequency test. In both cases it may help to sample random numbers (points) that are not consecutive, but are some (random) distance apart in the generation scheme. Clearly, this serial correlation test is an empirical version of lattice tests. It can be extended to three and higher dimensions, but then the sample size will have to increase exponentially with the dimension to ensure that the expected frequency is at least five in each cell.

2.4.3 Other empirical tests

There is no limit to the number and type of empirical tests that can be devised, and it will always be possible to construct a test that will yield a statistically significant result in respect of some aspect of dependence. This must be the case since, among other reasons, the stream is the result of a deterministic recurrence. The important point is that 'gross' forms of nonrandomness are not present. It may be asked what constitutes 'gross'? It might be defined as those forms of dependence (or nonuniformity) that are detrimental to a particular Monte Carlo application. For example, in a k-dimensional definite integration, it is the uniformity of k-tuples $(R_i, R_{i+1}, \ldots, R_{i+k-1}), i = 0, k, 2k, \ldots$, that is important. In practice, the user of random numbers rarely has the time or inclination to check these aspects. Therefore, one must rely on random number generators that have been thoroughly investigated in the literature and have passed a battery of theoretical and empirical tests. Examples of empirical (statistical) tests are the gap test, poker test, coupon collector's test, collision test, runs test, and test of linear dependence. These and a fuller description of generating and testing random numbers appear in Knuth (1998) and Dagpunar (1988a). The Internet server 'Plab' is devoted to research on random number generation at the Mathematics Department, Salzburg University. It is a useful source of generation methods and tests and is located at http://random.mat.sbg.ac.at/.

2.5 Combinations of generators

By combining the output from several independent generators it is hoped to (a) increase the period and (b) improve the randomness of the output. Aspect (a) is of relevance when working in high dimensions. A generator developed by Wichman and Hill (1982, 1984) combines the output of three congruential generators:

$$X_{i+1} = (171X_i) \bmod 30269,$$

$$Y_{i+1} = (172Y_i) \bmod 30307,$$

$$Z_{i+1} = (170Z_i) \bmod 30323.$$

Now define

$$R_{i+1} = \left(\frac{X_{i+1}}{30269} + \frac{Y_{i+1}}{30307} + \frac{Z_{i+1}}{30323} \right) \bmod 1 \qquad (2.17)$$

where $y \bmod 1$ denotes the fractional part of a positive real y. Thus R_{i+1} represents the fractional part of the sum of three uniform variates. It is not too difficult to show that $R_{i+1} \sim U(0, 1)$ (see Problem 10).

Since the three generators are maximum period prime modulus, their periods are 30268, 30306, and 30322 respectively. The period of the combined generator is the least common multiple of the individual periods, which is $30268 \times 30306 \times 30324/4 \approx 6.95 \times 10^{12}$. The divisor of 4 arises as the greatest common divisor of the three periods is 2. This is a reliable if rather slow generator. It is used, for example, in Microsoft Excel2003 (http://support.microsoft.com).

2.6 The seed(s) in a random number generator

The seed X_0 of a generator provides the means of reproducing the random number stream when this is required. This is essential when comparing different experimental policies and also when debugging a program. This reproducibility at run-time eliminates the need to store and access a long list of random numbers, which would be both slow and take up substantial memory.

In most cases, *independent* simulation runs (replications, realizations) are required and therefore nonoverlapping sections of the random number sequence should be used. This is most conveniently done by using, as a seed, the last value used in the previous simulation run. If output observations *within* a simulation run are not independent, then this is not sufficient and a buffer of random numbers should be 'spent' before starting a subsequent run. In practice, given the long cycle lengths of many generators, an alternative to these strategies is simply to choose a seed randomly (for example by using the computer's internal clock) and hope that the separate sections of the sequence do not overlap. In that case, results will not be reproducible.

2.7 Problems

1. Generate by hand the complete cycle for the linear recurrences (a) to (f) below. State the observed period and verify that it is in agreement with theory.

 (a) $X_{i+1} = (5X_i + 3) \bmod 16$, $X_0 = 5$;

 (b) $X_{i+1} = (5X_i + 3) \bmod 16$, $X_0 = 7$;

 (c) $X_{i+1} = (7X_i + 3) \bmod 16$, $X_0 = 5$;

 (d) $X_{i+1} = (5X_i + 4) \bmod 16$, $X_0 = 5$;

 (e) $X_{i+1} = (5X_i) \bmod 64$, $X_0 = 3$;

 (f) $X_{i+1} = (5X_i) \bmod 64$, $X_0 = 4$.

2. Modify the procedure 'r1' in Section 2.1.1 to generate numbers in [0,1) using

 (a) $X_{i+1} = (7X_i) \bmod 61$, $X_0 = 1$;

 (b) $X_{i+1} = (49X_i) \bmod 61$, $X_0 = 1$.

 In each case observe the period and verify that it agrees with theory.

3. Verify the correctness of the procedure 'schrage' listed in Section 2.1.1.

4. Consider the maximum period prime modulus generator

$$X_{i+1} = 1000101X_i \bmod (10^{12} - 11)$$

with $X_0 = 53547507752$. Compute by hand $1000101X_0 \bmod (10^{12})$ and $\lfloor 1000101X_0/10^{12} \rfloor$. Hence find X_1 by hand calculation.

5. (a) Consider the multiplicative prime modulus generator

$$X_{i+1} = (aX_i) \bmod m$$

where a is a primitive root of m. Show that over the entire cycle

$$E(X_i) = \frac{m}{2},$$

$$\text{Var}(X_i) = \frac{m(m-2)}{12}.$$

[*Hint.* Use the standard results $\sum_{i=1}^{k} i = k(k+1)/2$ and $\sum_{i=1}^{k} i^2 = k(k+1)(2k+1)/6$.] Put $R_i = X_i/m$ and show that $E(R_i) = \frac{1}{2}\forall m$ and $\text{Var}(R_i) \to 1/12$ as $m \to \infty$.

(b) Let $f(r)$ denote the probability density function of R, a $U(0, 1)$ random variable. Then $f(r) = 1$ when $0 \le r \le 1$ and is zero elsewhere. Let μ and σ denote the mean and standard deviation of R. Show that

$$\mu = \int_0^1 f(r)\, dr = \frac{1}{2},$$

$$\sigma^2 = \int_0^1 (r - \mu)^2 f(r)\, dr = \frac{1}{12},$$

thereby verifying that over the *entire* sequence the generator in (a) gives numbers with the correct mean $\forall m$, and with almost the correct variance when m is large.

6. Show that 2 is a primitive root of 13. Hence find all multiplicative linear congruential generators with modulus 13 and period 12.

7. Consider the multiplicative generator

$$X_{i+1} = (aX_i) \bmod 2^b$$

where $a = 5 \bmod 8$. This has a cycle length $m/4 = 2^{b-2}$. The random numbers may be denoted by $R_i = X_i/2^b$. Now consider the mixed full period generator

$$X_{i+1}^* = (aX_i^* + c) \bmod 2^{b-2}$$

where $c = (X_0 \bmod 4)[(a-1)/4]$. Denote the random numbers by $R_i^* = X_i^*/2^{b-2}$. It is shown in Theorem 2.1 that

$$R_i = R_i^* + \frac{X_0 \bmod 4}{2^b}.$$

Verify this result for $b = 5$, $a = 13$, and $X_0 = 3$ by generating the entire cycles of $\{R_i\}$ and $\{R_i^*\}$.

8. The linear congruential generator obtains X_{i+1} from X_i. A *Fibonacci* generator obtains X_{i+1} from X_i and X_{i-1} in the following way:

$$X_{i+1} = (X_i + X_{i-1})\bmod m$$

where X_0 and X_1 are specified.

(a) Without writing out a complete sequence, suggest a good upper bound for the period of the generator in terms of m.

(b) Suppose $m = 5$. Only two cycles are possible. Find them and their respective periods and compare with the bound in (a). [Note that an advantage of the Fibonacci generator is that no multiplication is involved – just the addition modulo m. However, the output from such a generator is not too random as all the triples (X_{i-1}, X_i, X_{i+1}) lie on just two planes, $X_{i+1} = X_i + X_{i-1}$ or $X_{i+1} = X_i + X_{i-1} - m$. *Shuffling* the output from such a generator can considerably improve its properties.]

9. (a) Obtain the full cycle of numbers from the generators $X_{i+1} = (7X_i)\bmod 13$ and $Y_{i+1} = (Y_i + 5)\bmod 16$. Using suitable plots, compare the two generators in respect of uniformity of the overlapping pairs $\{(X_i, X_{i+1})\}$ and $\{(Y_i, Y_{i+1})\}$ for $i = 1, 2, \ldots$. What are the periods of the two generators?

(b) Construct a combined $U(0, 1)$ generator from the two generators in (a). The combined generator should have a period greater than either of the two individual generators. When $X_0 = 1$ and $Y_0 = 0$, use this generator to calculate the next two $U(0, 1)$ variates.

10. (a) If U and V are independently distributed random variables that are uniformly distributed in $[0,1)$ show that $(U + V)\bmod 1$ is also $U[0, 1)$. Hence justify the assertion that $R_{i+1} \sim U[0, 1)$ in equation (2.17).

(b) A random number generator in $[0, 1)$ is designed by putting

$$R_n = \left(\frac{X_n}{8} + \frac{Y_n}{7}\right)\bmod 1$$

where $X_0 = 0$, $Y_0 = 1$, $X_{n+1} = (9X_n + 3)\bmod 8$, and $Y_{n+1} = (3Y_n)\bmod 7$ for $n = 0, 1, \ldots$. Calculate $R_0, R_1, \ldots, R_5$. What is the period of the generator, $\{R_n\}$?

11. A $U(0, 1)$ random sequence $\{U_n, n = 0, 1, \ldots\}$ is

0.69	0.79	0.10	0.02	0.43	0.61	0.76	0.66	0.58	....

The pseudo-code below gives a method for shuffling the order of numbers in a sequence, where the first five numbers are entered into the array $\{T\{j\}, j = 0, \ldots, 4\}$.

Output T (4) into shuffled sequence
$N := \lfloor 4T(4) \rfloor$
$T(4) := T(N)$
Input next U from unshuffled sequence
$T(N) := U$

Obtain the first six numbers in the shuffled sequence.

12. Consider the generator $X_{n+1} = (5X_n + 3) \bmod 16$. Obtain the full cycle starting with $X_0 = 1$. Shuffle the output using the method described in Section 2.3 with $k = 6$. Find the first 20 integers in the shuffled sequence.

13. A frequency test of 10 000 supposed $U(0, 1)$ random numbers produced the following frequency table:

Interval	0–0.1	0.1–0.2	0.2–0.3	0.3–0.4	0.4–0.5	0.5–0.6
Frequency	1023	1104	994	993	1072	930

Interval	0.6–0.7	0.7–0.8	0.8–0.9	0.9–1.0
Frequency	1104	969	961	850

What are your conclusions?

3

General methods for generating random variates

In this chapter some general principles will be given for sampling from arbitrary univariate continuous and discrete distributions. Specifically, a sequence is required of independent realizations (random variates) $x_1, x_2, \ldots$ of a random variable X. If X is continuous then it will have a probability density function (p.d.f.) f_X or simply f. In the discrete case X will have a probability mass function (p.m.f.) p_x. In either case the cumulative distribution function is denoted by F_X or just F.

3.1 Inversion of the cumulative distribution function

If X is a continuous random variable with cumulative distribution function F and $R \sim U(0, 1)$, then the random variable $F^{-1}(R)$ has a cumulative distribution function that is F. To see this, let x denote any real value belonging to the support of f. Then

$$P\left[F^{-1}(R) \leq x\right] = P[0 \leq R \leq F(x)]$$

since F is strictly monotonic increasing on the support of f. Since $R \sim U(0, 1)$,

$$P[0 \leq R \leq F(x)] = F(x),$$

giving the required result.

Example 3.1 *Derive a method based on inversion for generating variates from a distribution with density*

$$f(x) = \lambda e^{-\lambda x}$$

on support $[0, \infty)$.

Simulation and Monte Carlo: With applications in finance and MCMC J. S. Dagpunar
© 2007 John Wiley & Sons, Ltd

Solution.   The distribution function is

$$F(x) = \int_0^x \lambda e^{-\lambda u} \, du = 1 - e^{-\lambda x}$$

for $x \geq 0$. Therefore

$$1 - e^{-\lambda X} = R$$

and so

$$X = F^{-1}(R) = -\frac{1}{\lambda} \ln(1 - R). \tag{3.1}$$

Equation (3.1) shows how to transform a uniform random number into a negative exponentially distributed random variable. Since R has the same distribution as $1 - R$ we could equally well use

$$X = -\frac{1}{\lambda} \ln(R). \tag{3.2}$$

This result will be used frequently. For example, to sample a random variate $x \sim \text{Exp}(\lambda)$ from a negative exponential distribution with $\lambda = \frac{1}{3}$,

$$x = -3 \ln R.$$

Therefore, if the next random number is $R = 0.1367$, the exponential variate is

$$x = -3 \ln(0.1367) = 5.970.$$

Inversion is somewhat limited in that many standard distributions do not have closed forms for F^{-1}. For example, applying inversion to a standard normal density yields

$$\int_{-\infty}^X \frac{1}{\sqrt{2\pi}} e^{-u^2/2} \, du = R.$$

This cannot be inverted analytically. It is true that X can be solved numerically. However, this approach is to be avoided if at all possible. It is likely to be much slower computationally, compared with other methods that will be developed. This is important as perhaps millions of such variates will be needed in a simulation. Table 3.1 shows the common standard continuous distributions that do have a closed form for F^{-1}.

Table 3.1   Continuous distributions

Name	$f(x)$	Parameters	Support
Weibull	$\alpha \lambda^\alpha x^{\alpha-1} \, e^{-(\lambda x)^\alpha}$	$\alpha > 0, \lambda > 0$	$[0, \infty)$
Logistic	$\dfrac{\lambda e^{-\lambda x}}{\left(1 + e^{-\lambda x}\right)^2}$	$\lambda > 0$	$(-\infty, \infty)$
Cauchy	$\dfrac{1}{\pi\left(1 + x^2\right)}$		$(-\infty, \infty)$

Turning now to discrete distributions, suppose that X is a discrete random variable with support $\{0, 1, \dots\}$ and cumulative distribution function F. Let $R \sim U(0, 1)$ and $W = \min\{x : R < F(x), x = 0, 1, \dots\}$. Then W has a cumulative distribution function F. To see this, note that $W = x$ $(x = 0, 1, \dots)$ if and only if $F(x - 1) \leq R < F(x)$. This happens with probability $F(x) - F(x - 1) = p_x$, as required.

Example 3.2 *Suppose* $p_x = \theta^{x-1}(1 - \theta), x = 1, 2, \dots,$ *where* $0 < \theta < 1$. *Derive a method of generating variates from this (geometric) distribution.*

Solution. Now

$$F(x) = \sum_{i=1}^{x} p_i = 1 - \theta^x$$

and so

$$X = \min\{x : R < 1 - \theta^x, x = 1, 2, \dots\}$$

or

$$X = \min\left\{x : x > \frac{\ln(1 - R)}{\ln \theta}, x = 1, 2, \dots\right\}$$

Replacing $1 - R$ by R as before gives

$$X = \left\lfloor \frac{\ln R}{\ln \theta} + 1 \right\rfloor$$

where $\lfloor \quad \rfloor$ is the floor function.

There are very few discrete distributions that can be inverted analytically like this. In general, and particularly for empirical distributions, it is usually necessary to search an array of cumulative probabilities, $\{F(0), F(1), \dots\}$. The Maple procedure below shows how to do this for any discrete distribution with a finite support, say $\{0, \dots, k\}$. Note how the parameter 'cdf' is specified to be of type *list*.

```
> discinvert:=proc(cdf::list)local R,x;
  R:=evalf(rand()/10^12);
  x:= 1;
  do;
  if R<cdf[x] then return x − 1 else x:= x+1 end if;
  end do;
  end proc;
```

The parameter 'cdf' is a *list* of the cumulative probabilities. Thus, for example, with $F[0] = 0.1, F[1] = 0.3, \dots, F[7] = 1$, invoking the procedure generates the random variate 2 as shown below:

```
> discinvert([.1, .3, .5, .55, .8, .95, .99, 1.0]);
2
```

Note in the procedure listing that the value $x - 1$ is returned, since in Maple the smallest subscript for any list is 1, yet the support of the distribution here is $\{0, \dots, 7\}$.

3.2 Envelope rejection

We wish to sample variates from the density f that is proportional to some non-negative function h, that is

$$f(x) = \frac{h(x)}{\int_{-\infty}^{\infty} h(x)\,dx},$$

and it is supposed that there is no closed form for the inverse F^{-1}. Choose another (*proposal*) density,

$$\frac{g(y)}{\int_{y \in \text{support}(g)} g(y)\,dy},$$

from which it is easy to sample variates. Additionally, choose g such that $\text{support}(h) \subseteq \text{support}(g)$ and $g(x) \geq h(x)\ \forall x \in \text{support}(g)$. Now generate a prospective variate, y say, from the proposal density. Then *accept* y with probability $h(y)/g(y)$. If it is not accepted it is rejected, in which case repeat the process until a prospective variate is accepted. The idea of the method is to alter the relative frequency distribution of y values, through the probabilistic rejection step, in such a way that the accepted ones have precisely the required distribution f. Note how the function g majorizes h. If, in addition, the two functions are equal at one or more points, then the graph of g *envelopes* that of h.

The following algorithm will generate a variate from a density proportional to h.

Algorithm 3.1 *1. Sample independent variates $y \sim g(y)/\int_{y \in \text{support}(g)} g(y)\,dy$ and $R \sim U(0,1)$. If $R < h(y)/g(y)$ accept y, else goto 1.*

Proof. To show the validity of the algorithm let Y have a density proportional to g. Then

$$P(y < Y \leq y + dy \,|\, Y \text{ is accepted})$$

$$\propto P(Y \text{ is accepted} \,|\, y < Y \leq y + dy)\, \frac{g(y)\,dy}{\int_{y \in \text{support}(g)} g(y)\,dy} \qquad (3.3)$$

$$\propto P\left(R < \frac{h(y)}{g(y)}\right) g(y)\,dy. \qquad (3.4)$$

Now $h(y)/g(y) \leq 1$, so $P(R < h(y)/g(y))$ is just $h(y)/g(y)$. Substituting into (3.4) it is found that

$$P(y < Y \leq y + dy \,|\, Y \text{ is accepted}) \propto h(y)\,dy.$$

It follows that accepted y values have the density f, as required.

The algorithm is equally valid for discrete distributions, once the integrals are replaced by sums. It is clear that the efficiency of the algorithm will be improved if the overall probability of accepting y values is large, and this is achieved by choosing g to be similar in shape to h. Since

$$P(Y \text{ is accepted} \mid y < Y \le y + dy) = \frac{h(y)}{g(y)},$$

the overall probability of acceptance is

$$\int_{y \in \text{support}(g)} \frac{h(y)}{g(y)} \frac{g(y)\,dy}{\int_{u \in \text{support}(g)} g(u)\,du} = \frac{\int_{y \in \text{support}(h)} h(y)\,dy}{\int_{y \in \text{support}(g)} g(y)\,dy}. \tag{3.5}$$

One of the merits of the algorithm is that the density f is required to be known only up to an arbitrary multiplicative constant. As we shall see in chapter 8 this is particularly useful when dealing with Bayesian statistics.

Example 3.3 *Devise an envelope rejection scheme for generating variates from a standard normal density using a negative exponential envelope.*

Solution. It will suffice to generate variates from a folded standard normal distribution with density

$$f(x) = \sqrt{\frac{2}{\pi}} e^{-x^2/2}$$

on support $[0, \infty)$, and then to multiply the variate by -1 with probability $\frac{1}{2}$. In the notation introduced above let

$$h(x) = e^{-x^2/2}$$

on support $[0, \infty)$. Consider an envelope

$$g(x) = K e^{-\lambda x}$$

on support $[0, \infty)$, where $K > 0$ and $\lambda > 0$. As g should majorize h, for given λ we must have

$$K e^{-\lambda x} \ge e^{-x^2/2}$$

$\forall x \in [0, \infty)$. Efficiency is maximized by choosing

$$K = \min \left\{ k : k e^{-\lambda x} \ge e^{-x^2/2} \ \forall x \ge 0 \right\}$$

$$= \min \left\{ k : k \ge e^{-(x-\lambda)^2/2} e^{\lambda^2/2} \ \forall x \ge 0 \right\}$$

$$= e^{\lambda^2/2}$$

Now, the overall probability of acceptance is, by Equation (3.5),

$$\frac{\int_{y \in \text{support}(h)} h(y)\,dy}{\int_{y \in \text{support}(g)} g(y)\,dy} = \frac{\int_0^\infty e^{-y^2/2}\,dy}{\int_0^\infty e^{\lambda^2/2} e^{-\lambda y}\,dy} = \frac{\sqrt{\pi/2}}{e^{\lambda^2/2} \lambda^{-1}}. \tag{3.6}$$

This is maximized when $\lambda = 1$, showing that the best exponential envelope is

$$g(y) = e^{1/2}e^{-y}.$$

Figure 3.1 shows how g envelopes h.

It remains to generate variates from the proposal density that is proportional to $g(y)$. This is

$$\frac{g(y)}{\int_{y \in \text{support}(g)} g(y)\,dy} = e^{-y}.$$

Given a random number $R_1 \sim U(0, 1)$ we put

$$y = -\ln(R_1).$$

Given a second random number R_2, the acceptance condition is

$$R_2 < \frac{h(y)}{g(y)} = \frac{e^{-y^2/2}}{e^{1/2}e^{-y}} = e^{-(y-1)^2/2},$$

which can be rewritten as

$$-\ln(R_2) > \frac{1}{2}(y-1)^2$$

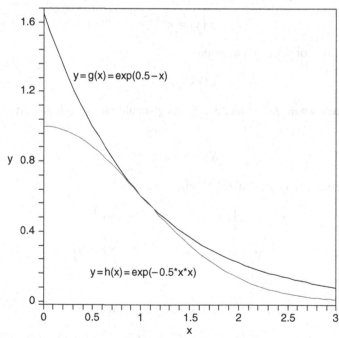

Envelope rejection for a folded normal
density with an exponential proposal density

$y = g(x) = \exp(0.5 - x)$

$y = h(x) = \exp(-0.5 \ast x \ast x)$

Figure 3.1   An exponential envelope for a folded standard normal density

Now $-\ln(R_2)$ and y are independent Exp (1) variates (i.e. from a negative exponential density with expectation 1). Writing these as E_2 and E_1 respectively, the condition is to deliver E_1 if and only if $E_2 > \frac{1}{2}(E_1 - 1)^2$. The Maple procedure below shows an implementation of the algorithm.

```
> stdnormal:=proc() local r1,r2,r3,E1,E2;
   do;
     r1:=evalf(rand()/10^12);
     r2:=evalf(rand()/10^12);
     E1:=-ln(r1);
     E2:=-ln(r2);
     if E2 > 0.5*(E1-1)^2 then break end if;
   end do;
   r3:=evalf(rand()/10^12);
   if r3 > 0.5 then E1:=-E1 end if;
   E1;
   end proc;
```

Note how a third random number, r3, is used to decide whether or not to negate the folded variate. A seed can now be specified for Maple's uniform generator 'rand'. The few lines of Maple below invoke the procedure 'stdnormal' to generate five variates:

```
> randomize(4765);
   for j from 1 to 5 do;
     stdnormal();
   end do;
```

```
4765
.4235427382
.1856983287
.6611634273
-.6739401269
-.9897380691
```

How good is this algorithm? From a probabilistic point of view it is fine. The accepted variates will have the required standard normal density, providing the uniform random number generator is probabilistically adequate. However, because a typical simulation requires so many variates, we are also concerned with the *speed* at which they can be generated. This depends upon various features of the algorithm. 'Expensive' calculations such as evaluating a logarithm or exponential are time consuming. To a lesser extent, so are multiplications and divisions. Also, the greater the number of uniform random numbers required to generate a random variate, the larger the execution time. Finally, if there are preliminary calculations associated with the parameters of the distribution, we would prefer this 'set-up' time to be small, or at least that a large number of variates be generated to justify this. For this standard normal density there is no set-up time. From Equation (3.6), the probability of acceptance is

$$\left.\frac{\sqrt{\pi/2}}{e^{\lambda^2/2}\lambda^{-1}}\right|_{\lambda=1} = \sqrt{\frac{\pi}{2e}} = 0.760.$$

Each prospective variate requires two random numbers and two logarithmic evaluations. Therefore, the expected number of logarithmic evaluations required per accepted variate is $2/0.760 = 2.63$ and the expected number of uniform random numbers is $2.63 + 1 = 3.63$, since one is required for unfolding the variate. It can be seen that such a procedure is rather inefficient compared with, for example, the inversion method for generating exponential variates. In that case each variate requires exactly one random number and one logarithmic evaluation. More efficient ways of generating standard normal variates will be discussed in Chapter 4.

Appendix 3.1 contains a procedure 'envelopeaccep' which computes the overall acceptance probability, given a target density (i.e. the density from which variates are to be sampled) proportional to $h(x)$ and an envelope proportional to $r(x)$.

3.3   Ratio of uniforms method

Suppose that there are two independent uniformly distributed random variables, $U \sim U(0, 1)$ and $V \sim U(-a, a)$, where a is a known positive constant. We can create a new random variable, $X = V/U$, from the ratio of these two uniforms. What is the distribution of X? The joint density of U and V is

$$f_{U,V}(u, v) = \frac{1}{2a}$$

over the support $0 < u < 1$ and $-a < v < a$. Now $v = xu$, where x is a realized value of X. Therefore the joint density of U and X is

$$f_{U,X}(u, x) = f_{X|U=u}(x, u)f_U(u)$$

$$= \left|\frac{\partial v}{\partial x}\right| f_{V|U=u}(xu, u)f_U(u)$$

$$= u f_{U,V}(u, xu)$$

$$= \frac{u}{2a}$$

over support $-a < xu < a$ and $0 < u < 1$. Integrating over u, the marginal density of X is

$$f_X(x) = \int_0^{\min(a/|x|,1)} \frac{u}{2a}du = \begin{cases} \dfrac{1}{4a} & (|x| < a), \\ \dfrac{a}{4x^2} & (|x| \geq a). \end{cases}$$

Such a density is bell-shaped, except that the top of the bell has been sliced off by a line that is parallel to the x axis.

Can a ratio of uniforms be used to obtain variates from any chosen distribution? The previous example selected a point (U, V) uniformly distributed over the rectangle $(0, 1) \times (-a, a)$. The following theorem due to Kinderman and Monahan (1977) modifies this by selecting (U, V) to be uniformly distributed over a subset of a rectangle.

Theorem 3.1 *Let h be any non-negative function with $\int_{x \in \text{support}(h)} h(x) \, dx < \infty$. Let $C = \left\{ (u, v) : 0 < u \leq \sqrt{h(v/u)}, v/u \in \text{support}(h) \right\}$. Let the point (U, V) be uniformly distributed over C and let $X = V/U$. Then the probability density function of X is $h(x) / \int_{-\infty}^{\infty} h(x) \, dx$.*

Proof. The joint density of U and V is

$$f_{U,V}(u, v) = A^{-1}$$

on support C, where $A = \int \int_C du \, dv$. Now put $X = V/U$. Then the joint density of U and X is

$$f_{U,X}(u, x) = u A^{-1}$$

on support $\left\{ (u, x) : 0 < u \leq \sqrt{h(x)}, x \in \text{support}(h) \right\}$. It follows that the marginal density of X is

$$f_X(x) = \int_0^{\sqrt{h(x)}} \frac{u \, du}{A} = \frac{h(x)}{2A}.$$

Since f is a density, we must have

$$1 = \frac{\int_{-\infty}^{\infty} h(x) \, dx}{2A} \tag{3.7}$$

and

$$f_X(x) = \frac{h(x)}{\int_{-\infty}^{\infty} h(x) \, dx},$$

as required.

The theorem leads directly to a method of generating random variates with the density proportional to h. The only technical problem is how to generate points uniformly over the region C. Usually, this is done by enclosing C within a rectangle D with sides parallel to the u and v axes. Generating points uniformly within D is easy. If such a point also falls in C it is accepted and V/U (a ratio of uniforms) is accepted; otherwise it is rejected.

Example 3.4 *Devise a ratio of uniforms algorithm for sampling from the density proportional to* $h(x) = \exp(-x)$ *on support* $[0, \infty)$.

Solution.   Let

$$
\begin{aligned}
C &= \left\{ (u, v) : 0 < u \le \sqrt{\exp\left(-\frac{v}{u}\right)}, \frac{v}{u} \in \text{support}(h) \right\} \\
&= \left\{ (u, v) : 0 < u \le \exp\left(-\frac{v}{2u}\right), \frac{v}{u} \in \text{support}(h) \right\} \\
&= \left\{ (u, v) : 0 < u \le 1, 0 \le v \le -2u \ln u \right\}
\end{aligned}
\tag{3.8}
$$

Note how C is the closed region bounded by the curves $v = 0, v = -2u \ln u$. We will enclose this region by the minimal rectangle $D = (0, u^+] \times (0, v^+]$, where $u^+ = 1$ and $v^+ = \max_{0<u\le 1}(-2u \ln u) = 2/e$. This leads to the algorithm

> *1. generate* $R_1, R_2 \sim U(0, 1)$
>
> $$ U := R_1, V := \frac{2}{e} R_2, X := \frac{V}{U} $$
>
> *If* $-2 \ln U \ge X$ *then deliver* X *else goto 1*

The algorithm generates a point (U, V) uniformly over D and accepts it only if it falls in C, that is if $0 < U \le \exp[-V/(2U)]$. The acceptance probability is

$$
\frac{\int \int_C du \, dv}{\int \int_D du \, dv} = \frac{A}{u^+ v^+}.
$$

From Equation (3.7) this is

$$
\frac{\frac{1}{2} \int_{-\infty}^{\infty} h(x) dx}{v^+ u^+} = \frac{\int_0^{\infty} \frac{1}{2} \exp(-x) dx}{v^+ u^+} = \frac{1/2}{2/e} = \frac{e}{4}.
$$

The following theorem allows a minimal enclosing rectangle to be found, given any non-negative valued function h.

Theorem 3.2   *Define* C *as in Theorem 3.1. Let*

$$ u^+ = \max_x \sqrt{h(x)}, $$

$$ v^+ = \max_{x>0} \left(x\sqrt{h(x)} \right), $$

$$ v^- = \min_{x\le 0} \left(x\sqrt{h(x)} \right). $$

Let $D = [0, u^+] \times [v^-, v^+]$. *Then* $C \subseteq D$.

Proof.   Suppose $(u, v) \in C$. Put $x = v/u$. Then $0 < u \le \sqrt{h(x)} \le u^+$. Now suppose $v > 0$. Then $x > 0$. Multiplying both sides of the inequality on u by v/u gives $v \le x\sqrt{h(x)} \le v^+$. Similarly, if $v \le 0$ then $x \le 0$ and $v \ge x\sqrt{h(x)} \ge v^-$. This shows that $(u, v) \in D$.

Example 3.5 *Devise a ratio of uniforms method for the standard normal distribution.*

Solution. Since $f(x) \propto \exp\left(-\frac{1}{2}x^2\right)$ we will take $h(x) = \exp\left(-\frac{1}{2}x^2\right)$. Therefore,

$$u^+ = \max_x \sqrt{\exp\left(-\frac{1}{2}x^2\right)} = 1,$$

$$v^+ = \max_{x>0}\left(x\sqrt{\exp\left(-\frac{1}{2}x^2\right)}\right) = \sqrt{\frac{2}{e}},$$

$$v^- = \min_{x\leq 0}\left(x\sqrt{\exp\left(-\frac{1}{2}x^2\right)}\right) = -\sqrt{\frac{2}{e}}.$$

Therefore the algorithm is

    1. *generate $R_1, R_2 \sim U(0,1)$.*

        $U =: R_1, V := \sqrt{2/e}\,(2R_2 - 1), X := V/U.$

        If $-\ln U \geq X^2/4$ *deliver X else goto 1.*

The acceptance probability is

$$\frac{\frac{1}{2}\int_{-\infty}^{\infty}\exp(-\frac{1}{2}x^2)dx}{2\sqrt{2/e}} = \frac{\sqrt{2\pi}}{4\sqrt{2/e}} = \frac{\sqrt{e\pi}}{4} = 0.731.$$

Figure 3.2 shows the region C in this case.

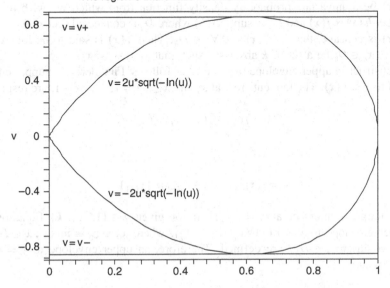

The region C in the ratio method for the standard normal

Figure 3.2 Ratio of uniforms: standard normal density

Note that any ratio algorithm requires knowledge only of the density up to a multiplicative constant, as in envelope rejection. However, in order to calculate the probability of acceptance, that constant must be known. Appendix 3.2 contains a procedure 'ratioaccep' which computes the overall acceptance probability for arbitrary p.d.f.s.

3.4   Adaptive rejection sampling

A feature of some simulations is that each time a random variate is required, it is from a *different* distribution. In envelope rejection we have seen that in order to sample from a density that is proportional to h an envelope function g must be found such that $g(x) \geq h(x) \forall x \in$ support (h). The (fixed) processing time to find such a function (it often involves numerical maximization of a function) can be considerably more than the extra time required to generate a successful proposal variate. In a sense, the 'fixed cost' of variate generation is high compared with the variable cost, and therefore the procedure is economic only if sufficient variates are generated from the distribution in question. Adaptive rejection sampling aims to make this fixed cost lower, which is of great benefit if just one variate is to be generated from the distribution in question. The method was derived by Gilks and Wild (1992) in response to variate generation in Gibbs sampling (see Chapter 8), where a single variate is required from each of a vast number of different nonstandard distributions.

The basic idea of adaptive rejection sampling is to use a piecewise exponential envelope $g(x) = \exp[u_k(x)]$, where $u_k(x)$ is a piecewise linear function based on k abscissae in the domain of h. The *adaptive* element refers to the insertion of extra pieces $(k := k+1)$ into the envelope, thereby making it a better fit to h. The usual form of the algorithm is applicable only to densities that are *log-concave*. Fortunately, many densities satisfy this condition.

Let $f(x)$ be a univariate probability density function from which we wish to sample variates. Let $h(x) \propto f(x)$ and $D_h \equiv$ support(h) where D_h is connected. Let $r(x) = \ln[h(x)]$ where $r(x)$ is concave, that is $r''(x) < 0 \; \forall x \in D_h$. Then $h(x)$ is said to be log-concave. Let $x_1, \ldots, x_k \in D_h$ be a set of k abscissae such that $x_1 < \cdots < x_k$.

Now construct an upper envelope to $y = e^{r(x)}$ as follows. First derive an upper envelope $y = u_k(x)$ to $y = r(x)$. The tangents to r at x_j and $x_{j+1}, j = 1, \ldots, k-1$, are respectively

$$y = r(x_j) + (x - x_j) r'(x_j) \tag{3.9}$$

and

$$y = r(x_{j+1}) + (x - x_{j+1}) r'(x_{j+1}). \tag{3.10}$$

Let these tangents intersect at $x = z_j$. Then for given $j \in \{1, \ldots, k\}$ Equation (3.9) is an upper envelope to $y = r(x) \forall x \in (z_{j-1}, z_j]$, where $x_0 \equiv z_0 \equiv \inf\{x : x \in D_h\}$ and $x_{k+1} \equiv z_k \equiv \sup\{x : x \in D_h\}$ are defined. Therefore, an upper envelope to $y = r(x)$ is $y = u_k(x)$, where

$$u_k(x) = r(x_j) + (x - x_j) r'(x_j), \quad x \in (z_{j-1}, z_j]$$

for $j = 1, \ldots, k$. From Equations (3.9) and (3.10),

$$z_j = \frac{r(x_j) - r(x_{j+1}) - x_j r'(x_j) + x_{j+1} r'(x_{j+1})}{r'(x_{j+1}) - r'(x_j)}$$

for $j = 1, \ldots, k - 1$. Figure 3.3 shows a typical plot of $y = u_2(x)$ and $y = r(x)$.

From this construction, $u_k(x) \geq r(x) \ \forall x \in D_h$, and it follows that $e^{r(x)} \leq e^{u_k(x)} \ \forall x \in D_h$. Now $u_k(x)$ is piecewise linear in x and so $y = e^{u_k(x)}$ is a piecewise exponential upper envelope to $y = h(x) = e^{r(x)}$ (see Figure 3.4).

A prospective variate Y is sampled from the density $e^{u_k(y)} / \int_{x_0}^{x_{k+1}} e^{u_k(x)} dx$, which is accepted with probability $e^{r(Y)} / e^{u_k(Y)} (\leq 1)$. Therefore, in the usual envelope rejection manner

$$P(y \leq Y < y + dy \text{ and } Y \text{ is accepted}) = \frac{e^{r(y)}}{e^{u_k(y)}} \frac{e^{u_k(y)} dy}{\int_{x_0}^{x_{k+1}} e^{u_k(x)} dx}$$

$$= \frac{e^{r(y)} dy}{\int_{x_0}^{x_{k+1}} e^{u_k(x)} dx},$$

showing that the density of accepted variates is proportional to $e^{r(y)}$, as required, and that the overall probability of acceptance is

$$\frac{\int_{x_0}^{x_{k+1}} e^{r(y)} dy}{\int_{x_0}^{x_{k+1}} e^{u_k(x)} dx} = \frac{\int_{x_0}^{x_{k+1}} h(y) \, dy}{\int_{x_0}^{x_{k+1}} e^{u_k(x)} dx}. \tag{3.11}$$

To illustrate, take a variate $X \sim$ gamma $(\alpha, 1)$ with a density proportional to

$$h(x) = x^{\alpha-1} e^{-x}$$

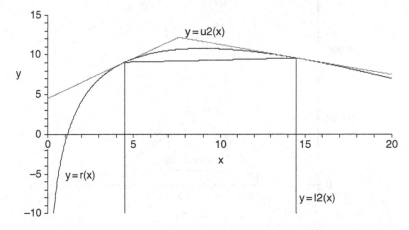

Adaptive rejection for standard gamma, alpha=10:
Piecewise linear upper and lower envelopes to r(x)
with two abscissae at x1 = 4.5, x2 = 14.5

Figure 3.3 Adaptive rejection with two abscissae : upper and lower envelopes to $r(x)$

on support $(0, \infty)$. It is easily verified that this is log-concave when $\alpha > 1$. Now,

$$r(x) = \ln[h(x)] = (\alpha - 1) \ln x - x.$$

Suppose $\alpha = 10$. Take two abscissae, one on either side of the mode $(x = 9)$, say at $x_1 = 4.5$ and $x_2 = 14.5$. The upper envelope $y = u_2(x)$ is constructed as shown in Figure 3.3.

Figure 3.4 shows the piecewise exponential envelope. Note how the overall probability of acceptance is given by a comparison of the areas under $\exp[u_2(x)]$ and $h(x)$ [see Equation (3.11)]. Suppose the prospective variate is $y = 11$ and that it is accepted. Then the procedure stops. Alternatively, if $y = 11$ is rejected, then a third abscissa, is introduced so that $x_1 = 4.5$, $x_2 = 11$, and $x_3 = 14.5$. To complete construction of the new envelope, $y = u_3(x)$, we will need $r(x_2) = r(11)$, but note that this has already been calculated in deciding to reject $y = 11$.

Figures 3.5 and 3.6 show the corresponding plots for the case of three abscissae. Note how the probability of acceptance in Figure 3.6 is now higher than in Figure 3.4. Experience with adaptive rejection algorithms shows that usually a variate is accepted with four or fewer abscissae. By that time the upper envelope is usually much closer to $h(x)$.

Note how the condition for acceptance is

$$R < \frac{\exp[r(y)]}{\exp[u_k(y)]}$$

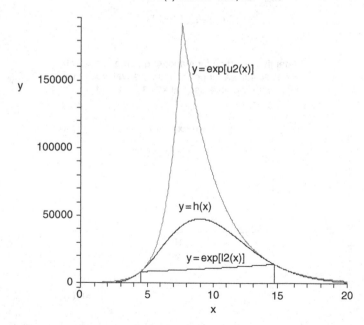

Adaptive rejection for standard gamma, alpha = 10:
Piecewise exponential upper and lower envelopes
to h(x) with x1 = 4.5, x2 = 14.5

Figure 3.4 Adaptive rejection with two abscissae : upper and lower envelopes to $\exp[r(x)]$

Adaptive rejection for standard gamma, alpha = 10:
The effect of a third abscissa. Piecewise linear upper and
lower envelopes to r(x) with x1 = 4.5, x2 = 11, x3 = 14.5

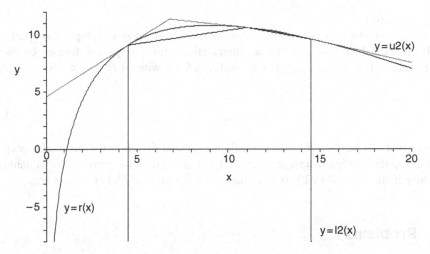

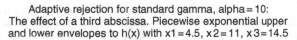

Figure 3.5 Adaptive rejection with three abscissae: upper and lower envelopes to $r(x)$

Adaptive rejection for standard gamma, alpha = 10:
The effect of a third abscissa. Piecewise exponential upper
and lower envelopes to h(x) with x1 = 4.5, x2 = 11, x3 = 14.5

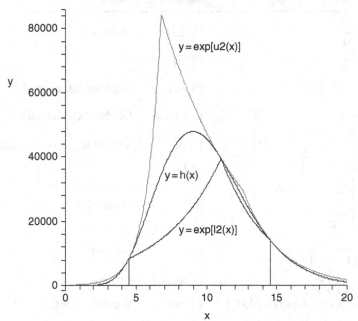

Figure 3.6 Adaptive rejection with three abscissae: upper and lower envelopes to $\exp[r(x)]$

where $R \sim U(0, 1)$. This reduces to

$$E > u_k(y) - r(y) \tag{3.12}$$

where $E \sim \text{Exp}(1)$.

A further refinement is to construct a piecewise linear *lower* envelope, $y = l_k(x)$, that is itself enveloped by $y = r(x)$. A convenient choice for $y = l_k(x)$ is formed by chords joining $(x_j, r(x_j))$ to $(x_{j+1}, r(x_{j+1}))$ for $j = 1, \ldots, k-1$ with $l_k(x) \equiv -\infty$ when $x < x_1$ or $x > x_k$. If

$$E > u_k(y) - l_k(y) \tag{3.13}$$

it follows that $E > u_k(y) - r(y)$ (as r envelopes l_k), and therefore the variate is accepted. This is computationally advantageous since $l_k(y)$ is often less expensive to evaluate than $r(y)$. Only if the *pre-test* (3.13) is not satisfied is the full test (3.12) carried out.

3.5 Problems

1. Use inversion of the cumulative distribution function to develop methods for sampling variates from each of the following probability density functions:

	$f(x)$	Support	Name
(a)	$\dfrac{1}{20}$	$[5, 25]$	Uniform
(b)	$\dfrac{x-2}{8}$	$(2, 6]$	
(c)	$\dfrac{1}{5}e^{-x/5}$	$[0, \infty)$	Exponential, mean 5
(d)	$\dfrac{1}{2}e^{-|x|}$	$(-\infty, \infty)$	Double exponential
(e)	$\dfrac{5}{2}\exp\left[-\dfrac{5}{2}(x-2)\right]$	$[2, \infty)$	Shifted negative exponential
(f)	$4x(1-x^2)$	$(0,1)$	
(g)	$\begin{cases} \dfrac{2x}{a} & (0 < x \le a) \\ \dfrac{2(1-x)}{1-a} & (a < x < 1) \end{cases}$	$(0,1)$	Triangular
(h)	$\dfrac{\exp(-x)}{[1+\exp(-x)]^2}$	$(-\infty, \infty)$	Logistic
(i)	$\dfrac{1}{\pi(1+x^2)}$	$(-\infty, \infty)$	Cauchy
(j)	$\alpha(\lambda x)^{\alpha-1}\lambda\exp[-(\lambda x)^\alpha]$	$(0, \infty)$	Weibull, $\lambda > 0, \alpha > 0$
(k)	$x\exp\left[-\dfrac{1}{2}(x^2-\theta^2)\right]$	(θ, ∞)	$\theta > 0$

2. Use inversion of the cumulative distribution function to develop a sampling method for the following discrete distributions:

   (a) $p_x = 1/(n+1)$ on support $\{0, 1, \ldots, n\}$, where n is a specified positive integer;

   (b)

p_x	0.1	0.2	0.4	0.2	0.1
x	1	2	3	4	5

   (c) binomial (3,0.9).

3. Derive an envelope rejection procedure for sampling variates from the density proportional to $h(x)$ where

$$h(x) = \frac{\exp\left(-x^2/2\right)}{1+|x|}$$

   on support $(-\infty, \infty)$. It may be assumed that there is access to a source of independent variates from a standard normal distribution. Comment on the efficiency of your method, given that $\int_{-\infty}^{\infty} h(x)dx = 1.5413$.

4. Derive envelope rejection algorithms for sampling from densities proportional to $h(x)$ below. In each case use an envelope $y = g(x)$ to the curve $y = h(x)$.

   (a) $h(x) = x(1-x)$ on support (0,1) using $g(x) = \text{constant}$;

   (b) $h(x) = \exp(-\frac{1}{2}x^2)$ on support $(-\infty, \infty)$ using $g(x) \propto (1+x^2)^{-1}$;

   (c) $h(x) = \exp\left(-\frac{1}{2}x^2\right)/(1+x^2)$ on support $(-\infty, \infty)$ using $g(x) \propto e^{-0.5x^2}$;

   (d) $h(x) = \exp\left(-\frac{1}{2}x^2\right)$ on support $[0, \infty)$ where $\theta > 0$, using $g(x) \propto x\exp\left[-\frac{1}{2}\left(x^2 - \theta^2\right)\right]$ as given in Problem 1, part (k).

5. A discrete random variable X has the probability mass function

$$P(X = x) = \frac{1/x}{\sum_{j=1}^{n} 1/j}$$

   on support $\{1, \ldots, n\}$, where n is a known integer ≥ 2.

   (a) Show that $X = \lfloor Y \rfloor$ where Y is a continuous random variable with density proportional to $h(y)$ on support $[1, n+1)$ and

$$h(y) = \frac{1}{\lfloor y \rfloor}.$$

   (b) Use an envelope g to h where

$$g(y) = \begin{cases} 1 & [1 \leq y < 2), \\ \frac{1}{y-1} & (2 \leq y < n+1), \end{cases}$$

   to establish the validity of the following Maple procedure for generating a value of the discrete random variable X.

```
> harmonic:=proc(n) local r1,r2,x,ix;
  do;
     r1:=evalf(rand()/10^12);
     if r1 < 1/(1+ln(n*1.0))
       then return 1
     else
       x := 1+(n*exp(1.0))^r1/exp(1.0);
       r2 := evalf(rand()/10^12);
       ix := floor(x);
       if r2 < (x−1)/ix then return ix end if;
     end if;
  end do;
  end proc;
```

(c) Show that a lower bound on the overall probability of acceptance is

$$\frac{1+\ln[(n+1)/2]}{1+\ln(n)}.$$

6. An envelope rejection procedure is required for generating variates from a density proportional to $h(x)$, where $h(x) = \cos^2(x)\exp(-\lambda x)$ on support $[0,\infty)$ and $\lambda > 0$.

   (a) Show that the optimal exponential envelope is $g(x) = \exp(-\lambda x)$ and that the overall probability of accepting a prospective variate is $(\lambda^2+2)/(\lambda^2+4)$.

   (b) Write a procedure in Maple suitable for sampling with any specified value of λ.

   (c) If inversion is to be used, show that, given $R \sim U(0,1)$, it is necessary to solve the equation

$$R = \frac{\left[\lambda^2\cos(2x) - 2\lambda\sin(2x) + \lambda^2 + 4\right]e^{-\lambda x}}{2(\lambda^2+2)}.$$

   for x. Which of the methods (b) or (c) do you prefer and why?

7. Envelope rejection can also be used with discrete distributions. Derive such a method for generating variates $\{x\}$ from a probability mass function proportional to $(1+x)\left(\frac{1}{2}\right)^x$ using an envelope proportional to $\left(\frac{2}{3}\right)^x$, both on support $\{0,1,2,\dots\}$. What is the probability of acceptance?

8. Devise ratio of uniforms algorithms for densities proportional to h below. In each case calculate the overall acceptance probability.

   (a) $h(x) = x^2(1-x)$ on support $(0,1)$. Can you suggest a method for improving the acceptance probability?

   (b) $h(x) = \sqrt{x}\exp(-x)$ on support $(0,\infty)$.

   (c) $h(x) = \left(1+\frac{x^2}{3}\right)^{-2}$ on support $(-\infty,\infty)$.

9. (a) Derive a ratio of uniform variates method for generating random variates from a density that is proportional to $h(x)$ where $h(x) = (1+x^2)^{-1}$.

   (b) Let U_1 and U_2 be independently distributed as a $U(-1, 1)$ density. Show from first principles that the density of U_2/U_1 subject to $U_1^2 + U_2^2 < 1$ is the Cauchy distribution with density

$$\frac{1}{\pi(1+x^2)}$$

   on support $(-\infty, \infty)$.

   (c) Let X and Y denote identically and independently distributed random variables with an $N(0, 1)$ density. Show from first principles that the density of Y/X is the Cauchy distribution of (b).

10. (This is a more difficult problem.) Let X have a density that is proportional to $h(x)$ and let

$$C = \left\{ (u, v) : 0 < u \le \sqrt{h\left(\frac{v}{u}\right)}, \frac{v}{u} \in \text{support}(h) \right\}.$$

Define a linear map,

$$T: \binom{u}{v} \mapsto A \binom{u}{v} \quad \text{where } A = \begin{pmatrix} 1 & 0 \\ -\tan\theta & 1 \end{pmatrix}$$

and define

$$C_\theta = \left\{ (r, s) : \binom{r}{s} = A \binom{u}{v}, (u, v) \in C \right\}.$$

   (a) Show that if (R, S) is uniformly distributed over C_θ then the density of S/R is proportional to $h(s/r + \tan\theta)$ and that consequently a variate x may be generated by taking $x = \tan\theta + s/r$.

   (b) Show that C_θ is obtained by rotating C through a clockwise angle θ about the origin and then applying a specified deformation.

   (c) Let $m = \text{mode}(X)$ and $\max\sqrt{h(x)} = u^+$. Show that the line $u = u^+$ touches the boundary of C at $(u^+, u^+ m)$.

   (d) Why might it be better to use the ratio method on C_θ rather than C?

   (e) If X has a symmetric distribution it may be conjectured that $\theta = \tan^{-1} m$ will maximize the overall probability of acceptance. However, for a nonsymmetric distribution this will not be generally true.

   If $h(x) = x^4(1-x)^{14}$ on support $(0,1)$ use Maple to compare the probabilities of acceptance using:

(i) the standard ratio method,

(ii) the relocated method where $Y = X - m = X - \frac{2}{9}$,

(iii) a numerical approximation to the optimal choice of θ.

11. Consider the following approach to sampling from a symmetric beta distribution with density

$$f_X(x) = \frac{\Gamma(2\alpha)}{\Gamma^2(\alpha)} x^{\alpha-1}(1-x)^{\alpha-1} \quad (0 \le x \le 1, \alpha > 1).$$

Put $Y = X - \frac{1}{2}$. Then the density of Y is proportional to $h(y)$ where

$$h(y) = \left(\frac{1}{4} - y^2\right)^{\alpha-1} \quad \left(-\frac{1}{2} \le y \le \frac{1}{2}, \alpha > 1\right).$$

Now show that the following algorithm will sample variates $\{x\}$ from the density $f_X(x)$. R_1 and R_2 are two independent $U(0, 1)$ random variables.

1. $U = \left(\frac{1}{2}\right)^{\alpha-1} R_1$

$$V = \frac{(1-\alpha^{-1})^{(\alpha-1)/2}(R_2 - \frac{1}{2})}{2^{\alpha-1}\sqrt{\alpha}}$$

$$Y = \frac{V}{U}$$

If $U < \left(\frac{1}{4} - Y^2\right)^{(\alpha-1)/2}$ *deliver* $Y + \frac{1}{2}$ *else goto* 1.

Using Maple to reduce the mathematical labour, show that the probability of acceptance is

$$\frac{\Gamma(\alpha)}{4\Gamma(\alpha+\frac{1}{2})}\sqrt{\pi}(\alpha-1)^{(1-\alpha)/2}\alpha^{\alpha/2}$$

Plot the probability of acceptance as a function of α and comment upon the efficiency of the algorithm. Show that for large α the acceptance probability is approximately

$$\frac{\sqrt{\pi e}}{4} = 0.731.$$

12. The adaptive rejection sampling method is applicable when the density function is log-concave. Examine whether or not the following densities are log-concave:

(a) normal;

(b) $f(x) \propto x^{\alpha-1}e^{-x}$ on support $(0, \infty)$, $\alpha > 0$;

(c) Weibull: $f(x) \propto \lambda^\alpha \alpha x^{\alpha-1} \exp[-(\lambda x)^\alpha]$ on support $(0, \infty)$, $\lambda > 0$, $\alpha > 0$;

(d) lognormal: $f(x) \propto 1/(\sigma x) \exp\left\{-\frac{1}{2}[(\ln x - \mu)/\sigma]^2\right\}$ on support $(0, \infty)$, $\sigma > 0$, $\mu \in \mathbb{R}$.

13. In adaptive rejection sampling from a density $f \propto h$, $r(x) = \ln[h(x)]$ must be concave. Given k ordered abscissae $x_0, \ldots, x_k \in \text{support}(h)$, the tangents to $y = r(x)$ at $x = x_j$ and $x = x_{j+1}$ respectively, intersect at $x = z_j$, for $j = 1, \ldots, k-1$. Let $x_0 \equiv z_0 \equiv \inf(x : x \in \text{support}(h))$ and $x_{k+1} \equiv z_k \equiv \sup(x : x \in \text{support}(h))$. Let $u_k(y)$, $x_0 < y < x_{k+1}$, be the piecewise linear hull formed from these tangents. Then u_k is an upper envelope to r. It is necessary to sample a prospective variate from the density

$$\phi(y) = \frac{\exp[u_k(y)]}{\int_{x_0}^{x_{k+1}} \exp[u_k(y)]dy}.$$

(a) Show that

$$\phi(y) = \sum_{j=1}^{k} p_j \phi_j(y)$$

where

$$\phi_j(y) = \begin{cases} \dfrac{\exp[u_k(y)]}{\int_{z_{j-1}}^{z_j} \exp[u_k(y)] \, dy}, & y \in (z_{j-1}, z_j], \\ 0, & y \notin (z_{j-1}, z_j], \end{cases}$$

and

$$p_j = \frac{\int_{z_{j-1}}^{z_j} \exp[u_k(y)]dy}{\int_{x_0}^{x_{k+1}} \exp[u_k(y)]dy}$$

for $j = 1, \ldots, k$.

(b) In (a), the density ϕ is represented as a *probability mixture*. This means that in order to sample a variate from ϕ a variate from the density ϕ_j is sampled with probability $p_j (j = 1, \ldots, k)$. Show that a variate Y from ϕ_j may be sampled by setting

$$Y = z_{j-1} + \frac{1}{r'(x_j)} \ln\left[1 - R + Re^{(z_j - z_{j-1})r'(x_j)}\right]$$

where $R \sim U(0, 1)$.

(c) Describe how you would randomly select in (b) the value of j so that a sample may be drawn from ϕ_j. Give an algorithm in pseudo-code.

14. In a single server queue with Poisson arrivals at rate $\lambda(< \mu)$ and service durations that are independently distributed as negative exponential with mean μ^{-1}, it can be shown that the distribution of waiting time W in the *queue*, when it has reached a steady state, is given by

$$P(W \geq w) = \frac{\lambda}{\mu} e^{-(\mu - \lambda)w}$$

when $w > 0$. Write a short Maple procedure to generate variates from this distribution (which is a mixture of discrete and continuous).

4

Generation of variates from standard distributions

4.1 Standard normal distribution

The standard normal distribution is so frequently used that it has its own notation. A random variable Z follows the standard normal distribution if its density is ϕ, where

$$\phi(z) = \frac{1}{\sqrt{2\pi}} e^{-z^2/2}$$

on support $(-\infty, \infty)$. The cumulative distribution is $\Phi(z) = \int_{-\infty}^{z} \phi(u) \, du$. It is easy to show that the expectation and variance of Z are 0 and 1 respectively.

Suppose $X = \mu + \sigma Z$ for any $\mu \in \mathbb{R}$ and $\sigma \geq 0$. Then X is said to be normally distributed with mean μ and variance σ^2. The density of X is

$$f_X(x) = \frac{1}{\sqrt{2\pi}\,\sigma} e^{-[(x-\mu)/\sigma]^2/2}$$

and for shorthand we write $X \sim N(\mu, \sigma^2)$.

Just two short algorithms will be described for generating variates. These provide a reasonable compromise between ease of implementation and speed of execution. The reader is referred to Devroye (1986) or Dagpunar (1988a) for other methods that may be faster in execution and more sophisticated in design.

4.1.1 Box–Müller method

The Box–Müller method is simple to implement and reasonably fast in execution. We start by considering two independent standard normal random variables, X_1 and X_2. The joint density is

$$f_{X_1, X_2}(x_1, x_2) = \frac{1}{2\pi} e^{-(x_1^2 + x_2^2)/2}$$

on support $\mathbb{R}^2$. Now transform to polars by setting $X_1 = R \cos \theta$ and $X_2 = R \sin \theta$. Then the joint density of R and θ is given by

$$f_{R,\theta}(r, \theta)\, dr\, d\theta = \frac{1}{2\pi} e^{-r^2/2} \begin{vmatrix} \dfrac{\partial x_1}{\partial r} & \dfrac{\partial x_1}{\partial \theta} \\ \dfrac{\partial x_2}{\partial r} & \dfrac{\partial x_2}{\partial \theta} \end{vmatrix} dr\, d\theta$$

$$= \frac{1}{2\pi} e^{-r^2/2} r\, dr\, d\theta$$

on support $r \in (0, \infty)$ and $\theta \in [0, 2\pi]$. It follows that R and θ are independently distributed with $\frac{1}{2} R^2 \sim \text{Exp}(1)$ and $\theta \sim U[0, 2\pi]$. Therefore, given two random numbers R_1 and R_2, on using inversion, $\frac{1}{2} R^2 = -\ln R_1$ or $R = \sqrt{-2 \ln R_1}$, and $\theta = 2\pi R_2$. Transforming back to the original Cartesian coordinates gives

$$X_1 = \sqrt{-2 \ln R_1} \cos (2\pi R_2),$$

$$X_2 = \sqrt{-2 \ln R_1} \sin (2\pi R_2).$$

The method delivers 'two for the price of one'. Although it is mathematically correct, it is not generally used in that form since it can produce fewer than expected tail variates (see Neave, 1973, and Problem 1). For example, using the sine form, an extreme tail value of X_2 is impossible unless R_1 is close to zero and R_2 is not. However, using a multiplicative linear congruential generator with a *small* multiplier will ensure that if R_1 is small then so is the next random number R_2. Of course, one way to avoid this problem is to shuffle the output from the uniform generator.

Usually a variant known as 'polar' Box–Müller is used without the problems concerning tail variates. This is now described. We recall that $X_1 = R \cos \theta$ and $X_2 = R \sin \theta$, where $\frac{1}{2} R^2 \sim \text{Exp}(1)$ and $\theta \sim U[0, 2\pi]$. Consider the point (U, V) uniformly distributed over the unit circle $C \equiv \{(u, v) : u^2 + v^2 \leq 1\}$. Then it is obvious that $\tan^{-1}(V/U) \sim U[0, 2\pi]$ and it is not difficult to show that $U^2 + V^2 \sim U(0, 1)$. Further, it is intuitive that these two random variables are independent (see Problem 2 for a derivation based on the Jacobian of the transformation). Therefore, $\tan^{-1}(V/U) = 2\pi R_2$ and $U^2 + V^2 = R_1$, where R_1 and R_2 are random numbers. Accordingly, (U, V) can be taken to be uniformly distributed over the square $D = \{(u, v) : -1 \leq u \leq 1, -1 \leq v \leq 1\}$. Subject to $U^2 + V^2 \leq 1$ we return two independent standard normal variates as

$$X_1 = \sqrt{-2 \ln (U^2 + V^2)} \frac{U}{\sqrt{U^2 + V^2}},$$

$$X_2 = \sqrt{-2 \ln (U^2 + V^2)} \frac{V}{\sqrt{U^2 + V^2}}.$$

A Maple procedure 'STDNORM' appears in Appendix 4.1.

4.1.2 An improved envelope rejection method

In Example 3.3 a rejection method was developed for sampling from a folded normal. Let us now see whether the acceptance probability of that method can be improved. In the usual notation for envelope rejection we let

$$h(x) = e^{-x^2/2}$$

on support $(-\infty, \infty)$ and use a majorizing function

$$g(x) = \begin{cases} 1 & (|x| \le c), \\ e^{-\lambda(|x|-c)} & (|x| > c), \end{cases}$$

for suitable $\lambda > 0$ and $c > 0$. Now

$$\frac{g(x)}{h(x)} = \begin{cases} e^{x^2/2} & (|x| \le c), \\ e^{x^2/2} - \lambda|x| + \lambda c & (|x| > c). \end{cases}$$

We require g to majorize h, so λ and c must be chosen such that $x^2/2 - \lambda|x| + \lambda c \ge 0 \; \forall |x| > c$. The probability of acceptance is

$$\frac{\int_{-\infty}^{\infty} h(x)\,dx}{\int_{-\infty}^{\infty} g(x)\,dx} = \frac{\sqrt{2\pi}}{2(c+1/\lambda)}$$

Therefore, we must minimize $c + 1/\lambda$ subject to $x^2/2 - \lambda|x| + \lambda c \ge 0 \forall |x| > c$. Imagine that c is given. Then we must maximize λ subject to $x^2/2 - \lambda|x| + \lambda c \ge 0 \; \forall \; |x| > c$. If $\lambda^2 > 2\lambda c$, that is if $\lambda > 2c$, then $x^2/2 - \lambda|x| + \lambda c < 0$ for some $|x| > c$. Since $x^2/2 - \lambda|x| + \lambda c \ge 0 \forall |x|$ when $\lambda \le 2c$, it follows that the maximizing value of λ, given c, is $\lambda = 2c$. This means that we must minimize $c + 1/(2c)$, that is set $c = \sqrt{2}/2$ and $\lambda = 2c = \sqrt{2}$, giving an acceptance probability of

$$\frac{\sqrt{2\pi}}{2\left(1/\sqrt{2} + 1/\sqrt{2}\right)} = \frac{\sqrt{\pi}}{2} = 0.88623.$$

Note that this is a modest improvement on the method due to Butcher (1961), where $c = 1$ and $\lambda = 2$ with an acceptance probability of $\sqrt{2\pi}/3 = 0.83554$, and also on that obtained in Example 3.3.

The prospective variate is generated by inversion from the proposal cumulative distribution function,

$$G(x) = \begin{cases} \dfrac{(1/\sqrt{2})e^{\sqrt{2}x+1}}{1/\sqrt{2} + 2(1/\sqrt{2}) + (1/\sqrt{2})} = \frac{1}{4}e^{\sqrt{2}x+1} & \left(x < -\dfrac{\sqrt{2}}{2}\right), \\[3ex] \dfrac{1/\sqrt{2} + x + 1/\sqrt{2}}{1/\sqrt{2} + 2(1/\sqrt{2}) + 1/\sqrt{2}} = \frac{\sqrt{2}x+2}{4} & \left(|x| \le \dfrac{\sqrt{2}}{2}\right), \\[3ex] \dfrac{4/\sqrt{2} - (1/\sqrt{2})e^{-\sqrt{2}x+1}}{1/\sqrt{2} + 2(1/\sqrt{2}) + 1/\sqrt{2}} = 1 - \frac{1}{4}e^{-\sqrt{2}x+1} & \left(x > \dfrac{\sqrt{2}}{2}\right). \end{cases}$$

Applying inversion, given a random number R_1, we have

$$x = \begin{cases} \dfrac{\sqrt{2}[\ln(4R_1) - 1]}{2} & \left(R_1 \leq \dfrac{1}{4}\right), \\[3mm] \sqrt{2}(2R_1 - 1) & \left(\dfrac{1}{4} < R_1 \leq \dfrac{3}{4}\right), \\[3mm] \dfrac{\sqrt{2}[1 - \ln\{4(1 - R_1)\}]}{2} & \left(\dfrac{3}{4} < R_1\right). \end{cases}$$

Given a second random number R_2, a prospective variate x is accepted if

$$R_2 < \begin{cases} e^{-x^2/2} & (|x| \leq c), \\ e^{-(x^2/2 - \lambda|x| + \lambda c)} & (|x| > c), \end{cases}$$

or, on putting $E = -\ln(R_2)$,

$$E > \begin{cases} x^2/2 & \left(|x| \leq \frac{1}{\sqrt{2}}\right), \\ x^2/2 - \sqrt{2}|x| + 1 = (x - \sqrt{2})^2/2 & \left(|x| > \frac{1}{\sqrt{2}}\right). \end{cases} \tag{4.1}$$

The acceptance probability is high, but there is still at least one (expensive) logarithmic evaluation, E, for each prospective variate. Some of these can be avoided by noting the bound

$$\frac{1}{R_2} - 1 \geq -\ln R_2 \geq 1 - R_2.$$

Let us denote the right-hand side of the inequality in (4.1) by W. Suppose that $1 - R_2 > W$. Then $-\ln R_2 > W$ and the prospective variate is accepted. Suppose $-\ln R_2 \not> W$ but $1/R_2 - 1 \leq W$. Then $-\ln R_2 \leq W$ and the prospective variate is rejected. Only if both these (inexpensive) pre-tests fail is the decision to accept or reject inconclusive. On those few occasions we test explicitly for $-\ln R_2 > W$. In the terminology of Marsaglia (1977), the function $-\ln R_2$ is *squeezed* between $1/R_2 - 1$ and $1 - R_2$.

4.2 Lognormal distribution

This is a much used distribution in financial mathematics. If $X \sim N(\mu, \sigma^2)$ and $Y = e^X$ then Y is said to have a lognormal distribution. Note that Y is always positive. To sample a Y value just sample an X value and set $Y = e^X$. It is useful to know that the expectation and standard deviation of Y are

$$\mu_Y = e^{\mu + \sigma^2/2}$$

and

$$\sigma_Y = E(Y)\sqrt{e^{\sigma^2} - 1}$$

respectively. Clearly

$$P(Y < y) = P(X < \ln y)$$

$$= \Phi \left(\frac{\ln y - \mu}{\sigma} \right)$$

so that

$$f_Y(y) = \frac{1}{\sigma y} \phi \left(\frac{\ln y - \mu}{\sigma} \right)$$

$$= \frac{1}{\sqrt{2\pi}\sigma y} e^{-[(\ln y - \mu)/\sigma]^2/2}$$

on support $(0, \infty)$.

4.3 Bivariate normal density

Suppose $X \sim N(\mu_1, \sigma_1^2)$ and $Y \sim N(\mu_2, \sigma_2^2)$, and the conditional distribution of Y given that $X = x$ is

$$N \left(\mu_2 + \rho \frac{\sigma_2}{\sigma_1}(x - \mu_1), \sigma_2^2(1 - \rho^2) \right) \tag{4.2}$$

where $-1 \le \rho \le 1$. Then the correlation between X and Y is ρ and the conditional distribution of X given Y is

$$N \left(\mu_1 + \rho \frac{\sigma_1}{\sigma_2}(y - \mu_2), \sigma_1^2(1 - \rho^2) \right).$$

The vector (X, Y) is said to have a bivariate normal distribution. In order to generate such a vector two *independent* standard normal variates are needed, $Z_1, Z_2 \sim N(0, 1)$. Set

$$x = \mu_1 + \sigma_1 Z_1$$

and (from 4.2)

$$y = \mu_2 + \rho \sigma_2 Z_1 + \sigma_2 \sqrt{1 - \rho^2} \, Z_2.$$

In matrix terms,

$$\begin{pmatrix} x \\ y \end{pmatrix} = \begin{pmatrix} \mu_1 \\ \mu_2 \end{pmatrix} + \begin{pmatrix} \sigma_1 & 0 \\ \rho\sigma_2 & \sigma_2\sqrt{1 - \rho^2} \end{pmatrix} \begin{pmatrix} Z_1 \\ Z_2 \end{pmatrix}. \tag{4.3}$$

Later it will be seen how this lower triangular structure for the matrix in the right-hand side of Equation (4.3) also features when generating $n - variate$ normal vectors, where $n \ge 2$.

4.4   Gamma distribution

The gamma distribution with *shape parameter* α and *scale parameter* λ has the density

$$f_Z(z) = \frac{(\lambda z)^{\alpha-1}\lambda e^{-\lambda z}}{\Gamma(\alpha)} \qquad (z > 0) \tag{4.4}$$

where $\lambda > 0$, $\alpha > 0$ and $\Gamma(\alpha)$ is the *gamma function* given by

$$\Gamma(\alpha) = \int_0^\infty v^{\alpha-1}e^{-v}dv.$$

This has the property that $\Gamma(\alpha) = (\alpha-1)\Gamma(\alpha-1)$, $\alpha > 1$, and $\Gamma(\alpha) = (\alpha-1)!$ when α is an integer. The notation $Z \sim$ gamma (α, λ) will be used.

The density (4.4) may be reparameterized by setting $X = \lambda Z$. Therefore,

$$f_X(x) = \frac{x^{\alpha-1}e^{-x}}{\Gamma(\alpha)} \tag{4.5}$$

and $X \sim$ gamma$(\alpha, 1)$. Thus we concentrate on sampling from Equation (4.5) and set $Z = X/\lambda$ to deliver a variate from Equation (4.4). The density (4.5) is monotonically decreasing when $\alpha \leq 1$, and has a single mode at $x = \alpha - 1$ when $\alpha > 1$. The case $\alpha = 1$ describes a negative exponential distribution. The density (4.5) therefore represents a family of distributions and is frequently used when we wish to model a *non-negative* random variable that is positively skewed.

When α is *integer* the distribution is known as the *special Erlang* and there is an important connection between X and the negative exponential density with mean one. It turns out that

$$X = E_1 + E_2 + \cdots + E_\alpha \tag{4.6}$$

where $E_1, \ldots, E_\alpha$ are independent random variables with density $f_{E_i}(x) = e^{-x}$. Since the mean and variance of such a negative exponential are both known to be unity, from (4.6)

$$E(X) = \text{Var}(X) = \alpha$$

and therefore, in terms of the original gamma variate Z,

$$E(Z) = \frac{\alpha}{\lambda}$$

and

$$\text{Var}(Z) = \text{Var}\left(\frac{X}{\lambda}\right) = \frac{\alpha}{\lambda^2}$$

It can easily be shown that these are also the moments for nonintegral α. The representation (4.6) allows variates (for the α integer) to be generated from unit negative exponentials. From Equation (3.2),

$$X = \sum_{i=1}^{\alpha} -\ln R_i = -\ln\left(\prod_{i=1}^{\alpha} R_i\right). \tag{4.7}$$

Result (4.7) is very convenient to implement for small integer α, but is clearly inefficient for large α. Also, in that case, we must test that the product of random numbers does not underflow; that is the result must not be less than the smallest real that can be represented on the computer. However, that will not be a problem with Maple.

When α is *noninteger* a different method is required (which will also be suitable for integer α).

4.4.1 Cheng's log-logistic method

An algorithm will be constructed to sample from a density proportional to $h(x)$ where $h(x) = x^{\alpha-1}e^{-x}$, $x \geq 0$. The envelope rejection method of Cheng (1977) will be used, which samples prospective variates from a distribution with density proportional to $g(x)$ where

$$g(x) = \frac{Kx^{\lambda-1}}{(\mu + x^\lambda)^2} \tag{4.8}$$

where $\mu = \alpha^\lambda$, $\lambda = \sqrt{2\alpha - 1}$ when $\alpha \geq 1$, and $\lambda = \alpha$ when $\alpha < 1$. Since g must envelope h, for the maximum probability of acceptance, K is chosen as

$$K = \max_{x \geq 0}\left\{x^{\alpha-1}e^{-x}\left[\frac{x^{\lambda-1}}{(\mu + x^\lambda)^2}\right]^{-1}\right\}. \tag{4.9}$$

It is easily verified that the right-hand side of Equation (4.9) is maximized at $x = \alpha$. Therefore K is chosen as

$$K = \alpha^{\alpha-1}e^{-\alpha}(\alpha^\lambda + \alpha^\lambda)^2 \alpha^{1-\lambda} = 4\alpha^{\alpha+\lambda}e^{-\alpha}.$$

According to Equation (3.5) the acceptance probability is

$$\frac{\int_0^\infty h(y)\,dy}{\int_0^\infty g(y)\,dy} = \frac{\Gamma(\alpha)}{K/(\lambda\mu)} = \frac{\lambda\mu\Gamma(\alpha)}{4\alpha^{\alpha+\lambda}e^{-\alpha}} = \frac{\lambda\Gamma(\alpha)e^\alpha}{4\alpha^\alpha}.$$

Figure 4.1 shows the acceptance probability as a function of α. It is increasing from $\frac{1}{4}$ (at $\alpha = 0^+$) to $2/\sqrt{\pi} = 0.88623$ as $\alpha \to \infty$ (use Stirling's approximation, $\Gamma(\alpha)e^{\alpha-1} \sim \sqrt{2\pi}(\alpha - 1)^{\alpha-1/2}$). Note that for $\alpha > 1$ the efficiency is uniformly high. An advantage of the method, not exhibited by many other gamma generators, is that it can still be used when $\alpha < 1$, even though the shape of the gamma distribution is quite different in such cases (the density is unbounded as $x \to 0$ and is decreasing in x).

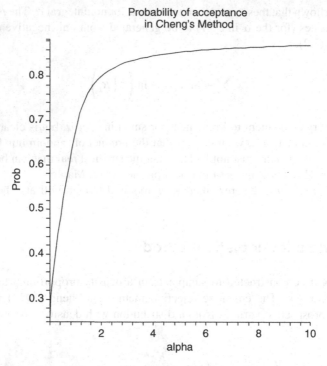

Figure 4.1   Gamma density: acceptance probability and shape parameter value

It remains to derive a method for sampling from the prospective distribution. From Equation (4.8) the cumulative distribution function is

$$G(x) = \frac{1/\mu - 1/(\mu + x^\lambda)}{1/\mu} = \frac{x^\lambda}{\mu + x^\lambda}$$

and so, given a random number R_1, the prospective variate is obtained by solving $G(x) = 1 - R_1$ (this leads to a tidier algorithm than R_1). Therefore,

$$x = \alpha \left(\frac{1 - R_1}{R_1} \right)^{1/\lambda}. \tag{4.10}$$

Given a random number R_2, the acceptance condition is

$$R_2 < \frac{h(x)}{g(x)} = \frac{x^{\alpha - \lambda} e^{-x} \left(\alpha^\lambda + x^\lambda \right)^2}{\alpha^{\alpha - \lambda} e^{-\alpha} (\alpha^\lambda + \alpha^\lambda)^2} = \left(\frac{x}{\alpha} \right)^{\alpha - \lambda} e^{\alpha - x} \left(\frac{1 + (x/\alpha)^\lambda}{2} \right)^2$$

or, on using Equation (4.10),

$$\ln \left(4 R_1^2 R_2 \right) < (\alpha - \lambda) \ln \left(\frac{x}{\alpha} \right) + \alpha - x.$$

4.5 Beta distribution

A continuous random variable X is said to have a beta distribution on support $(0,1)$ with shape parameters α (> 0) and β (> 0) if its probability density function is

$$f(x) = \frac{\Gamma(\alpha+\beta)x^{\alpha-1}(1-x)^{\beta-1}}{\Gamma(\alpha)\Gamma(\beta)} \quad (0 \le x \le 1).$$

A shorthand is $X \sim \text{beta}(\alpha, \beta)$. A method of generating variates $\{x\}$ that is easy to remember and to implement is to set

$$x = \frac{w}{w+y}$$

where $w \sim \text{gamma}(\alpha, 1)$ and $y \sim \text{gamma}(\beta, 1)$ with w and y independent. To show this, we note that the joint density of w and y is proportional to

$$e^{-(w+y)} w^{\alpha-1} y^{\beta-1}.$$

Therefore the joint density of W and X is proportional to

$$e^{-w/x} w^{\alpha-1} \left[\frac{w(1-x)}{x} \right]^{\beta-1} \left| \frac{\partial y}{\partial x} \right| = e^{-w/x} w^{\alpha+\beta-2}(1-x)^{\beta-1} x^{1-\beta} \left| \frac{-w}{x^2} \right|.$$

Integrating over w, the marginal density of X is proportional to

$$(1-x)^{\beta-1} x^{1-\beta-2} \int_0^\infty e^{-w/x} w^{\alpha+\beta-1} \, dw = (1-x)^{\beta-1} x^{-\beta-1} \int_0^\infty e^{-a}(ax)^{\alpha+\beta-1} x \, da$$

$$\propto (1-x)^{\beta-1} x^{\alpha-1};$$

that is X has the required beta density. Providing an efficient gamma generator is available (e.g. that described in Section 4.4.1), the method will be reasonably efficient.

4.5.1 Beta log-logistic method

This method was derived by Cheng (1978). Let $\rho = \alpha/\beta$ and let Y be a random variable with a density proportional to $h(y)$ where

$$h(y) = y^{\alpha-1}(1+\rho y)^{-\alpha-\beta} \quad (y \ge 0). \tag{4.11}$$

Then it is easy to show that

$$X = \frac{\rho y}{1+\rho y} \tag{4.12}$$

has the required beta density. To sample from Equation (4.11) envelope rejection with a majorizing function g is used where

$$g(y) = \frac{K y^{\lambda-1}}{[\mu + (\rho y)^\lambda]^2} \quad (y \ge 0),$$

$\mu = \rho^{\lambda}, \lambda = \min\ (\alpha, \beta)$ if $\min\ (\alpha, \beta) \leq 1$, and $\lambda = \sqrt{(2\alpha\beta - \alpha - \beta)/(\alpha + \beta - 2)}$ otherwise. Now choose the smallest K such that $g(y) \geq h(y)\ \forall y \in [0, \infty)$. Thus

$$K = \max_{y>0}\left[\frac{y^{\alpha-1}(1+\rho y)^{-\alpha-\beta}}{y^{\lambda-1}/(\mu+[\rho y]^{\lambda})^2}\right].$$

The maximum occurs at $y = 1$, and so given a random number R_2 the acceptance condition is

$$R_2 < \frac{\dfrac{y^{\alpha-1}(1+\rho y)^{-\alpha-\beta}}{y^{\lambda-1}/(1+y^{\lambda})^2}}{\dfrac{y^{\alpha-1}(1+\rho y)^{-\alpha-\beta}}{y^{\lambda-1}/(1+y^{\lambda})^2}\Big|_{y=1}}$$

$$= y^{\alpha-\lambda}\left(\frac{1+\rho y}{1+\rho}\right)^{-\alpha-\beta}\left(\frac{1+y^{\lambda}}{2}\right)^2.$$

Now, to generate a prospective variate from a (log-logistic) density proportional to g, we first obtain the associated cumulative distribution function as

$$G(y) = \frac{\displaystyle\int_0^y \frac{v^{\lambda-1}dv}{[\mu+(\rho v)^{\lambda}]^2}}{\displaystyle\int_0^\infty \frac{v^{\lambda-1}dv}{[\mu+(\rho v)^{\lambda}]^2}}$$

$$= \frac{-[\mu+(\rho v)^{\lambda}]^{-1}\Big|_0^y}{-[\mu+(\rho v)^{\lambda}]^{-1}\Big|_0^\infty}$$

$$= 1-\left(1+y^{\lambda}\right)^{-1}.$$

Setting $1 - R_1 = G(y)$ gives

$$y = \left(\frac{1-R_1}{R_1}\right)^{1/\lambda}.$$

Therefore the acceptance condition becomes

$$4R_1^2 R_2 < y^{\alpha-\lambda}\left(\frac{1+\rho}{1+\rho y}\right)^{\alpha+\beta}.$$

The method is unusual in that the probability of acceptance is bounded from below by $\frac{1}{4}$ for all $\alpha > 0$ and $\beta > 0$. When both parameters are at least one the probability of acceptance is at least $e/4$. A Maple procedure 'beta' appears in Appendix 4.2.

4.6 Chi-squared distribution

A random variable χ_n^2 has a chi-squared distribution with n degrees of freedom if its p.d.f. is

$$f_{\chi_n^2}(x) = \frac{e^{-x/2}x^{n/2-1}}{2^{n/2}\Gamma(n/2)} \quad (x \geq 0)$$

where n is a positive integer. It is apparent that $\chi_n^2 \sim$ gamma $(n/2, 1/2)$. It is well known that

$$\chi_n^2 = \sum_{i=1}^{n} Z_i^2$$

where $Z_1, \ldots, Z_n$ are i.i.d. $N(0, 1)$. However, that is unlikely to be an efficient way of generation unless n is small and odd-valued. When n is small and even-valued $\chi_n^2 = -2 \ln (R_1 \cdots R_{n/2})$ could be used. For all other cases a gamma variate generator is recommended.

A *non-central* chi-squared random variable arises in the following manner. Suppose $X_i \sim N(\mu_i, 1)$ for $i = 1, \ldots, n$, and these random variables are independent. Let $Y = \sum_{i=1}^{n} X_i^2$. By obtaining the moment generating function of Y it is found that the density depends only upon n and $\theta = \sum_{i=1}^{n} \mu_i^2$. For given θ, one parameter set is $\mu_i = 0$ for $i = 1, \ldots, n-1$ and $\mu_n = \sqrt{\theta}$, but then $Y = \sum_{i=1}^{n-1} Z_i^2 + X_n^2$. Therefore, a convenient way of generating $\chi_{n,\theta}^2$, a chi-squared random variable with n degrees of freedom and a noncentrality parameter θ, is to set

$$\chi_{n,\theta}^2 = \chi_{n-1}^2 + \left(Z + \sqrt{\theta}\right)^2$$

where χ_{n-1}^2 and $Z \sim N(0, 1)$ are independent.

4.7 Student's *t* distribution

This is a family of continuous symmetric distributions. They are similar in shape to a standard normal density, but have fatter tails. The density of a random variable, T_n, having a Student's *t* distribution with n $(n = 1, 2, \ldots)$ degrees of freedom is proportional to $h_n(t)$ where

$$h_n(t) = \left(1 + \frac{t^2}{n}\right)^{-(n+1)/2}$$

on support $(-\infty, \infty)$. It is a standard result that $T_n = Z/\sqrt{\chi_n^2/n}$ where Z is independently $N(0, 1)$. This provides a convenient but perhaps rather slow method of generation.

A faster method utilizes the connection between a t-variate and a variate from a shifted symmetric beta density (see, for example, Devroye 1996). Suppose X has such a distribution with density proportional to $r(x)$ where

$$r(x) = \left(\frac{1}{2}+x\right)^{n/2-1} \left(\frac{1}{2}-x\right)^{n/2-1}$$

on support $\left(-\frac{1}{2}, \frac{1}{2}\right)$ where $n = 1, 2, \ldots$. Let

$$Y = \frac{\sqrt{n}X}{\sqrt{\frac{1}{4}-X^2}}. \tag{4.13}$$

Then Y is monotonic in X and therefore the density of Y is proportional to

$$\left|\frac{dx}{dy}\right| r\left(x[y]\right).$$

Now, $|dx/dy| = (n/2)(n+y^2)^{-3/2}$ and $r\left(x[y]\right)$ is proportional to $\left(1+y^2/n\right)^{1-n/2}$. Therefore, the density of Y is proportional to $h_n(y)$, showing that Y has the same distribution as T_n.

It only remains to generate a variate x from the density proportional to $r(x)$. This is done by subtracting $\frac{1}{2}$ from a beta $(n/2, n/2)$ variate. The log-logistic method of Section 4.5.1 leads to an efficient algorithm. Alternatively, for $n > 1$, given two uniform random numbers R_1 and R_2, we can use a result due to Ulrich (1984) and set

$$X = \frac{\sqrt{1 - R_1^{2/(n-1)}} \cos(2\pi R_2)}{2}. \tag{4.14}$$

On using Equation (4.13) it follows that

$$T_n = \frac{\sqrt{n} \cos(2\pi R_2)}{\sqrt{\left(1 - R_1^{2/(n-1)}\right)^{-1} - \cos^2(2\pi R_2)}}. \tag{4.15}$$

The case $n = 1$ is a Cauchy distribution, which of course can be sampled from via inversion. A Maple procedure, 'tdistn', using Equation (4.15) and the Cauchy appears in Appendix 4.3.

A *noncentral* Student's t random variable with n degrees of freedom and noncentrality parameter δ is defined by

$$T_{n,\delta} = \frac{X}{\sqrt{\chi_n^2/n}}$$

where $X \sim N(\delta, 1)$, independent of χ_n^2.

A *doubly* noncentral Student's t random variable with n degrees of freedom and noncentrality parameters δ and θ is defined by

$$T_{n,\delta,\theta} = \frac{X}{\sqrt{\chi_{n,\theta}^2/n}}.$$

Since $\chi^2_{n,\theta} = \chi^2_{n+2j}$ with probability $e^{-\theta/2} (\theta/2)^j /j!$ (see, for example, Johnson *et al.*, 1995, pp. 435–6) it follows that

$$T_{n,\delta,\theta} = \frac{X}{\sqrt{\chi^2_{n+2j}/n}}$$

$$= \frac{X\sqrt{n/(n+2j)}}{\sqrt{\chi^2_{n+2j}/(n+2j)}}$$

with the same probability, and therefore

$$T_{n,\delta,\theta} = \sqrt{\frac{n}{n+2J}} T_{n+2J,\delta}$$

where $J \sim$ Poisson $(\theta/2)$.

4.8 Generalized inverse Gaussian distribution

Atkinson (1982) has described this as an enriched family of gamma distributions. The random variable $Y \sim$ gig (λ, ψ, χ) (where gig is a generalized inverse gaussian distribution) if it has a p.d.f. proportional to

$$y^{\lambda-1} \exp\left[-\frac{1}{2}\left(\psi y + \frac{\chi}{y}\right)\right] \quad (y \geq 0) \tag{4.16}$$

where $\chi > 0, \psi \geq 0$ if $\lambda < 0, \chi > 0, \psi > 0$ if $\lambda = 0$, and $\chi \geq 0, \psi > 0$ if $\lambda > 0$. It is necessary to consider only the case $\lambda \geq 0$, since gig $(-\lambda, \psi, \chi) = 1/\text{gig}(\lambda, \chi, \psi)$. Special cases of note are

$$\text{gig } (\lambda, \psi, 0) = \text{gamma}\left(\lambda, \frac{\psi}{2}\right) \quad (\lambda > 0)$$

$$\text{gig }\left(-\frac{1}{2}, \psi, \chi\right) = \text{ig } (\psi, \chi)$$

where ig is an inverse Gaussian distribution. If $W \sim N\left(\mu + \phi\sigma^2, \sigma^2\right)$ where $\sigma^2 \sim$ gig $(\lambda, \theta^2 - \phi^2, \chi)$, where $\theta > |\phi| \geq 0$, then the resulting mixture distribution for W belongs to the generalized hyperbolic family (Barndorff-Nielson, 1977). Empirical evidence suggests that this provides a better fit than a normal distribution to the return (see Section 6.5) on an asset (see, for example, Eberlein, 2001).

Atkinson (1982) and Dagpunar (1989b) devised sampling methods for the generalized inverse Gaussian. However, the following method is extremely simple to implement and gives good acceptance probabilities over a wide range of parameter values. We exclude the cases $\chi = 0$ or $\psi = 0$, which are gamma and reciprocal gamma variates respectively.

Reparameterizing (4.16), we set $\alpha = \sqrt{\psi/\chi}$ and $\beta = \sqrt{\psi\chi}$ and define $X = \alpha Y$. Then the p.d.f. of X is proportional to $h(x)$ where

$$h(x) = x^{\lambda-1} \exp\left[-\frac{\beta}{2}\left(x+\frac{1}{x}\right)\right] \quad (x \geq 0).$$

Now select an envelope $g(x) \propto r(x)$ where

$$r(x) = x^{\lambda-1} \exp\left[-\frac{\gamma x}{2}\right] \quad (x \geq 0)$$

and $\gamma < \beta$. A suitable envelope is $g(x) = Kr(x)$ where

$$K = \max_{x \geq 0} \left\{\frac{h(x)}{r(x)}\right\}$$

$$= \max_{x \geq 0} \left\{\exp\left[-\frac{\beta}{2}\left(x+\frac{1}{x}\right)+\frac{\gamma x}{2}\right]\right\}$$

$$= \exp\left[-\sqrt{\beta(\beta-\gamma)}\right].$$

Therefore an algorithm is:

1. Sample $X \sim$ gamma $(\lambda, \frac{\gamma}{2}), R \sim U(0,1)$
   If $-\ln R > \frac{(\beta-\gamma)X}{2} + \frac{\beta}{2X} - \sqrt{\beta(\beta-\gamma)}$ then deliver X else goto 1

It follows from Equation (3.5) that the overall acceptance probability is

$$\frac{\int_0^\infty h(x)\,dx}{\int_0^\infty g(x)\,dx} = \frac{\int_0^\infty x^{\lambda-1}\exp[-(\beta/2)(x+1/x)]\,dx}{K\int_0^\infty x^{\lambda-1}\exp(-\gamma x/2)\,dx}$$

$$= \frac{\int_0^\infty x^{\lambda-1}\exp[-(\beta/2)(x+1/x)]\,dx}{\exp\left[-\sqrt{\beta(\beta-\gamma)}\right]\Gamma(\lambda)(2/\gamma)^\lambda}$$

$$= \frac{\exp\left[\sqrt{\beta(\beta-\gamma)}\right](\gamma/2)^\lambda \int_0^\infty x^{\lambda-1}\exp[-(\beta/2)(x+1/x)]\,dx}{\Gamma(\lambda)}$$

This is maximized, subject to $\gamma < \beta$ when

$$\gamma = \frac{2\lambda^2\left(\sqrt{1+\beta^2/\lambda^2}-1\right)}{\beta}$$

A Maple procedure, 'geninvgaussian', using this optimized envelope appears in Appendix 4.4. Using this value of γ, the acceptance probabilities for specimen values of λ and β are shown in Table 4.1. Given the simplicity of the algorithm, these are adequate apart from when λ is quite small or β is quite large. Fortunately, complementary behaviour is exhibited in Dagpunar (1989b), where the ratio of uniforms method leads to an algorithm that is also not computer intensive.

Table 4.1 Acceptance probabilities for sampling from generalized inverse Gaussian distributions

$\lambda\backslash\beta$	0.2	0.4	1	5	10	20
0.1	0.319	0.250	0.173	0.083	0.060	0.042
0.5	0.835	0.722	0.537	0.264	0.189	0.135
1	0.965	0.909	0.754	0.400	0.287	0.205
5	0.9995	0.998	0.988	0.824	0.653	0.482
20	0.99997	0.9999	0.9993	0.984	0.943	0.837

4.9 Poisson distribution

The random variable $X \sim \text{Poisson}(\mu)$ if its p.d.f. is

$$f(x) = \frac{\mu^x e^{-\mu}}{x!} \quad (x = 0, 1, 2, \dots)$$

where $\mu > 0$. For all but large values of μ, *unstored inversion* of the c.d.f. is reasonably fast. This means that a partial c.d.f. is computed each time the generator is called, using the recurrence

$$f(x) = \frac{\mu f(x-1)}{x}$$

for $x = 1, 2, \dots$ where $f(0) = e^{-\mu}$. A variant using *chop-down search*, a term conceived of by Kemp (1981), eliminates the need for a variable in which to store a cumulative probability. The algorithm below uses such a method.

$W := e^{-\mu}$
Sample $R \sim U(0, 1)$
$X := 0$
While $R > W$ do
$R := R - W$
$X := X + 1$
$W := \dfrac{W\mu}{X}$
End do
Deliver X

A Maple procedure, 'ipois1' appears in Appendix 4.5. For large values of μ a rejection method with a logistic envelope due to Atkinson (1979), as modified by Dagpunar (1989a), is faster. The exact switching point for μ at which this becomes faster than inversion depends upon the computing environment, but is roughly 15/20 if μ is fixed between successive calls of the generator and 20/30 if μ is reset.

4.10 Binomial distribution

A random variable $X \sim$ binomial (n, p) if its p.m.f. is

$$f(x) = \binom{n}{x} p^x q^{n-x} \quad (x = 0, 1, \ldots, n)$$

where n is a positive integer, $0 \le p \le 1$, and $p + q = 1$. Using unstored inversion a generation algorithm is

If $p > 0.5$ then $p := 1 - p$
$W := q^n$
Sample $R \sim U(0, 1)$
$X := 0$
While $R > W$ do
$R := R - W$
$X := X + 1$
$W := \dfrac{W(n - X + 1)p}{qX}$
End do
If $p \le 0.5$ deliver X else deliver $n - X$

The expected number of comparisons to sample one variate is $E(X + 1) = 1 + np$. That is why the algorithm returns n – binomial (n, q) if $p > 0.5$. A Maple procedure, 'ibinom', appears in Appendix 4.6. Methods that have a bounded execution time as $n \min(p, q) \to \infty$ include those of Fishman (1979) and Ahrens and Dieter (1980). However, such methods require significant set-up time and therefore are not suitable if the parameters of the binomial are reset between calls, as, for example, in Gibbs sampling (Chapter 8).

4.11 Negative binomial distribution

A discrete random variable $X \sim$ negbinom (k, p) if its p.m.f. is f where

$$f(x) = \frac{\Gamma(x + k) p^x q^k}{\Gamma(k)\Gamma(x + 1)} \quad (x = 0, 1, 2, \ldots)$$

and where $0 < p < 1$, $p + q = 1$, and $k > 0$. If k is an integer then $X + k$ represents the number of Bernoulli trials to the kth 'success' where q is the probability of success on any trial. Therefore, X is the sum of k independent geometric random variables, and so could be generated using

$$X = \sum_{i=1}^{k} \left\lfloor \frac{\ln(R_i)}{\ln p} \right\rfloor$$

where $R_i \sim U(0, 1)$. However, this is likely to be slow for all but small k. When k is real valued then unstored inversion is fast unless the expectation of X, kp/q, is large.

A procedure, 'negbinom', appears in Appendix 4.7. An extremely convenient, but not necessarily fast, method uses the result that

$$\text{negbinom}(k, p) = \text{Poisson}\left[\frac{p}{q}\text{gamma}(k, 1)\right].$$

A merit of this is that since versions of both Poisson and gamma generators exist with bounded execution times, the method is robust for all values of k and p.

4.12 Problems

1. (This is a more difficult problem.) Consider the Box–Müller generator using the sine formula only, that is

$$x = \sqrt{-2\ln R_1}\sin(2\pi R_2).$$

Further, suppose that R_1 and R_2 are obtained through a maximum period prime modulus generator

$$R_1 = \frac{j}{m}$$

$$R_2 = \frac{(aj)\bmod m}{m}$$

for any $j \in [1, m-1]$, where $a = 131$ and $m = 2^{31} - 1$. Show that the largest positive value of X that can be generated is approximately $\sqrt{2\ln 524} = 3.54$ and that the largest negative value is approximately $-\sqrt{2\ln[(4 \times 131)/3]} = -3.21$. In a random sample of 2^{30} such variates, corresponding to the number of variates produced over the entire period of such a generator, calculate the expected frequency of random variates that take values (i) greater than 3.54 and (ii) less than −3.21 respectively. Note the deficiency of such large variates in the Box–Müller sample, when used with a linear congruential generator having a small multiplier. This can be rectified by increasing the multiplier.

2. Let (U, V) be uniformly distributed over the unit circle $C \equiv \{(u, v) : u^2 + v^2 \leq 1\}$. Let $Y = U^2 + V^2$ and $\Theta = \tan^{-1}(V/U)$. Find the joint density of Y and Θ. Show that these two random variables are independent. What are the marginal densities?

3. Write a Maple procedure for the method described in Section 4.1.2.

4. Under standard assumptions, the price of a share at time t is given by

$$X(t) = X(0)\exp\left[\left(\mu - \frac{1}{2}\sigma^2\right)t + \sigma\sqrt{t}Z\right],$$

where $X(0)$ is known, μ and σ are the known growth rates and volatility, and $Z \sim N(0, 1)$.

(a) Use the standard results for the mean and variance of a lognormally distributed random variable to show that the mean and standard deviation of $X(t)$ are given by

$$E[X(t)] = X(0)e^{\mu t}$$

and

$$\sigma[X(t)] = X(0)e^{\mu t}\sqrt{\exp(\sigma^2 t) - 1}.$$

(b) Write a Maple procedure that samples 1000 values of $X(0.5)$ when $X(0) = 100$ pence, $\mu = 0.1$, and $\sigma = 0.3$. Suppose a payoff of $\max[0, 110 - X(0.5)]$ is received at $t = 0.5$. Estimate the expected payoff.

(c) Suppose the prices of two shares A and B are given by

$$X_A(t) = 100\exp\left[\left(0.1 - \frac{1}{2}0.09\right)t + 0.3\sqrt{t}\,Z_A\right]$$

and

$$X_B(t) = 100\exp\left[\left(0.08 - \frac{1}{2}0.09\right)t + 0.3\sqrt{t}\,Z_B\right]$$

where Z_A and Z_B are standard normal variates with correlation 0.8. A payoff at time 0.5 is $\max[0, 320 - X_A(0.5) - 2X_B(0.5)]$. Estimate the expected payoff, using randomly sampled values of the payoff.

5. Derive an envelope rejection procedure for a gamma $(\alpha, 1)$ distribution using a negative exponential envelope. What is the best exponential envelope and how does the efficiency depend upon α? Is the method suitable for large α? Why can the method not be used when $\alpha < 1$?

6. To generate variates $\{x\}$ from a standard gamma density where the shape parameter α is less than one, put $W = X^\alpha$. Show that the density of W is proportional to $\exp(-w^{1/\alpha})$ on support $(0, \infty)$. Design a Maple algorithm for generating variates $\{w\}$ using the minimal enclosing rectangle. Show that the probability of acceptance is $[e/(2\alpha)]^\alpha [\Gamma(\alpha+1)/2]$. How good a method is it? Could it be used for shape parameters of value at least one?

7. Let $\alpha > 0$ and $\beta > 0$. Given two random numbers R_1 and R_2, show that conditional upon $R_1^{1/\alpha} + R_2^{1/\beta} \leq 1$, the random variable $R_1^{1/\alpha}/\left(R_1^{1/\alpha} + R_2^{1/\beta}\right)$ has a beta density with parameters α and β. This method is due to Jöhnk (1964).

8. Suppose that $0 < \alpha < 1$ and $X = WY$ where W and Y are independent random variables that have a negative exponential density with expectation one and a beta density with shape parameters α and $1 - \alpha$ respectively. Show that $X \sim gamma(\alpha, 1)$. This method is also due to Jöhnk (1964).

9. Let $\alpha > 0$, $\beta > 0$, $\rho = \alpha/\beta$, and Y be a random variable with a density proportional to $h(y)$ where

$$h(y) = y^{\alpha-1}(1+\rho y)^{-\alpha-\beta}$$

on support $(0, \infty)$. Prove the result in Equation (4.12), namely that the random variable $X = \rho y/(1+\rho y)$ has a beta density with parameters α and β.

10. (This is a more difficult problem.) Let Θ and R be two independent random variables distributed respectively as $U(0, 2\pi)$ and density f on domain $(0, \infty)$. Let $X = R\cos\Theta$ and $Y = R\sin\Theta$. Derive the joint density of X and Y and hence or otherwise show that the marginal densities of X and Y are

$$\int_0^\infty \frac{f\left(\sqrt{x^2+y^2}\right)}{\pi\sqrt{x^2+y^2}}\,dy \quad \text{and} \quad \int_0^\infty \frac{f\left(\sqrt{y^2+x^2}\right)}{\pi\sqrt{y^2+x^2}}\,dx$$

respectively. This leads to a generalization of the Box–Müller method (Devroye, 1996).

(a) Hence show that if $f(r) = r\exp\left(-r^2/2\right)$ on support $(0, \infty)$ then X and Y are independent standard normal random variables.

(b) Show that if $f(r) \propto r\left(1-r^2\right)^{c-1}$ on support $(0, 1)$, then the marginal densities of X and Y are proportional to $\left(1-x^2\right)^{c-1/2}$ and $\left(1-y^2\right)^{c-1/2}$ respectively on support $(-1, 1)$. Now put $c = (n-1)/2$ where n is a positive integer. Show that the random variables $\sqrt{n}X/\sqrt{1-X^2}$ and $\sqrt{n}Y/\sqrt{1-Y^2}$ are distributed as Student's t with n degrees of freedom. Are these two random variables independent? Given two random numbers R_1 and R_2 derive the result (4.15).

11. Refer to Problem 8(b) in Chapter 3. Derive a method of sampling variates from a density proportional to $h(x)$ where

$$h(x) = \sqrt{x}\,e^{-x}$$

on support $(0, \infty)$. The method should not involve a rejection step.

5

Variance reduction

It is a basic fact of life that in any sampling experiment (including Monte Carlo) the standard error is proportional to $1/\sqrt{n}$ where n is the sample size. If *precision* is defined as the reciprocal of standard deviation, it follows that the amount of computational work required varies as the square of the desired precision. Therefore, design methods that alter the constant of proportionality in such a way as to increase the precision for a given amount of computation are well worth exploring. These are referred to as variance reduction methods. Among the methods that are in common use and are to be discussed here are those involving *antithetic variates, importance sampling, stratified sampling, control variates*, and *conditional Monte Carlo*. Several books contain a chapter or more on variance reduction and the reader may find the following useful: Gentle (2003), Hammersley and Handscomb (1964), Law and Kelton (2000), Ross (2002), and Rubinstein (1981).

5.1 Antithetic variates

Suppose that $\widehat{\theta}_1$ is an unbiased estimator of some parameter θ where $\widehat{\theta}_1$ is some function of a known number m of random numbers $R_1, \ldots, R_m$. Since $1 - R_i$ has the same distribution as R_i (both $U(0, 1)$) another unbiased estimator can be constructed simply by replacing R_i by $1 - R_i$ in $\widehat{\theta}_1$ to give a second estimator $\widehat{\theta}_2$. Call the two simulation runs giving these estimators the *primary* and *antithetic* runs respectively. Now take the very particular case that $\widehat{\theta}_1$ is a linear function of $R_1, \ldots, R_n$. Then

$$\widehat{\theta}_1 = a_0 + \sum_{i=1}^{m} a_i R_i \tag{5.1}$$

and

$$\widehat{\theta}_2 = a_0 + \sum_{i=1}^{m} a_i (1 - R_i).$$

Now consider the combined estimator

$$\widehat{\theta} = \frac{\widehat{\theta}_1 + \widehat{\theta}_2}{2}.$$

Then

$$\widehat{\theta} = a_0 + \frac{1}{2}\sum_{i=1}^{m} a_i, \tag{5.2}$$

which is clearly the value of θ and does not depend upon $\{R_i\}$. The variance of the estimator has been reduced to zero!

In practice, of course, the estimator is unlikely to be linear (and, if it were, simulation would be unnecessary), but the above shows that if the response from a simulation is approximately linear in the random numbers, then some good variance reduction can be expected. For any function (not just linear ones) the combined estimator $\widehat{\theta}$ is clearly unbiased and has variance

$$\text{var}(\widehat{\theta}) = \frac{1}{4}(\sigma^2 + \sigma^2 + 2\rho\sigma^2)$$

$$= \frac{1}{2}\sigma^2(1+\rho) \tag{5.3}$$

where σ^2 is the common variance of $\widehat{\theta}_1$ and $\widehat{\theta}_2$ and ρ is the correlation between them. Putting $\rho = 0$ in Equation (5.3), the variance of the average of two *independent* estimators is simply $\frac{1}{2}\sigma^2$. The variance ratio of 'naive' Monte Carlo to one employing this variance reduction device is therefore $1 : (1+\rho)$, and the corresponding quotient will be referred to as the *variance reduction ratio* (v.r.r.). Therefore,

$$\text{v.r.r.} = \frac{1}{1+\rho}.$$

The hope is that the use of antithetic variates will induce negative correlation between the responses in primary and antithetic runs, leading to a variance reduction ratio greater than 1. A correlation of -1 gives a variance reduction ratio of infinity, and clearly this is the case for linear response functions, as in Equation (5.1).

By way of an example, let us return to the Monte Carlo estimate of θ where

$$\theta = \int_0^\infty x^{0.9} e^{-x}\, dx$$

previously met in Section 1.1. Given a random number $R_i \sim U(0, 1)$, an unbiased estimator of θ is

$$[-\ln(R_i)]^{0.9}. \tag{5.4}$$

An antithetic estimator is

$$[-\ln(1 - R_i)]^{0.9}, \tag{5.5}$$

so an unbiased combined estimator is

$$\frac{1}{2}\left\{[-\ln(R_i)]^{0.9}+[-\ln(1-R_i)]^{0.9}\right\}. \tag{5.6}$$

In Appendix 5.1 the Maple procedure 'theta1_2' samples m independent values of (5.4). By replacing R_i by $1-R_i$ in the procedure and using the *same* seed, m further values of (5.5) are sampled. Alternatively, procedure 'theta_combined' samples m independent values of (5.6), performing the two sets of sampling in one simulation run. Each procedure returns the sample mean (i.e. an estimate of θ) and the estimated standard error (e.s.e.) of the estimate. With $m = 1000$, using the same random number stream (note the use of 'randomize( )') it is found that

$$\widehat{\theta}_1 = 0.9260$$

$$\text{e.s.e.}\left(\widehat{\theta}_1\right) = 0.0280$$

$$\widehat{\theta}_2 = 1.0092$$

$$\text{e.s.e.}\left(\widehat{\theta}_2\right) = 0.0277$$

$$\widehat{\theta} = 0.9676$$

$$\text{e.s.e.}\left(\widehat{\theta}\right) = 0.0107$$

The true value of θ obtained through numerical integration is 0.9618 ($e1$ in the Maple worksheet), so the antithetic design has improved the estimate of both the primary and antithetic runs. Using the individual values within the primary and antithetic runs, the Maple command 'describe[linearcorrelation]' shows the sample correlation coefficient to be $\widehat{\rho} = -0.70353$, giving an estimated variance reduction ratio of $(1-0.7035)^{-1} = 3.373$. Another way to estimate the v.r.r. is to note that an estimate of $(1+\rho)^{-1}$ from Equation (5.3) is

$$\widehat{\text{v.r.r.}} = \frac{\frac{1}{4}\left\{\widehat{\text{Var}(\widehat{\theta}_1)}+\widehat{\text{Var}\left(\widehat{\theta}_1\right)}\right\}}{\widehat{\text{Var}\left(\widehat{\theta}\right)}} \tag{5.7}$$

$$= 3.373 \tag{5.8}$$

The numerator in (5.7) is merely an unbiased estimate of $\sigma^2/2$, the variance using a single naive simulation of 2000 variates (note that (5.8) will be slightly biased as it is a ratio estimate). Finally, using numerical integration, the worksheet shows the exact value of ρ to be -0.71061 and hence the true v.r.r. $= 3.4555$. Using this value, the *efficiency* of the method of antithetic variates is

$$\frac{\text{v.r.r.} \times t_1}{t_2} = \frac{1.047 \times 3.4555}{0.625} = 5.79$$

where t_1 is the time to generate 2000 variates using naive Monte Carlo and t_2 the time using the combined method. This means that using this variance reduction device, on average, provides similar precision answers to those obtained using naive Monte Carlo in approximately one-sixth of the time.

How effective is the method of antithetic variates in more complicated simulations? To answer, we should note the factors that are associated with a failure to induce large negative correlation. These are:

(i) Several sampling activities. For example, in a queueing system it is important to reserve one random number stream for interarrival times and another one for service durations. If this is not done, then a particular random number used for sampling interarrival times on the primary run might be used for a service duration on the antithetic run. The contribution of these variates to the primary and antithetic responses (e.g. waiting times) would then be positively correlated, which would tend to increase variance!

(ii) Use of rejection sampling rather than inversion of the cumulative distribution function. If rejection sampling is employed, then the one-to-one mapping of a random number to a variate is lost and the variance reduction will be attenuated.

(iii) The primary and antithetic runs use a *random* number of uniform numbers. This will certainly be the case in rejection sampling, and can also be the case where the end of the simulation run is marked by the passage of a predetermined amount of time (e.g. in estimating the total waiting time of all customers arriving during the next 10 hours, rather than the total waiting time for the next 25 customers, say).

(iv) More complicated systems with more interacting component parts are more likely to generate nonlinear responses.

Finally, it should be noted that it is always easy to generate antithetic variates from a symmetric distribution. For example, if $X \sim N(\mu, \sigma^2)$ then the antithetic variate is $2\mu - X$. If the antithetic variate is generated at the same time as the primary variate (as in the procedure 'theta_combined') this saves computation time. If it is generated within a separate run, then providing one-to-one correspondence between uniform random numbers and variates can be maintained, the variate generated will be $2\mu - X$. If the response is a linear function of symmetrically distributed random variables then use of antithetic variates will result in a zero variance estimator. Of course, simulation would not be used in such a situation as the expectation of response could be written down as a linear combination of the expectations of the random variables. However, if the response function is near-linear, good variance reduction can still be expected.

5.2 Importance sampling

Suppose we wish to evaluate the multiple integral

$$\theta = \int_{\mathbb{R}^n} h(x) f(x) \, dx$$

where $x' = (x_1, \dots, x_n)$ is a vector, $f(x)$ is a multivariate probability density function, and $h(x)$ is any function such that the integral exists. Then

$$\theta = E_f [h(\mathbf{X})]. \tag{5.9}$$

Now let $g(x)$ be any probability density function such that support $(hf) \subseteq$ support (g). (Recall that the support of a function is the set of those x values for which the function value is nonzero.) Then

$$\theta = \int_{x \in \text{support}(g)} \frac{h(x)f(x)}{g(x)} g(x) \, dx$$

$$= E_g \left[\frac{h(\mathbf{X})f(\mathbf{X})}{g(\mathbf{X})} \right]. \tag{5.10}$$

Using Equation (5.9), an unbiased estimator of θ is

$$\widehat{\theta} = \frac{1}{m} \sum_{i=1}^{m} h(\mathbf{X}_i)$$

where $\mathbf{X}_1, \ldots, \mathbf{X}_m$ are identically and independently distributed with density f. From Equation (5.10) an alternative unbiased estimator is

$$\widehat{\theta} = \frac{1}{m} \sum_{i=1}^{m} \frac{h(\mathbf{X}_i)f(\mathbf{X}_i)}{g(\mathbf{X}_i)}$$

where this time the variates are distributed with density g. The variance of this second estimator is

$$\text{Var}_g \left(\widehat{\theta} \right) = \frac{1}{m} \text{Var}_g \left(\frac{h(\mathbf{X})f(\mathbf{X})}{g(\mathbf{X})} \right).$$

It is clear that if $h(x) > 0 \; \forall x \in$ support (hf), and we choose $g(x) \propto h(x)f(x)$, then $h(x)f(x)/g(x)$ is constant and so $\text{Var}_g \left(\widehat{\theta} \right) = 0$. In this case

$$g(x) = \frac{h(x)f(x)}{\int_{u \in \text{support}(hf)} h(u)f(u) \, du}. \tag{5.11}$$

However, the denominator in Equation (5.11) is just θ, which is what we require to find, so this ideal is not achievable in practice. Nevertheless, this observation does indicate that a good choice of *importance sampling distribution*, $g(x)$, is one that is similar in shape to $h(x)f(x)$. A good choice will sample heavily from those values x for which $h(x)f(x)$ is large, and only infrequently where $h(x)f(x)$ is small – hence the name 'importance sampling'.

Suppose the choice of $g(x)$ is restricted to members of a family of distributions $\{g(x, \alpha)\}$ parameterized on α which of course is not necessarily of the same dimension as x. Then we wish to minimize with respect to α, the function

$$\text{Var}_g \left(\frac{h(x)f(x)}{g(x, \alpha)} \right) = E_g \left[\left(\frac{h(x)f(x)}{g(x, \alpha)} \right)^2 \right] - \theta^2$$

$$= \int_{x \in \text{support}(g)} \frac{h^2(x)f^2(x)}{g(x, \alpha)} \, dx - \theta^2. \tag{5.12}$$

Note that if $h(x) > 0 \; \forall x \in \text{support}\,(hf)$ then

$$\int_{x \in \text{support}(g)} \frac{h^2(x)f^2(x)}{g(x, \boldsymbol{\alpha})} \, dx$$

$$< \max_{x \in \text{support}(g)} \left\{ \frac{h(x)f(x)}{g(x, \boldsymbol{\alpha})} \right\} \int_{x \in \text{support}(g)} h(x)f(x) \, dx.$$

Therefore an upper bound on the variance is given by

$$\text{Var}_g \left(\frac{h(x)f(x)}{g(x, \boldsymbol{\alpha})} \right) < \theta \left\{ M\,(\boldsymbol{\alpha}) - \theta \right\} \tag{5.13}$$

where

$$M\,(\boldsymbol{\alpha}) = \max_{x \in \text{support}(g)} \left\{ \frac{h(x)f(x)}{g(x, \boldsymbol{\alpha})} \right\}.$$

This upper bound is minimized when $M\,(\boldsymbol{\alpha})$ is minimized with respect to $\boldsymbol{\alpha}$. The resulting density g happens to be the optimal envelope were an envelope rejection algorithm for generating variates sought from a probability density proportional to hf. Although this g is unlikely to be the best importance sampling density (parameterized on $\boldsymbol{\alpha}$), this observation does indicate that g is providing a good fit to hf, albeit with a different objective to the one of minimizing variance.

To fix ideas, let us evaluate by Monte Carlo

$$\theta = \int_a^\infty x^{\alpha-1} e^{-x} \, dx$$

where $\alpha > 1$ and $a > 0$. A naive sampling design identifies $f(x) = g(x) = e^{-x}$ on support $(0, \infty)$ and $h(x) = x^{\alpha-1} 1_{x>a}$, where $1_{x>a} = 1, 0$ according to whether $x > a$ or otherwise. Then

$$\theta = E_g\,(h(X))$$

$$= E_g\left(X^{\alpha-1} 1_{X>a} \right)$$

and the associated estimator is

$$\frac{1}{m} \sum_{i=1}^m X_i^{\alpha-1} 1_{X_i>a}$$

where $X_1, \ldots, X_m$ are sampled from f. This will be very inefficient for large a, since the density $g(x) = e^{-x}$ on support $(0, \infty)$ is quite dissimilar in shape to the function $h(x)f(x) = e^{-x} x^{\alpha-1} 1_{x>a}$.

It would be better to choose an importance sampling distribution that has support $[a, \infty)$, just as $h(x)f(x)$ does. We might choose

$$g(x) = \lambda e^{-\lambda(x-a)} \tag{5.14}$$

on support $[a, \infty)$ where λ is to be determined. Since Equation (5.14) is just a shifted exponential density, the random variable $X = a - \lambda^{-1} \ln R$ where $R \sim U(0, 1)$. An importance sampling estimator based on a single realization of X is

$$\widehat{\theta} = \frac{h(X)f(X)}{g(X)}$$

$$= \frac{X^{\alpha-1}e^{-X}}{\lambda e^{-\lambda(X-a)}}.$$

From Equation (5.12),

$$\mathrm{Var}_g \left(\widehat{\theta} \right) = \int_a^\infty \frac{x^{2\alpha-2}e^{-2x}}{\lambda e^{-\lambda(x-a)}} \, dx - \theta^2.$$

Minimization of this with respect to λ requires an evaluation of an integral, which is of a similar type to the one we are trying to evaluate. However, we might try minimizing the bound in (5.13). In that case we must find

$$\min_{0<\lambda<1} \max_{x>a} \frac{x^{\alpha-1}e^{-x}}{\lambda e^{-\lambda(x-a)}} \tag{5.15}$$

($\lambda < 1$, otherwise the inner maximization yields infinity). Therefore, we seek the unique saddle point in $(a, \infty) \times (0, 1)$ of the function in (5.15). This occurs at (x^*, λ^*) where

$$\frac{\alpha - 1}{x^*} - 1 + \lambda^* = 0,$$

$$-\frac{1}{\lambda^*} + x^* - a = 0,$$

that is at

$$\lambda^* = \frac{a - \alpha + \sqrt{(a-\alpha)^2 + 4a}}{2a}$$

$$x^* = a + \frac{1}{\lambda}.$$

For example, when $\alpha = 3$ and $a = 5$,

$$\lambda^* = \frac{1 + \sqrt{6}}{5} = 0.690,$$

$$x^* = 4 + \sqrt{6},$$

and

$$\mathrm{Var}_g \left(\widehat{\theta} \right) < 0.0050159.$$

The actual variance (rather than an upper bound) for this value of λ is 0.00028119. The minimum variance (rather than the minimum value of this upper bound on variance) is achieved when $\lambda = 0.72344$, giving a variance of 0.00015739. Using λ^* as computed above, the variance reduction ratio is

$$\frac{\int_a^\infty x^{2\alpha-2}e^{-x}\,dx - \theta^2}{0.00028119} = 37376,$$

showing that the method is highly effective.

5.2.1 Exceedance probabilities for sums of i.i.d. random variables

Suppose $X_i, i = 1, \ldots, n$, are identically and independently distributed random variables with the probability density function $f(x)$. We wish to estimate the exceedance probability

$$\theta = P\left(\sum_{i=1}^n X_i > a\right) \tag{5.16}$$

for values of a that make θ small. For example, X_i might represent the wear (a non-negative random variable) that a component receives in week i. Suppose the component is replaced at the first week end for which the cumulative wear exceeds a specified threshold, a. Let T represent the time between replacements. Then, for any positive integer n,

$$P(T \le n) = P\left(\sum_{i=1}^n X_i > a\right) = \theta.$$

For some densities $f(x)$, an easily computed distribution for $\sum_{i=1}^n X_i$ is not available (normal, gamma, binomial, and negative binomial are exceptions), so Monte Carlo is an option. However, when θ is small the precision will be low as the coefficient of variation of an estimate based on a sample of size m, say, is

$$\frac{\sqrt{\theta(1-\theta)/m}}{\theta} \approx \frac{1}{\sqrt{m\theta}}.$$

This motivates a design using an importance sampling distribution that samples more heavily from large values of X_i.

In another example, this time drawn from project management, suppose that a project consists of n similar activities in series. Let Y_i denote the duration of the ith activity. Further, suppose that the $\{Y_i\}$ are identically distributed as a beta density with shape parameters α and β, but on support (A, B) rather than $(0, 1)$ as described in Section 4.5. Then $Y_i = A + (B - A) X_i$ where X_i is beta distributed on support $(0, 1)$. Consequently, the problem of finding the (perhaps small) probability that the project duration exceeds a target duration $d (\le nB)$ is given by Equation (5.16) where $a = (d - nA)/(B - A) \le n$. Now define

$$f(x) = \prod_{i=1}^n f(x_i)$$

and an importance sampling n variate density

$$g(x) = \prod_{i=1}^{n} g(x_i)$$

where

$$g(x) = \gamma x^{\gamma-1}$$

on support $(0, 1)$ and where $\gamma \geq 1$. Then

$$\theta = E_f \left(1_{\sum_{i=1}^{n} X_i > a} \right)$$

where $1_P = 1$ if P is true; otherwise $1_P = 0$. It follows that

$$\theta = E_g \left(1_{\sum_{i=1}^{n} X_i > a} \prod_{i=1}^{n} \frac{\Gamma(\alpha+\beta) X_i^{\alpha-1} (1-X_i)^{\beta-1}}{\Gamma(\alpha) \Gamma(\beta) \gamma X_i^{\gamma-1}} \right).$$

Using the upper bound on variance (5.13), we hope to find a good value of γ by minimizing $M(\gamma)$, where

$$M(\gamma) = \max_{x \in (0,1)^n} \left(1_{\sum_{i=1}^{n} x_i > a} \prod_{i=1}^{n} \frac{\Gamma(\alpha+\beta) x_i^{\alpha-1} (1-x_i)^{\beta-1}}{\Gamma(\alpha) \Gamma(\beta) \gamma x_i^{\gamma-1}} \right)$$

$$= \max_{x \in (0,1)^n, \sum_{i=1}^{n} x_i > a} \left(\prod_{i=1}^{n} \frac{\Gamma(\alpha+\beta) x_i^{\alpha-\gamma} (1-x_i)^{\beta-1}}{\Gamma(\alpha) \Gamma(\beta) \gamma} \right).$$

To determine $M(\gamma)$ note that $x_i^{\alpha-\gamma}(1-x_i)^{\beta-1}$ is decreasing in $(0, 1)$ when $\alpha - \gamma < 0$ and $\beta - 1 \geq 0$. In such cases, the constraint $\sum_{i=1}^{n} x_i > a$ is active and so the maximum is at $x_i = a/n$, $i = 1, \ldots, n$. Consequently,

$$M(\gamma) = K \left\{ \frac{(a/n)^{\alpha-\gamma}}{\gamma} \right\}^n$$

where K is a constant independent of γ. $M(\gamma)$ is minimized when

$$\gamma = \frac{1}{\ln(n/a)}$$

and so this minimizes the upper bound on variance, providing $\beta \geq 1$ and $\gamma > \alpha$. The latter condition gives

$$n \geq a > n \exp\left(-\frac{1}{\alpha}\right).$$

The following algorithm estimates θ based upon a sample of size m:

Input a, α, β, n, m

$$\gamma := \frac{1}{\ln(n/a)}$$

If $a \leq n \exp\left(-\frac{1}{\alpha}\right)$ or $\beta < 1$ then stop

For $j = 1, \ldots, m$ do

$s := 0$

$p := 1$

For $i = 1, \ldots, n$ do

$X := R^{1/\gamma}$ where $R \sim U(0, 1)$

$s := s + X$

$p := p X^{\alpha - \gamma}(1 - X)^{\beta - 1}$

end do

If $s > a$ then $\theta_j := p \left(\dfrac{\Gamma(\alpha + \beta)}{\Gamma(\alpha)\Gamma(\beta)\gamma}\right)^n$ else $\theta_j := 0$ end if

end do

$$\widehat{\theta} := \frac{1}{m}\sum_{j=1}^{m}\theta_j$$

$$\text{e.s.e.} := \sqrt{\frac{1}{m(m-1)}\sum_{j=1}^{m}\left(\theta_j - \widehat{\theta}\right)^2}$$

A procedure, 'impbeta', appears in Appendix 5.2. For given α, β, n, and a, a simulation sampled $\{\theta_j, j = 1, \ldots, 5000\}$. Table 5.1 shows the resulting $\widehat{\theta}$ and e.s.e. $\left(\widehat{\theta}\right)$. The standard error for a naive Monte Carlo simulation is $\sqrt{\theta(1-\theta)/5000}$ and this may be estimated using $\widehat{\theta}$ obtained from 'impbeta'. The resulting estimated variance reduction ratio is therefore

$$\widehat{\text{v.r.r.}} = \frac{\widehat{\theta}\left(1 - \widehat{\theta}\right)/5000}{\left[\text{e.s.e.}\left(\widehat{\theta}\right)\right]^2}.$$

In all cases, comparison with a normal approximation using a central limit approximation is interesting. None of these lie within two standard deviations of the Monte Carlo estimate. Clearly, such an approximation needs to be used with caution when estimating small tail probabilities.

Table 5.1 Exceedance probability estimates for the sum of beta random variables

Parameter values				impbeta			Tilted		
α	β	n	a	$\widehat{\theta}$	e.s.e. $\left(\widehat{\theta}\right)$	$\widehat{\text{v.r.r.}}$	$\widehat{\theta}$	e.s.e. $\left(\widehat{\theta}\right)$	$\widehat{\text{v.r.r.}}$
1.5	2.5	12	6.2	0.01382	0.00092	3	0.01399	0.00031	29
1.5	2.5	12	6.5	0.00474	0.00032	9	0.00498	0.00012	70
2.5	1.5	12	9	0.01937	0.00057	12	0.01872	0.00039	24
2.5	1.5	24	18	0.00176	0.00008	50	0.00156	0.00004	214
1.5	2.5	12	5.5	$a < n\exp(-1/\alpha)$		—	0.09503	0.00178	5

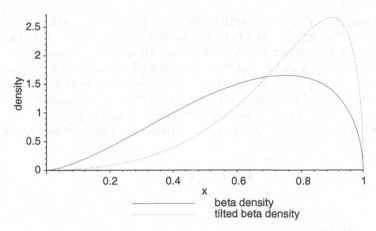

Figure 5.1   Beta and tilted beta densities

Another approach to this estimation is to use a *tilted* importance sampling distribution. This modifies the original beta density f to give an importance sampling density

$$g(x) = \frac{e^{tx} f(x)}{M_f(t)}$$

where $M_f(t)$ is the moment generating function of f. By making t positive, g samples more heavily from large values of x than does f. Of course, this is what is required in estimating exceedance probabilities. Further, a good value of t can be determined by minimizing the bound in (5.13). Figure 5.1 shows f and g for the case $\alpha = 2.5, \beta = 1.5, n = 12, a = 9$, and $t = 3.192$. This is discussed in more detail in problem 4(b) at the end of this chapter. The reader is encouraged to carry out the computations suggested there and to verify the conclusion to be drawn from Table 5.1, namely that this provides even better variance reduction. In the table, the two sets of estimated standard errors under naive Monte Carlo are calculated using the $\widehat{\theta}$ obtained from the corresponding variance reduction experiment.

5.3   Stratified sampling

The idea of stratified sampling is borrowed from a theory developed in the design of sample surveys. Suppose Monte Carlo is used to estimate a scalar parameter θ. A scalar statistic Y is simulated where it is known that $\theta = E(Y)$. Y is a function of random variables $\{W_j\}$. Often $\{W_j\}$ might be a set of uniform random numbers $\{R_j\}$. A stratified design works as follows. Select a *stratification variable* X which is another statistic realized during the simulation and which is also a function of the driving variables $\{W_j\}$. A good stratification variable is one for which:

 (i) There is a strong dependence (not necessarily linear) between X and Y.

 (ii) The density of X is known analytically and is easy to sample from.

(iii) It is easy to sample from the conditional distribution of Y given $X = x$.

First the range of X is split into disjoint strata $\{S^{(i)}, i = 1, .., M\}$. Then we fix in advance the number, n_i, of pairs (X, Y) for which $X \in S^{(i)}, i = 1, \ldots, M$. The required sample size from each stratum is then simulated. Of course, one way to do this is to sample $\{W_j\}$, calculate X and Y, and throw away those for which $X \notin S^{(i)}$. Repeat this for all i. This would be hopelessly inefficient. A better way is to sample n_i values of X from the density of X subject to $X \in S^{(i)}$ and then sample from the conditional distribution of $\{W_j\}$ for each such X $(= x$, say). Finally, calculate Y given $\{W_j\}$.

Let θ_i and $\sigma_i^2 (i = 1, \ldots, M)$ denote the ith stratum mean and variance respectively of Y, that is

$$\theta_i = E(Y \mid X \in S^{(i)}),$$
$$\sigma_i^2 = \text{Var}(Y \mid X \in S^{(i)}),$$

for $i = 1, \ldots, M$. Let

$$p_i = P(X \in S^{(i)}),$$
$$\sigma^2 = \text{Var}(Y),$$

and let (X_{ij}, Y_{ij}) denote the jth sample in the ith stratum, where $j = 1, \ldots, n_i$. The stratum sample means for the statistic of interest are

$$\overline{Y}_i = \frac{1}{n_i} \sum_{j=1}^{n_i} Y_{ij}$$

and a natural stratified estimator for θ is

$$\widehat{\theta}_{\text{ST}} = \sum_{i=1}^{M} p_i \overline{Y}_i.$$

This is clearly unbiased and since $Y_{i1}, \ldots, Y_{in_i}$ are identically and independently distributed then

$$\text{Var}(\widehat{\theta}_{\text{ST}}) = \sum_{i=1}^{M} p_i^2 \frac{\sigma_i^2}{n_i}. \tag{5.17}$$

A common scheme is *proportional stratified sampling* where we set $n_i = N p_i$ and N is the total sample size. In that case Equation (5.17) becomes

$$\text{Var}(\widehat{\theta}_{\text{PS}}) = \frac{1}{N} \sum_{i=1}^{M} p_i \sigma_i^2. \tag{5.18}$$

How does this compare with a naive estimator? In that case

$$\text{Var}(\widehat{\theta}) = \frac{\sigma^2}{N} = \frac{E(Y^2) - \theta^2}{N} \tag{5.19}$$

$$= \frac{1}{N} \left[\sum_{i=1}^{M} p_i E(Y^2 \mid X \in S^{(i)}) - \theta^2 \right] \tag{5.20}$$

$$= \frac{1}{N} \left[\sum_{i=1}^{M} p_i \left(\sigma_i^2 + \theta_i^2 \right) - \theta^2 \right]$$

$$= \frac{1}{N} \sum_{i=1}^{M} p_i \sigma_i^2 + \frac{1}{N} \sum_{i=1}^{M} p_i \left(\theta_i - \theta \right)^2. \tag{5.21}$$

Comparing Equations (5.18) and (5.21) gives

$$\mathrm{Var}\left(\widehat{\theta} \right) - \mathrm{Var}\left(\widehat{\theta}_{\mathrm{PS}} \right) = \frac{1}{N} \sum_{i=1}^{M} p_i \left(\theta_i - \theta \right)^2,$$

which is the amount of variance that has been removed through proportional stratified sampling.

In theory we can do better than this. If Equation (5.17) is minimized subject to $\sum_{i=1}^{M} n_i = N$, it is found that the optimum number to select from the ith stratum is

$$n_i^* = \frac{N p_i \sigma_i}{\sum_{i=1}^{M} p_i \sigma_i},$$

in which case the variance becomes

$$\mathrm{Var}\left(\widehat{\theta}_{\mathrm{OPT}} \right) = \frac{1}{N} \left(\sum_{i=1}^{M} p_i \sigma_i \right)^2 = \frac{\overline{\sigma}^2}{N}, \tag{5.22}$$

say. However,

$$\sum_{i=1}^{M} p_i \left(\sigma_i - \overline{\sigma} \right)^2 = \sum_{i=1}^{M} p_i \sigma_i^2 - \overline{\sigma}^2. \tag{5.23}$$

Therefore, from Equations (5.18) and (5.22),

$$\mathrm{Var}\left(\widehat{\theta}_{\mathrm{PS}} \right) - \mathrm{Var}\left(\widehat{\theta}_{\mathrm{OPT}} \right) = \frac{1}{N} \sum_{i=1}^{M} p_i \left(\sigma_i - \overline{\sigma} \right)^2.$$

Now the various components of the variance of the naive estimator can be shown:

$$\mathrm{Var}\left(\widehat{\theta} \right) = \frac{1}{N} \left[\sum_{i=1}^{M} p_i \left(\theta_i - \theta \right)^2 + \sum_{i=1}^{M} p_i \left(\sigma_i - \overline{\sigma} \right)^2 + \overline{\sigma}^2 \right] \tag{5.24}$$

The right-hand side of Equation (5.24) contains the variance removed due to use of the proportional $\{n_i\}$ rather than the naive estimator, the variance removed due to use of the optimal $\{n_i\}$ rather than the proportional $\{n_i\}$, and the residual variance respectively. Now imagine that very fine stratification is employed (i.e. $M \to \infty$). Then the outcome, $X \in S^{(i)}$, is replaced by the actual value of X and so from Equation (5.21)

$$\mathrm{Var}\left(\widehat{\theta} \right) = \frac{1}{N} \left\{ \mathrm{Var}_X \left[E(Y|X) \right] + E_X \left[\sigma^2(Y|X) \right] \right\} \tag{5.25}$$

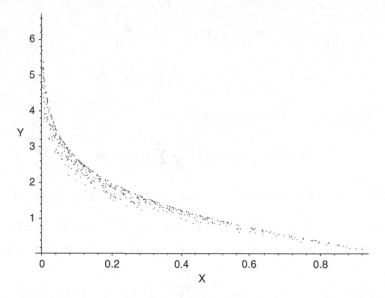

Figure 5.2   An example where X is a good stratification but poor control variable

The first term on the right-hand side of Equation (5.25) is the amount of variance removed from the naive estimator using proportional sampling. The second term is the residual variance after doing so. If proportional sampling is used (it is often more convenient than optimum sampling which requires estimation of the stratum variances $\{\sigma_i^2\}$ through some pilot runs), then we choose a stratification variable that tends to minimize the residual variance or equivalently one that tends to maximize $\mathrm{Var}_X[E(Y|X)]$.

Equation (5.25) shows that with a fine enough proportional stratification, *all* the variation in Y that is due to the variation in $E(Y|X)$ can be removed, leaving only the residual variation $E_X[\sigma^2(Y|X)]$. This is shown in Figure 5.2 where a scatter plot of 500 realizations of (X, Y) demonstrates that most of the variability in Y will be removed through fine stratification. It is important to note that it is not just variation in the linear part of $E(Y|X)$ that is removed, but all of it.

5.3.1   A stratification example

Suppose we wish to estimate

$$\theta = E\left(W_1 + W_2\right)^{5/4}$$

where W_1 and W_2 are independently distributed Weibull variates with density

$$f(x) = \frac{3}{2}x^{1/2}\exp(-x^{3/2})$$

on support $(0, \infty)$. Given two uniform random numbers R_1 and R_2,

$$W_1 = (-\ln R_1)^{2/3},$$
$$W_2 = (-\ln R_2)^{2/3},$$

and so a naive Monte Carlo estimate of θ is

$$Y = \left[(-\ln R_1)^{2/3} + (-\ln R_2)^{2/3}\right]^{5/4}$$

A naive simulation procedure, 'weibullnostrat' in Appendix 5.3.1, was called to generate 20 000 independent realizations of Y (seed $= 639\,156$) with the result

$$\widehat{\theta} = 2.15843$$

and

$$\text{e.s.e.} \left(\widehat{\theta}\right) = 0.00913. \tag{5.26}$$

For a stratified Monte Carlo, note that Y is monotonic in both R_1 and R_2, so a reasonable choice for a stratification variable is

$$X = R_1 R_2.$$

This is confirmed by the scatter plot (Figure 5.2) of 500 random pairs of (X, Y). The joint density of X and R_2 is

$$f_{X,R_2}(x, r_2) = f_{R_1,R_2}\left(\frac{x}{r_2}, r_2\right) \left|\frac{\partial r_1}{\partial x}\right|$$

$$= \frac{1}{r_2}$$

on support $0 < x < r_2 < 1$. Therefore

$$f_X(x) = \int_x^1 \frac{\mathrm{d}r_2}{r_2}$$

$$= -\ln x,$$

and the cumulative distribution is

$$F_X(x) = x - x \ln x. \tag{5.27}$$

on $(0,1)$. The conditional density of R_2 given X is

$$f_{R_2|X}(r_2, x) = \frac{f_{X,R_2}(x, r_2)}{f_X(x)} = -\frac{1}{r_2 \ln x}$$

on support $0 < x < r_2 < 1$, and the cumulative conditional distribution function is

$$F_{R_2|X}(r_2, x) = \int_x^{r_2} \frac{\mathrm{d}u}{-u \ln x} = 1 - \frac{\ln r_2}{\ln x}. \tag{5.28}$$

N realizations of (X, Y) will be generated with N strata where $p_i = 1/N$ for $i = 1, \ldots, N$. With this design and under proportional stratified sampling there is exactly one pair

(X, Y) for which $X \in S^{(i)}$ for each i. Let U_i, V_i be independently distributed as $U(0, 1)$. Using Equation (5.27) we generate X_i from the ith stratum through

$$X_i - X_i \ln X_i = \frac{i - 1 + U_i}{N}. \qquad (5.29)$$

Using Equation (5.28),

$$\frac{\ln R_2^{(i)}}{\ln X_i} = V_i,$$

that is

$$R_2^{(i)} = X_i^{V_i}.$$

Therefore

$$R_1^{(i)} = \frac{X_i}{R_2^{(i)}}.$$

Note that Equation (5.29) will need to be solved numerically, but this can be made more efficient by observing that $X_i \in (X_{i-1}, 1)$. The ith response is

$$Y_i = \left[\left(-\ln R_1^{(i)} \right)^{2/3} + \left(-\ln R_2^{(i)} \right)^{2/3} \right]^{5/4}$$

and the estimate is

$$\widehat{\theta}_{PS} = \sum_{i=1}^{N} p_i Y_i = \frac{1}{N} \sum_{i=1}^{N} Y_i.$$

To estimate $\mathrm{Var}\left(\widehat{\theta}_{PS}\right)$ we cannot simply use $[1/(N-1)] \sum_{i=1}^{N} \left(Y_i - \widehat{\theta}_{PS} \right)^2 / N$ as the $\{Y_i\}$ are from different strata and are therefore not identically distributed. One approach is to simulate K independent realizations of $\widehat{\theta}$ as in the algorithm below:

For $j = 1, \ldots, K$ do
For $i = 1, \ldots, N$ do
generate $u, v \sim U(0, 1)$
solve: $x - \ln x = \dfrac{i - 1 + u}{N}$
$r_2 := x^v$
$r_1 := \dfrac{x}{r_2}$
$y_i := \left[(-\ln r_1)^{2/3} + (-\ln r_2)^{2/3} \right]^{5/4}$
end do
$\overline{y}_j := \frac{1}{N} \sum_{i=1}^{N} y_i$
end do
$\widehat{\theta}_{PS} := \frac{1}{K} \sum_{j=1}^{K} \overline{y}_j$
$\mathrm{Var}\left(\widehat{\theta}_{PS}\right) := \frac{1}{K(K-1)} \sum_{j=1}^{K} \left(\overline{y}_j - \widehat{\theta}_{PS} \right)^2$

Using procedure 'weibullstrat' in Appendix 5.3.2 with $N = 100$ and $K = 200$ (and with the same seed as in the naive simulation), the results were

$$\widehat{\theta}_{PS} = 2.16644$$

and

$$\text{e.s.e.} \left(\widehat{\theta}_{PS}\right) = 0.00132.$$

Comparing this with Equation (5.26), stratification produces an estimated variance reduction ratio,

$$\widehat{\text{v.r.r.}} = 48. \tag{5.30}$$

The efficiency must take account of both the variance reduction ratio and the relative computer processing times. In this case stratified sampling took 110 seconds and naive sampling 21 seconds, so

$$\text{Efficiency} = \frac{21 \times 47.71}{110}$$

$$\approx 9$$

Three points from this example are worthy of comment:

(i) The efficiency would be higher were it not for the time consuming numerical solution of Equation (5.29). Problem 5 addresses this.

(ii) A more obvious design is to employ *two* stratification variables, R_1 and R_2. Accordingly, the procedure 'grid' in Appendix 5.3.3 uses 100 equiprobable strata on a 10×10 grid on $(0, 1)^2$, with exactly one observation in each stratum. Using $N = 200$ replications (total sample size $= 20\,000$ as before) and the same random number stream as before, this gave

$$\widehat{\theta} = 2.16710,$$

$$\text{e.s.e.} \left(\widehat{\theta}\right) = 0.00251,$$

$$\widehat{\text{v.r.r.}} = 13,$$

and

$$\text{Efficiency} \approx 13.$$

Compared with the improved stratification method suggested in point (i), this would not be competitive. Moreover, this approach is very limited, as the number of strata increases exponentially with the dimension of an integral.

(iii) In the example it was fortuitous that it was easy to sample from both the distribution of the stratification variable X and from the conditional distribution of Y given X. In fact, this is rarely the case. However, the following method of *post stratification* avoids these problems.

5.3.2 Post stratification

This refers to a design in which the number of observations in each stratum is counted *after* naive sampling has been performed. In this case $\{n_i\}$ will be replaced by $\{N_i\}$ to emphasize that $\{N_i\}$ are now random variables (with expectation $\{Np_i\}$). A naive estimator is

$$\widehat{\theta} = \sum_{i=1}^{M} \frac{N_i}{N} \overline{Y}_i,$$

but this takes no account of the useful information available in the $\{p_i\}$. Post (after) stratification uses

$$\widehat{\theta}_{AS} = \sum_{i=1}^{M} p_i \overline{Y}_i,$$

conditional on *no empty strata*. The latter is easy to arrange with sufficiently large $\{Np_i\}$. The naive estimator assigns equal weight $(1/N)$ to each realization of the response Y, whereas $\widehat{\theta}_{AS}$ assigns more weight (p_i/N_i) to those observations in strata that have been undersampled $(N_i < Np_i)$ and less to those that have been oversampled $(N_i > Np_i)$. Cochran (1977, p. 134) suggests that if $E(N_i) > 20$ or so for all i, then the variance of $\widehat{\theta}_{AS}$ differs little from that of $\widehat{\theta}_{PS}$ obtained through proportional stratification with fixed $n_i = Np_i$. Of course, the advantage of post stratification is that there is no need to sample from the conditional distribution of Y given X, nor indeed from the marginal distribution of X. Implementing post stratification requires only that cumulative probabilities for X can be calculated. Given there are M equiprobable strata, this is needed to calculate $j = \lfloor MF_X(x) + 1 \rfloor$, which is the stratum number in which a pair (x, y) falls.

This will now be illustrated by estimating

$$\theta = E(W_1 + W_2 + W_3 + W_4)^{3/2}$$

where $W_1, \ldots, W_4$ are independent Weibull random variables with cumulative distribution functions $1 - \exp(-x^2)$, $1 - \exp(-x^3)$, $1 - \exp(-x^4)$, and $1 - \exp(-x^5)$ respectively on support $(0, \infty)$. Bearing in mind that a stratification variable is a function of other random variables, that it should have a high degree of dependence upon the response $Y = (W_1 + W_2 + W_3 + W_4)^{3/2}$ and should have easily computed cumulative probabilities, it will be made a linear combination of standard normal random variables. Accordingly, define $\{z_i\}$ by

$$F_{W_i}(w_i) = \Phi(z_i)$$

for $i = 1, \ldots, 4$ where Φ is the cumulative normal distribution function. Then

$$\theta = \int_{(0,\infty)^4} \left(\sum_{i=1}^{4} w_i \right)^{3/2} \prod_{i=1}^{4} f_{W_i}(w_i) \, dw_i$$

$$= \int_{(-\infty,\infty)^4} \left(\sum_{i=1}^{4} F_{W_i}^{-1}[\Phi(z_i)] \right)^{3/2} \prod_{i=1}^{4} \phi(z_i) \, dz_i$$

$$= E_{Z \sim N(0,\mathbf{I})} \left(\sum_{i=1}^{4} F_{W_i}^{-1}[\Phi(Z_i)] \right)^{3/2}$$

where ϕ is the standard normal density. Note that an unbiased estimator is

$$\left(\sum_{i=1}^{4} F_{W_i}^{-1} [\Phi (Z_i)] \right)^{3/2}$$

where the $\{Z_i\}$ are independently $N(0, 1)$, that is the vector $\mathbf{Z} \sim N(\mathbf{0}, \mathbf{I})$, where the covariance matrix is the identity matrix $\mathbf{I}$. Now

$$\sum_{i=1}^{4} F_{W_i}^{-1} [\Phi (Z_i)] = \sum_{i=1}^{4} \{- \ln [1 - \Phi (Z_i)]\}^{1/\alpha_i} \tag{5.31}$$

where $\alpha_1 = 2, \alpha_2 = 3, \alpha_3 = 4, \alpha_4 = 5$. Using Maple a linear approximation to Equation (5.31) is found by expanding as a Taylor series about $z = 0$. It is

$$X' = a_0 + \sum_{i=1}^{4} a_i z_i$$

where $a_0 = 3.5593$, $a_1 = 0.4792$, $a_2 = 0.3396$, $a_3 = 0.2626$, and $a_4 = 0.2140$. Let

$$X = \frac{X' - a_0}{\sqrt{\sum_{i=1}^{4} a_i^2}} \sim N(0, 1).$$

Since X is monotonic in X' the same variance reduction will be achieved with X as with X'. An algorithm simulating K independent realizations, each comprising N samples of $\left(\sum_{i=1}^{4} F_{W_i}^{-1} [\Phi (Z_i)] \right)^{3/2}$ on M equiprobable strata, is shown below:

For $k = 1, \ldots, K$ do
For $j = 1, \ldots, M$ do $s_j := 0$ and $n_j := 0$ end do
For $n = 1, \ldots, N$
generate $z_1, z_2, z_3, z_4 \sim N(0, 1)$
$x := \dfrac{\sum_{i=1}^{4} a_i z_i}{\sqrt{\sum_{i=1}^{4} a_i^2}}$
$y := \left(\sum_{i=1}^{4} F_{W_i}^{-1} [\Phi (z_i)] \right)^{3/2}$
$j := \lfloor M \Phi (x) + 1 \rfloor$
$n_j := n_j + 1$
$s_j := s_j + y$
end do
$\overline{y}_k := \frac{1}{M} \sum_{j=1}^{M} \frac{s_j}{n_j}$
end do
$\widehat{\theta} := \frac{1}{K} \sum_{k=1}^{K} \overline{y}_k$
$\widehat{\mathrm{Var} (\widehat{\theta})} := \frac{1}{K(K-1)} \sum_{k=1}^{K} \left(\overline{y}_k - \widehat{\theta} \right)^2$

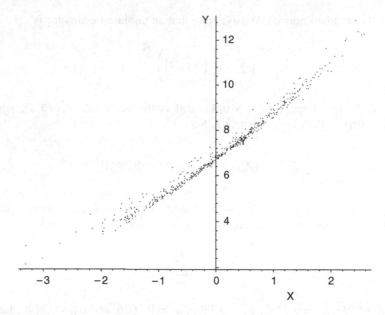

Figure 5.3 An example where X is both a good stratification and control variable

Using $K = 50$, $N = 400$, $M = 20$ (seed $= 566309$) it is found that

$$\widehat{\theta}_{AS} = 6.93055 \tag{5.32}$$

and

$$\text{e.s.e.}\left(\widehat{\theta}_{AS}\right) = 0.00223.$$

Using naive Monte Carlo with the same random number stream,

$$\text{e.s.e.}\left(\widehat{\theta}\right) = 0.01093$$

and so the estimated variance reduction ratio is

$$\widehat{\text{v.r.r.}} = 24. \tag{5.33}$$

A scatter plot of 500 random pairs of (X, Y) shown in Figure 5.3 illustrates the small variation about the regression curve $E(Y|X)$. This explains the effectiveness of the method.

5.4 Control variates

Whereas stratified sampling exploits the *dependence* between a response Y and a stratification variable X, the method of control variates exploits the *correlation* between a response and one or more control variables. As before, there is a response Y from a simulation and we wish to estimate $\theta = E(Y)$ where $\sigma^2 = \text{Var}(Y)$. Now suppose that in the same simulation we collect additional statistics $\mathbf{X}' = \left(X^{(1)}, \ldots, X^{(d)}\right)$ having *known* mean $\mu'_{\mathbf{X}} = \left(\mu^{(1)}, \ldots, \mu^{(d)}\right)$ and that the covariance matrix for $(\mathbf{X}, Y)'$ is

$$\begin{pmatrix} \Sigma_{\mathbf{XX}} & \Sigma_{\mathbf{X}Y} \\ \Sigma'_{\mathbf{X}Y} & \sigma^2 \end{pmatrix}.$$

The variables $X^{(1)}, \ldots, X^{(d)}$ are *control variables*. A control variate estimator

$$\widehat{\theta}_b = Y - b' \left(\mathbf{X} - \mu_{\mathbf{X}}\right)$$

is considered for any known vector $b' = (b_1, \ldots, b_d)$. Now, $\widehat{\theta}_b$ is unbiased and

$$\text{Var}\left(\widehat{\theta}_b\right) = \sigma^2 + b'\Sigma_{\mathbf{XX}}b - 2b'\Sigma_{\mathbf{X}Y}.$$

This is minimized when

$$b = b^* = \Sigma_{\mathbf{XX}}^{-1}\Sigma_{\mathbf{X}Y} \tag{5.34}$$

leading to a variance of

$$\text{Var}\left(\widehat{\theta}_{b^*}\right) = \sigma^2 - \Sigma'_{\mathbf{X}Y}\Sigma_{\mathbf{XX}}^{-1}\Sigma_{\mathbf{X}Y} = \left(1 - R^2\right)\sigma^2$$

where R^2 is the proportion of variance removed from the naive estimator $\widehat{\theta} = Y$. In practice the information will not be available to calculate Equation (5.34) so it may be estimated as follows. Typically, there will be a sample of independent realizations of $(\mathbf{X}_k, Y_k), k = 1, .., N$. Then

$$Y = \overline{Y} = \frac{1}{N}\sum_{k=1}^{N} Y_k,$$

$$\mathbf{X} = \overline{\mathbf{X}} = \frac{1}{N}\sum_{k=1}^{N}\mathbf{X}_k.$$

Let X_{ik} denote the ith element of column vector $\mathbf{X}_k$. Then an unbiased estimator of b^* is

$$\widehat{b}^* = \mathbf{S}_{\mathbf{XX}}^{-1}\mathbf{S}_{\mathbf{X}Y}$$

where the ijth element of $\mathbf{S}_{\mathbf{XX}}$ is

$$\frac{\sum_{k=1}^{N}\left(X_{ik} - \overline{X}_i\right)\left(X_{jk} - \overline{X}_j\right)}{N - 1}$$

and the ith element of $\mathbf{S}_{\mathbf{X}Y}$ is

$$\frac{\sum_{k=1}^{N}\left(X_{ik} - \overline{X}_i\right)\left(Y_k - \overline{Y}\right)}{N - 1}.$$

Now the estimator

$$\widehat{\theta}_{\widehat{b}^*} = \overline{Y} - \widehat{b}^{*'}\left(\overline{\mathbf{X}} - \mu_{\mathbf{X}}\right) \tag{5.35}$$

can be used. Since $\widehat{b}^*$ is a function of the data, $E\left(\widehat{b}^{*\prime}\left(\overline{\mathbf{X}}-\mu_{\mathbf{X}}\right)\right)\neq 0$, and so the estimator is biased. Fortunately, the bias is $O\left(1/N\right)$. Given that the standard error is $O\left(1/\sqrt{N}\right)$, the bias can be neglected providing N is large enough. If this method is not suitable, another approach is to obtain $\widehat{b}^*$ from a shorter pilot run (it is not critical that it deviates slightly from the unknown b^*) and then to use this in a longer independent simulation run to obtain $\widehat{\theta}_{\widehat{b}^*}$. This is unbiased for all N. It is worth noting that if $E\left(Y\,|\mathbf{X}\right)$ is linear then there is no question of any bias when there is no separate pilot run, even for small sample sizes.

A nice feature of the control variate method is its connection with linear regression. A regression of Y upon $\mathbf{X}$ takes the form

$$Y_k = \beta_0 + \boldsymbol{\beta}'\mathbf{X}_k + \varepsilon_k$$

where $\{\varepsilon_k\}$ are identically and independently distributed with zero mean. The *predicted value* (in regression terminology) at $\mathbf{X}^*$ is the (unbiased) estimate of $E\left(Y\,|\mathbf{X}^*\right)$ and is given by

$$Y^* = \overline{Y} + \widehat{\boldsymbol{\beta}}'\left(\mathbf{X}^* - \overline{\mathbf{X}}\right) \tag{5.36}$$

where

$$\widehat{\boldsymbol{\beta}} = \mathbf{S}_{\mathbf{X}\mathbf{X}}^{-1}\mathbf{S}_{\mathbf{X}Y}.$$

However, this is just $\widehat{b}^*$. This means that variance reduction can be implemented using multiple controls with standard regression packages. Given $(\mathbf{X}_k, Y_k)$, $k = 1, \ldots, N$, the control variate estimator is obtained by comparing Equations (5.35) and (5.36). It follows that $\widehat{\theta}_{\widehat{b}^*}$ is the predicted value Y^* at $\mathbf{X}^* = \mu_{\mathbf{X}}$.

Let us investigate how the theory may be applied to the simple case where there is just one control variable X $(d = 1)$. In this case

$$\widehat{b}^* = \frac{\sum_{k=1}^{N}\left(x_k - \overline{x}\right)\left(y_k - \overline{y}\right)}{\sum_{k=1}^{N}\left(x_k - \overline{x}\right)^2},$$

$\mathbf{X}^* = \mu_X$, and

$$\widehat{\theta}_{\widehat{b}^*} = \overline{Y} + \widehat{b}^*\left(\mu_X - \overline{X}\right).$$

An obvious instance where $d = 1$ is when a stratification variable X is used as a control variable. In the example considered in Section 5.3.1 a response was defined as

$$Y = \left[\left(-\ln R_1\right)^{2/3} + \left(-\ln R_2\right)^{2/3}\right]^{5/4}$$

and a stratification variable as

$$X = R_1 R_2.$$

Accordingly, a control variate estimator is

$$\widehat{\theta}_{\widehat{b}^*} = \overline{Y} + \widehat{b}^* \left(\frac{1}{4} - \overline{X} \right).$$

The effectiveness of this is given by R^2, which is simply the squared correlation between X and Y. A sample of 500 pairs $\{(X, Y)\}$ produced the scatter plot in Figure 5.2 and gave a sample correlation of -0.8369. So $\widehat{R}^2 = (-0.8369)^2 = 0.700$. Therefore, the proportion of variance that is removed through the use of this control variable is approximately 0.7 and the variance reduction ratio is approximately $(1 - 0.7)^{-1} = 3.3$. Although this is a useful reduction in variance, it does not compare well with the estimated variance reduction ratio of 48 given in result (5.30), obtained through fine stratification of X. The reason for this lies in the scatter plot (Figure 5.2), which shows that the regression $E(Y|X)$ is highly nonlinear. A control variable removes only the *linear* part of the variation in Y.

In contrast, using the stratification variable X as a control variate in the post stratification example considered in Section 5.3.2 will produce a variance reduction ratio of approximately $(1 - 0.9872^2)^{-1} = 40$, 0.9872 being the sample correlation of 500 pairs of (X, Y). Now compare this with the estimated variance reduction ratio of 24 given in result (5.33) using stratification. The control variate method is expected to perform well in view of the near linear dependence of Y upon X (Figure 5.3). However, the apparently superior performance of the control variate seems anomalous, given that fine stratification of X will always be better than using it as a control variate. Possible reasons for this are that $M = 20$ may not equate to fine stratification. Another is that $K = 50$ is a small sample as far as estimating the standard error is concerned, which induces a large estimated standard error on the variance reduction ratio. This does not detract from the main point emerging from this example. It is that if there is strong *linear* dependence between Y and X, little efficiency is likely to be lost in using a control variate in preference to stratification.

5.5 Conditional Monte Carlo

Conditional Monte Carlo works by performing as much as possible of a multivariate integration by analytical means, before resorting to actual sampling. Suppose we wish to estimate θ where

$$\theta = E_{g(x,y)} \{f(x,y)\}$$

where g is a multivariate probability density function for a random vector that can be partioned as the row vector $(\mathbf{X}', \mathbf{Y}')$. Suppose, in addition, that by analytical means the value of

$$E_{r(x|y)} \{f(x,y)\}$$

is known where r is the conditional density of $\mathbf{X}$ given that $\mathbf{Y} = y$. Then if $h(y)$ is the marginal density of $\mathbf{Y}$,

$$\theta = E_{h(y)} \left[E_{r(x|y)} \{f(x,y)\} \right].$$

Accordingly, a conditional Monte Carlo estimate of θ is given by sampling n variates y from h in the algorithm below:

For $i = 1, \ldots, n$
Sample $y_i \sim h(y)$
$\theta_i := E_{r(x|y_i)}\{f(x, y_i)\}$
end do
$\widehat{\theta} := \frac{1}{n} \sum_{i=1}^{n} \theta_i$

e.s.e. $\left(\widehat{\theta}\right) := \sqrt{\dfrac{\sum_{i=1}^{n} \left(\theta_i - \widehat{\theta}\right)^2}{n(n-1)}}$

For example, suppose a construction project has duration X where $X \sim N(\mu, \sigma^2)$ and where the distribution of the parameters μ and σ are independently $N(100,16)$ and exponential, mean 4, respectively. The company undertaking the project must pay £1000 for each day (and pro rata for part days) that the project duration exceeds K days. What is the expected cost C of delay? A naive simulation would follow the algorithm (note that $(X - K)^+ = \max(0, X - K)$):

For $i = 1, \ldots, n$ do
Sample $R \sim U(0, 1)$ and $Z \sim N(0, 1)$
$\sigma := -4 \ln(R)$
$\mu := 100 + 4Z_1$
Sample $X \sim N(\mu, \sigma^2)$
$C_i := 1000(X - K)^+$
end do
$\widehat{C} := \frac{1}{n} \sum_{i=1}^{n} C_i$

e.s.e. $\left(\widehat{C}\right) := \sqrt{\dfrac{\sum_{i=1}^{n} \left(C_i - \widehat{C}\right)^2}{n(n-1)}}$

Alternatively, using conditional Monte Carlo gives

$$\theta = E_{\sigma \sim \text{Exp}(1/4), \mu \sim N(100,16)} \left\{ E_{X \sim N(\mu, \sigma^2)} \left[1000(X - K)^+ \right] \right\}.$$

Let

$$C(\mu, \sigma^2) = E_{X \sim N(\mu, \sigma^2)} \left[1000(X - K)^+ \right]$$

$$= 1000 \int_{K}^{\infty} (x - K) \frac{1}{\sqrt{2\pi}\sigma} \exp\left[-\frac{1}{2} \left(\frac{x - \mu}{\sigma} \right)^2 \right] dx$$

$$= 1000 \int_{(K-\mu)/\sigma}^{\infty} (\sigma v + \mu - K) \frac{1}{\sqrt{2\pi}} \exp\left(-\frac{1}{2} v^2 \right) dv$$

$$= 1000 \left| \frac{-\sigma e^{-v^2/2}}{\sqrt{2\pi}} \right|_{(K-\mu)/\sigma}^{\infty} + 1000 (\mu - K) \, \Phi \left(\frac{\mu - K}{\sigma} \right)$$

$$= 1000 \left[\sigma \phi \left(\frac{K - \mu}{\sigma} \right) + (\mu - K) \, \Phi \left(\frac{\mu - K}{\sigma} \right) \right]$$

Accordingly, an algorithm for conditional Monte Carlo is

For $i = 1, \ldots, n$

Sample $R \sim U(0, 1)$ and $Z \sim N(0, 1)$

$\sigma := -4 \ln(R)$

$\mu := 100 + 4 Z_1$

$C_i := 1000\sigma \left[\phi \left(\frac{K-\mu}{\sigma} \right) - \left(\frac{K-\mu}{\sigma} \right) \Phi \left(\frac{\mu-K}{\sigma} \right) \right]$

end do

$\widehat{C} := \frac{1}{n} \sum_{i=1}^{n} C_i$

e.s.e. $\left(\widehat{C} \right) := \sqrt{\dfrac{\sum_{i=1}^{n} \left(C_i - \widehat{C} \right)^2}{n(n-1)}}$

This should give a good variance reduction. The reader is directed to Problem 8.

5.6 Problems

1. Consider the following single server queue. The interarrival times for customers are independently distributed as $U(0, 1)$. On arrival, a customer either commences service if the server is free or waits in the queue until the server is free and then commences service. Service times are independently distributed as $U(0, 1)$. Let A_i, S_i denote the interarrival times between the $(i-1)th$ and ith customers and the service time of the ith customer respectively. Let W_i denote the waiting time (excluding service time) in the queue for the ith customer. The initial condition is that the first customer in the system has just arrived at time zero. Then

$$W_i = \max(0, W_{i-1} + S_{i-1} - A_i)$$

for $i = 2, \ldots, 5$ where $W_1 = 0$. Write a procedure to simulate 5000 realizations of the total waiting time in the queue for the first five customers, together with 5000 antithetic realizations.

(a) Using a combined estimator from the primary and antithetic realizations, estimate the expectation of the waiting time of the five customers and its estimated standard error. Estimate the variance reduction ratio.

(b) Now repeat the experiment when the service duration is $U(0, 2)$. Why is the variance reduction achieved here much better than that in (a)?

2. In order to estimate $\theta = \int_b^\infty x^{\alpha-1}e^{-x}dx$ where $\alpha \le 1$ and $b > 0$, an importance sampling density $g(x) = e^{-(x-b)}1_{x>b}$ is used. (The case $\alpha > 1$ is considered in Section 5.2). Given $R \sim U(0,1)$, show that an unbiased estimator is $\widehat{\theta} = X^{\alpha-1}e^{-b}$ where $X = b - \ln R$ and that $\mathrm{Var}\left(\widehat{\theta}\right) < \theta\left(b^{\alpha-1}e^{-b} - \theta\right)$.

3. (This is a more difficult problem.) If $X \sim N\left(\mu, \sigma^2\right)$ then $Y = \exp(X)$ is lognormally distributed with mean $\exp(\mu + \sigma^2/2)$ and variance $\exp\left(2\mu + \sigma^2\right)\left[\exp\left(\sigma^2\right) - 1\right]$. It is required to estimate the probability that the sum of n such identically and independently lognormal distributed random variables exceeds a. A similar type of problem arises when considering Asian financial options (see Chapter 6). Use an importance sampling density that shifts the lognormal such that $X \sim N\left(\Delta, \sigma^2\right)$ where $\Delta > \mu$. (Refer to Section 5.2.1 which describes the i.i.d. beta distributed case.)

   (a) Show that when $a > n\exp\left(\mu\right)$ the upper bound on variance developed in result (5.13) is minimized when $\Delta = \ln\left(a/n\right)$.

   (b) Now suppose the problem is to estimate

   $$\theta = E_f\left(\sum_{i=1}^n e^{X_i} - a\right)^+$$

   where f is the multivariate normal density $N\left(\mu, \sigma^2 I\right)$ and $x^+ = \max\left(0, x\right)$. Show that the corresponding value of $\Delta\left(\ge \mu + \sigma^2/n\right)$ satisfies

   $$\Delta = \ln\left(\frac{a}{n}\right) - \ln\left[1 - \frac{\sigma^2}{n\left(\Delta - \mu\right)}\right].$$

   Run some simulations using this value of Δ (solve numerically using Maple). Does the suboptimal use of $\Delta = \ln\left(a/n\right)$ decrease the variance reduction ratio appreciably?

4. (This is a more difficult problem.) Where it exists, the moment generating function of a random variable having probability density function $f(x)$ is given by

   $$M\left(t\right) = \int_{\text{support}(f)} e^{tx}f(x)\, dx.$$

   In such cases a *tilted density*

   $$g\left(x\right) = \frac{e^{tx}f(x)}{M\left(t\right)}$$

   can be constructed. For $t > 0$ g may be used as an importance sampling distribution that samples more frequently from larger values of X than f.

   (a) Consider the estimation of $\theta = P\left(\sum_{i=1}^n \exp\left(X_i\right) > a\right)$ where the $\{X_i\}$ are independently $N\left(\mu, \sigma^2\right)$. Show that the tilted distribution is $N\left(\mu + \sigma^2 t, \sigma^2\right)$.

Show that when $a > n \exp(\mu)$ the value of t that minimizes the bound on variance given in result (5.13) is

$$t = \frac{\ln(a/n) - \mu}{\sigma^2}$$

and that therefore the method is identical to that described in Problem 3(a).

(b) Consider the estimation of $\theta = P\left(\sum_{i=1}^{n} X_i > a\right)$ where the $\{X_i\}$ are independent and follow a beta distribution with shape parameters $\alpha\,(>1)$ and $\beta\,(>1)$ on support $(0, 1)$. Show that the corresponding value of t here is the one that minimizes

$$e^{-at/n} M(t)$$

   (i) Use symbolic integration and differentiation within Maple to find this value of t when $n = 12$, $a = 6.2$, $\alpha = 1.5$, and $\beta = 2.5$.

   (ii) Write a Maple procedure that estimates θ for any $\alpha\,(>1)$, $\beta\,(>1)$, n, and $a\,(>0)$. Run your simulation for the parameter values shown in Table 5.1 of Section 5.2.1 and verify that the variance reduction achieved is of the same order as shown there.

5. In Section 5.3.1, Equation (5.29) shows how to generate, from a cumulative distribution function, $x - x\ln(x)$ on support $(0, 1)$, subject to x lying in the *ith* of N equiprobable strata. This equation has to be solved numerically, which accounts for the stratified version taking approximately four times longer than the naive Monte Carlo version. Derive an efficient envelope rejection method that is faster than this inversion of the distribution function. Use this to modify the procedure 'weibullstrat' in Appendix 5.3.2. Run the program to determine the improvement in efficiency.

6. Write a Maple procedure for the post stratification algorithm in Section 5.3.2. Compare your estimate with the one obtained in result (5.32).

7. Suggest a suitable stratification variable for the queue simulation in Problem 1. Write a Maple program and investigate the variance reduction achieved for different parameter values.

8. Write procedures for naive and conditional Monte Carlo simulations to estimate the expected cost for the example in Section 5.5. How good is the variance reduction?

9. Revisit Problem 4(b). Suggest and implement a variance reduction scheme that combines the tilted importance sampling with post stratification.

10. Use Monte Carlo to estimate $\int \cdots \int_D \left[\sum_{j=1}^{m} (m - j + 1)^2 x_j\right] dx$ where m is a positive integer and

$$D = \{x_j : j = 1, \ldots, m; 0 < x_1 < \ldots < x_m < 1\}.$$

For the case $m = 10$, simulate $10\,000$ points lying in D, and hence find a 95% confidence interval for the integral.

6

Simulation and finance

A derivative is a tradeable asset whose price depends upon other underlying variables. The variables include the prices of other assets. Monte Carlo methods are now used routinely in the pricing of financial derivatives. The reason for this is that apart from a few 'vanilla' options, most calculations involve the evaluation of high-dimensional definite integrals. To see why Monte Carlo may be better than standard numerical methods, suppose we wish to evaluate

$$I = \int_{[0,1]} f(x) \, dx$$

where $f(x)$ is integrable. Using the composite trapezium rule a subinterval length of h is chosen such that $(m-1)h = 1$ and then f is evaluated at m equally spaced points in $[0, 1]$. The error in this method is $O(h^2) = O(1/m^2)$. Now compare this with Monte Carlo where f is evaluated at m values of X where $X \sim U(0, 1)$. Here, the standard error is a measure of the accuracy, so if $\sigma^2 = \text{Var}_{X \sim U(0,1)}[f(X)]$, the error in the estimate of I is $\sigma/\sqrt{m} = O(1/\sqrt{m})$. Therefore, for large sample sizes, it is better to use the trapezium rule. Now suppose that the integration is over the unit cube in d dimensions. The trapezium rule will require m function evaluations to be made over a regular lattice covering the cube. If h is again the subinterval length along any of the d axes, then $m h^d \simeq 1$. The resulting error is $O(h^2) = O(1/m^{2/d})$. However, using Monte Carlo, the error is still $O(1/\sqrt{m})$. Therefore, for $d > 4$ and for sufficiently large m, Monte Carlo will be better than the trapezium rule. This advantage increases exponentially with increasing dimension. As will be seen, in financial applications a value of $d = 100$ is not unusual, so Monte Carlo is the obvious choice.

This chapter provides an introduction to the use of Monte Carlo in financial applications. For more details on the financial aspects there are many books that can be consulted, including those by Hull (2006) and Wilmott (1998). For a state-of-the-art description of Monte Carlo applications Glasserman (2004) is recommended.

The basic mathematical models that have been developed in finance assume an underlying geometric Brownian motion. First the main features of a Brownian motion, also known as a Wiener process, will be described.

Simulation and Monte Carlo: With applications in finance and MCMC   J. S. Dagpunar
© 2007 John Wiley & Sons, Ltd

6.1 Brownian motion

Consider a continuous state, continuous time stochastic process $\{B(t), t \geq 0\}$ where

$$dB = B(dt) = B(t+dt) - B(t) \sim N(0, dt) = Z_{(t,t+dt)}\sqrt{dt} \qquad (6.1)$$

for all t. Here $Z_{(t,t+dt)} \sim N(0, 1)$. Suppose further that the process has *independent* increments. This means that if (u, v) and (t, s) are nonoverlapping intervals then $B(v) - B(u)$ and $B(t) - B(s)$ are independently distributed. Further assume that $B(0) = 0$. Then the solution to Equation (6.1) is

$$B(t) \sim N(0, t)$$

or

$$B(t) = \sqrt{t}\, W_t$$

where $W_t \sim N(0, 1)$. The process $\{B(t), t \geq 0, B(0) = 0\}$ is called a *standard Brownian motion*.

Since the process has independent increments, for any $t \geq s \geq 0$

$$B(t) = \sqrt{s}\, W_1 + \sqrt{t-s}\, W_2 \qquad (6.2)$$

where W_1 and W_2 are independently $N(0, 1)$. Therefore, such a process may be simulated in the interval $[0, T]$ by dividing it into a large number, n, of subintervals, each of length h so that $T = nh$. Then according to Equation (6.2),

$$B(jh) = B([j-1]h) + Z_j\sqrt{h}$$

for $j = 1, \ldots, n$ where $Z_1, \ldots, Z_n$ are independently distributed as $N(0, 1)$. This provides a realization that is a discrete approximation to the continuous time process. It is exact at times $j = 0, h, \ldots, nh$. If a choice is made to interpolate at intermediate times it is an approximation. By choosing n large enough the resulting error can be made arbitrarily small. Now refer to Appendix 6.1. There is a procedure 'STDNORM' for a Box–Müller standard normal generator (it is used in preference to 'STATS[random,normald](1)' as it is somewhat faster), together with a Maple procedure, 'Brownian'. These are used to plot three such discrete approximations ($n = 10000$) to $\{B(t), 100 \geq t \geq 0, B(0) = 0\}$.

Now suppose that

$$dX(t) = \mu\, dt + \sigma\, dB(t). \qquad (6.3)$$

The parameter μ gives the Brownian motion a *drift* and the parameter $\sigma (> 0)$ scales $B(t)$. Given that $X(0) = x(0)$ the solution to this is obviously

$$X(t) = x(0) + \mu t + \sigma B(t).$$

The process $\{X(t), t \geq 0\}$ is called a Brownian motion (or *Wiener process*) with drift μ and variance parameter σ^2. It also has independent increments. For any $t \geq s$,

$$X(t) - X(s) \sim N\left(\mu[t-s], \sigma^2[t-s]\right)$$

and the probability density of $X(t), t > 0$, given that $X(0) = x(0)$ is

$$u(x, t) = \frac{1}{\sqrt{2\pi\sigma^2 t}} \exp\left\{-\frac{1}{2}\left[\frac{x - x(0) - \mu t}{\sigma\sqrt{t}}\right]^2\right\}$$

which is a solution to a *diffusion equation*

$$\frac{\partial u}{\partial t} = -\mu\frac{\partial u}{\partial x} + \frac{1}{2}\sigma^2\frac{\partial^2 u}{\partial x^2}.$$

6.2 Asset price movements

Suppose we wish to model the price movements over time of an asset such as a share, interest rate, or commodity. If $X(t)$ represents the price at time t, the most frequently used model in finance is

$$\frac{dX}{X} = \mu\, dt + \sigma\, dB \tag{6.4}$$

where $\sigma \geq 0$. Note that the left-hand side represents the proportional change in the price in the interval $(t, t+dt)$. If $\sigma = 0$ then the solution to this is $X(t) = x(s)\exp[\mu(t-s)]$, for $t \geq s$, where $x(s)$ is the known asset price at time s. In that case it is said that the *return* in (t, s) is $\mu(t-s)$ and that the *growth rate* is μ.

Equation (6.4) is an example of an *Itô stochastic differential equation* of the form

$$dX = a(X, t)\, dt + b(X, t)\, dB. \tag{6.5}$$

For $t \geq 0$,

$$X(t) - X(0) = \int_0^t a(X(u), u)\, du + \int_0^t b(X(u), u)\, dB(u) \tag{6.6}$$

where the second integral is known as an *Itô stochastic integral*. For more details on this, see, for example, Allen (2003, Chapter 8). Now suppose that $G[X(t), t]$ is some function of $X(t)$ and t, where $\partial G/\partial X$, $\partial G/\partial t$, and $\partial^2 G/\partial X^2$ all exist. Then *Itô's lemma* states that the change in G in $(t, t+dt)$ is given by

$$dG = \frac{\partial G}{\partial X}dX + \frac{\partial G}{\partial t}dt + \frac{1}{2}b^2\frac{\partial^2 G}{\partial X^2}dt. \tag{6.7}$$

An easy way to remember this is to imagine a Taylor series expansion about $(X(t), t)$,

$$\delta G = \frac{\partial G}{\partial X}\delta X + \frac{\partial G}{\partial t}\delta t + \frac{1}{2}\frac{\partial^2 G}{\partial X^2}(\delta X)^2 + \frac{1}{2}\frac{\partial^2 G}{\partial t^2}(\delta t)^2 + \frac{\partial^2 G}{\partial t\partial X}\delta t\,\delta X + \dots. \tag{6.8}$$

From Equation (6.5),

$$\delta X \approx a\,\delta t + b\,\delta B \approx a\,\delta t + bZ_{(t,t+\delta t)}\sqrt{\delta t}$$

where $Z_{(t,t+\delta t)} \sim N(0,1)$. So $E\left[(\delta X)^2\right] = b^2\delta t E\left[Z^2_{(t,t+\delta t)}\right] + o(\delta t) = b^2\delta t + o(\delta t)$, and similarly $\text{Var}\left((\delta X)^2\right) = o(\delta t)$. So, in the limit, $(\delta X)^2$ is nonstochastic and equals $b^2\,dt$. Similarly, the last two terms on the right-hand side of Equation (6.8) are $o(\delta t)$. Putting these together gives Equation (6.7).

To solve Equation (6.4), rewrite as (6.5) where $a(X,t) = \mu X$ and $b(X,t) = \sigma X$. Let $G = \ln X$. Then $\partial G/\partial X = 1/X$, $\partial^2 G/\partial X^2 = -1/X^2$, and $\partial G/\partial t = 0$. Using Itô's lemma,

$$dG = \frac{\partial G}{\partial X}dX + \frac{\partial G}{\partial t}dt + \frac{1}{2}b^2\frac{\partial^2 G}{\partial X^2}dt$$

$$= \frac{dX}{X} - \frac{\sigma^2 X^2 dt}{2X^2}$$

$$= \frac{\mu X dt + \sigma X dB}{X} - \frac{\sigma^2 dt}{2}$$

$$= \left(\mu - \frac{\sigma^2}{2}\right)dt + \sigma dB.$$

Comparing this with Equation (6.3), it can be seen that $\{G(t), t \geq 0\}$ is a Wiener process with drift $\mu - \sigma^2/2$ and variance parameter σ^2. Therefore, because any Wiener process has independent increments, then for $t \geq s$,

$$G(t) - G(s) = \ln X(t) - \ln X(s)$$

$$= \ln\left[\frac{X(t)}{X(s)}\right]$$

$$\sim N\left[\left(\mu - \frac{1}{2}\sigma^2\right)(t-s), \sigma^2(t-s)\right]. \tag{6.9}$$

Suppose now that the asset price is known to be $x(s)$ at time s. Then at a later time t, the price will be, from Equation (6.9),

$$X(t) = x(s)e^Y \tag{6.10}$$

where $Y \sim N\left[\left(\mu - \frac{1}{2}\sigma^2\right)(t-s), \sigma^2(t-s)\right]$. Y is the *return* during (s,t). Therefore, given $X(s) = x(s)$, $X(t)$ is lognormally distributed. Using standard results for the expectation of a lognormal random variable,

$$E[X(t)\,|\,X(s) = x(s)] = x(s)E\left(e^Y\right)$$

$$= x(s)\exp\left[\left(\mu - \frac{1}{2}\sigma^2\right)(t-s) + \frac{1}{2}\sigma^2(t-s)\right]$$

$$= x(s)e^{\mu(t-s)}.$$

Therefore μ can be interpreted as the *expected growth rate*, as in the deterministic model.

It is fortuitous that the model (6.4) can be solved analytically. Frequently, stochastic differential equations cannot. In such cases, one remedy is to simulate a sample of paths $\{\{X_i(t), T \geq t \geq 0\}, i = 1, \ldots, n\}$ for Equation (6.5) and make inferences about the distribution of $X(t)$, or perhaps other functionals of $\{X(t), T \geq t \geq 0\}$, from such a sample. A discrete approximation to one such path is obtained using Euler's method. The resulting difference equation is

$$X([j+1]h) = X(jh) + ha(X(jh), jh) + b(X(jh), jh)Z_{j+1}\sqrt{h} \qquad (6.11)$$

for $j = 0, \ldots, m - 1$, where $mh = T$, and $\{Z_j\}$ are independently $N(0, 1)$. In addition to errors resulting from the Euler method, it is also an approximation in the sense that it gives the behaviour of the path at discrete times only, whereas the model is in continuous time.

For model (6.4), an Euler approximation is unnecessary, and if we wish to see the entire path (rather than just the terminal value $X(T)$) Equation (6.10) would be used, giving the difference equation

$$X([j+1]h) = X(jh)\exp\left[\left(\mu - \frac{1}{2}\sigma^2\right)h + \sigma\sqrt{h}Z_{j+1}\right].$$

The stochastic process (6.4) is called a *geometric* Brownian motion. In Appendix 6.2 there is a procedure, 'GeometricBrownian', which is used to plot three independent realizations of $\{X(t), 10 \geq t \geq 0\}$ where $m = 2000$, $X(0) = 100$, $\mu = 0.1$, and $\sigma = 0.3$. Each realization shows how the price of an asset subject to geometric Brownian motion, and initially costing £100, changes over the next 10 years. The asset has an expected growth rate of 10 % per annum and a volatility (i.e. the standard deviation of return in a year) of 30 %. In the second part of Appendix 6.2 three further plots are shown for a similar asset, but with volatilities of 2 %, 4 %, and 8 % respectively.

6.3 Pricing simple derivatives and options

A *derivative* is a contract that depends in some way on the price of one or more underlying assets. For example, a *forward* contract is a derivative where one party promises to pay the other a specified amount for underlying assets at some specified time. An *option* is a derivative where the two parties have certain rights, which they are not obliged to enforce. The simplest type of options are European call and put options.

A *European call* option gives the holder the right (but not the obligation) to buy an asset at a specified time T (the *expiration or exercise date*) for a specified price K (the *exercise* or *strike* price). Let X be the asset price at expiry. The *payoff* for a European call is therefore $\max(0, X - K)$ which is written as $(X - K)^+$. This follows, since if $X > K$ then it pays the holder of the call to enforce the right to buy the asset at K and immediately sell it in the market for X, making a profit of $X - K$. If $X \leq K$ then exercising the option would result in a loss of $K - X$. In that case the holder of the call option does nothing, giving zero profit. A *put* option gives the holder the right to sell the asset at the exercise price and the payoff is therefore $(K - X)^+$.

Let $V(X(t), t)$ be the price at time t of a derivative (what type has not yet been specified) on an underlying asset with price $X(t)$, where $dX(t)/X(t) = \mu\, dt + \sigma\, dB$. $V(X(t), t)$ is derived by constructing a portfolio, whose composition will be changed dynamically with time by the holder of the portfolio in such a way that its return is equal to the return on a risk-free investment. Consider such a portfolio consisting of one derivative and $-\Delta$ units of the asset (i.e. 'short' in the asset). The value of the portfolio is therefore π where

$$\pi(X(t), t) = V(X(t), t) - \Delta X(t).$$

The change in portfolio value during $(t, t + dt)$ is

$$d\pi = d(V - \Delta X).$$

Using Equations (6.4) and (6.7),

$$d\pi = \frac{\partial(V - \Delta X)}{\partial X} dX + \frac{\partial(V - \Delta X)}{\partial t} dt + \frac{1}{2}\frac{\partial^2(V - \Delta X)}{\partial X^2}\sigma^2 X^2\, dt$$

$$= \left(\frac{\partial V}{\partial X} - \Delta\right) dX + \frac{\partial V}{\partial t} dt + \frac{1}{2}\frac{\partial^2 V}{\partial X^2}\sigma^2 X^2\, dt.$$

By setting $\Delta = \partial V/\partial X$ the risky component of $d\pi$ can be removed! In that case

$$d\pi = \frac{\partial V}{\partial t} dt + \frac{1}{2}\frac{\partial^2 V}{\partial X^2}\sigma^2 X^2\, dt. \tag{6.12}$$

However, this must equal the interest on a riskless asset otherwise investors could make a risk-free profit (an *arbitrage*). One of the key assumptions in derivative pricing models is that arbitrage is not possible. The argument is that if it were possible, then market prices would immediately adapt to eliminate such possibilities. Let r denote the risk-free growth rate. Then

$$d\pi = r\pi\, dt = r(V - \Delta X)\, dt. \tag{6.13}$$

Equating (6.12) and (6.13) gives the *Black–Scholes* differential equation

$$\frac{\partial V}{\partial t} + \frac{1}{2}\frac{\partial^2 V}{\partial X^2}\sigma^2 X^2 = r\left(V - \frac{\partial V}{\partial X} X\right). \tag{6.14}$$

It is an understatement to say that a nice feature of this equation is that it does not contain μ, the expected growth rate of the asset. This is excellent since μ is unknown. The theory was developed by Black, Merton, and Scholes (Black and Scholes, 1973; Merton, 1973), and earned Merton and Scholes a Nobel prize in 1997 (Black died in 1993). The equation has to be solved subject to the boundary conditions specific to the derivative. Note that since the derivative price will change with t, in response to changes in the asset price, the *hedging parameter* Δ will have to be updated continuously. This balances the portfolio to produce a riskless return.

The major assumptions in the Black–Scholes model are that no arbitrage is possible, that the asset price follows a geometric Brownian motion, that there are no transaction costs, that the portfolio can be continually rebalanced, that the risk-free interest rate is known during the life of the option, and that the underlying asset does not generate an income such as a dividend (this last one is easily relaxed).

6.3.1 European call

Suppose we are dealing with a European call with exercise time T and exercise price K, and that the price of the asset at time t is known to be $x(t)$. Then the terminal condition is $V(X(T), T) = (X(T) - K)^+$. The solution of Equation (6.14) subject to this boundary condition turns out to be

$$V(x(t), t) = e^{-r(T-t)} \int_{-\infty}^{\infty} (x - K)^+ f_{X_r(T)|X(t)=x(t)}(x, x(t)) dx \qquad (6.15)$$

where $f_{X_r(T)|x(t)}$ is the density of the asset price X at expiration time T, given that the current price is known to be $x(t)$ and *taking the expected growth rate to be r*, the risk-free interest rate. This can be verified by direct substitution into Equation (6.14). It is worth noting that $f_{X_r(T)|X(T)=x(T)}(x, x(T)) = \delta(x - x(T))$, a delta function, and so from Equation (6.15) $V(X(T), T) = (X(T) - K)^+$, as expected (at time T the option will be exercised if and only if $X(T) > K$, making the value of the option at that time $(X(T) - K)^+$). For $t < T$ it is known that the density is lognormal (see result (6.10)). Also, note the discount factor $e^{-r(T-t)}$, which makes the right-hand side of Equation (6.15) equal to the *present value of the expected payoff at expiration, assuming the asset has an expected growth rate of r*. This can be referred to as the present value of the expected value of the payoff in a *risk-neutral world*.

Fortunately, Equation (6.15) can be obtained in closed form as follows. From Equation (6.10), given that $X(t) = x(t)$,

$$X(T) = x(t) e^{(r - \sigma^2/2)(T-t) + Z\sigma\sqrt{T-t}}.$$

Therefore,

$$V(x(t), t) = e^{-r(T-t)} E_\phi \left(x(t) \, e^{(r - \sigma^2/2)(T-t) + z\sigma\sqrt{T-t}} - K \right)^+ \qquad (6.16)$$

where ϕ is the standard normal density. Let

$$z_0 = \frac{\ln[K/x(t)] - (r - \sigma^2/2)(T - t)}{\sigma\sqrt{T - t}}.$$

Then

$$V(x(t), t) = e^{-r(T-t)} \int_{z_0}^{\infty} \left(x(t) \, e^{(r - \sigma^2/2)(T-t) + Z\sigma\sqrt{T-t}} - K \right) \frac{e^{-z^2/2}}{\sqrt{2\pi}} \, dz$$

$$= \int_{z_0}^{\infty} \frac{x(t) \, e^{-(z - \sigma\sqrt{T-t})^2/2}}{\sqrt{2\pi}} \, dz - K \, e^{-r(T-t)} \Phi(-z_0)$$

$$= x(t) \Phi\left(\sigma\sqrt{T-t} + z_0\right) - K \, e^{-r(T-t)} \Phi(-z_0)$$

$$= x(t) \Phi(d) - K \, e^{-r(T-t)} \Phi\left(d - \sigma\sqrt{T-t}\right)$$

where

$$d = \frac{(r + \sigma^2/2)(T - t) + \ln(x(t)/K)}{\sigma\sqrt{T - t}}.$$

$V(x(t), t)$ (or just $c(t)$ for short) is the price of a call option at time t when the current price is known to be $x(t)$. Now refer to Appendix 6.3. The built-in 'blackscholes' procedure (part of the Maple finance package) is used to calculate the price of a European call on a share that currently at time $t = 23/252$ years (there are 252 trading days in a year) has a price of £100. The original life of the option is 6 months, so $T = 126/252$ years. The risk-free interest rate is 5 % per annum, the volatility is 20 % per annum, and the strike price is £97. The solution is

$$V\left(100, \frac{23}{252}\right) = £7.84 \qquad (6.17)$$

In practice, no one uses simulation to price simple (*vanilla*) options such as a European call. Nevertheless, it will be instructive to write a procedure that does this, as a prelude to simulating more complex (*exotic*) options, where closed-form expressions are not available. From Equation (6.16) it is clear that

$$c_i = e^{-r(T-t)}\left(x(t)\ e^{(r-\sigma^2/2)(T-t)+Z_i\sigma\sqrt{T-t}} - K\right)^+ \qquad (6.18)$$

is an unbiased estimator of the call price at time t, $c(t)$. Given that $\{Z_i, i = 1, \dots, m\}$ are independently $N(0, 1)$, let $\widehat{c}$ and s denote the sample mean and sample standard deviation of $\{c_i, i = 1, \dots, m\}$. Then, for m sufficiently large, a 95 % confidence interval is $(\widehat{c} - 1.96\,s/\sqrt{m}, \widehat{c} + 1.96\,s/\sqrt{m})$. In Appendix 6.4, the procedure 'BS' provides just such an estimate, (7.75, 7.92), which happily includes the exact value from Equation (6.17).

Note how 'BS' uses antithetic variates as a variance reduction device. In this case replacing Z_i by $-Z_i$ in Equation (6.18) also gives an unbiased estimator of $c(t)$. How effective is the use of antithetic variates here? The correlation with the primary estimate will be large and negative if $\left(x(t)\ e^{(r-\sigma^2/2)(T-t)+Z_i\sigma\sqrt{T-t}} - K\right)^+$ is well approximated by a linear function of Z. This is the case if its value is usually positive (that is true when $x(t)\ e^{(r-\sigma^2/2)(T-t)} - K$ is sufficiently large, in which case the option is said to be *deep in the money*) and when $\sigma\sqrt{T-t}$, the standard deviation of return in $[t, T]$ is small. An example is shown in Appendix 6.4.

6.3.2 European put

For a European put, let $p(t)$ denote the price of the option at time t. Now consider a portfolio consisting at time t of one put plus one unit of the underlying asset. The value of this at time T is $\max(X(T), K)$. This is the same as the value at time T of a portfolio that at time t consisted of one call option (value $c(t)$) plus an amount of cash equal to $K\exp(-r[T-t])$. Therefore, the values of the two portfolios at time t must be equal otherwise arbitrage would be possible. It follows that

$$c(t) + Ke^{-r(T-t)} = p(t) + X(t)$$

This result is known as *put-call parity*.

6.3.3 Continuous income

How are these results modified when the underlying asset earns a continuous known income at rate r_f? Recall that the dynamically hedged portfolio consists of $-\partial V/\partial X$ units of the underlying asset. Therefore the interest earned in $(t, t + dt)$ is $-Xr_f \partial V/\partial X \, dt$. Therefore Equation (6.12) becomes

$$d\pi = \frac{\partial V}{\partial t} \, dt + \frac{1}{2} \frac{\partial^2 V}{\partial X^2} \sigma^2 X^2 \, dt - \frac{\partial V}{\partial X} Xr_f \, dt$$

$$= r\left(V - X\frac{\partial V}{\partial X} \right) dt$$

and the corresponding Black–Scholes equation is

$$\frac{\partial V}{\partial t} + \frac{1}{2} \frac{\partial^2 V}{\partial X^2} \sigma^2 X^2 = rV - (r - r_f) X\frac{\partial V}{\partial X}.$$

For a European call, the boundary condition is

$$c(T) = V(X(T), T)$$
$$= (X(T) - K)^+$$

and the solution is

$$c(t) = e^{-r(T-t)} \int_{-\infty}^{\infty} (x - K)^+ f_{X_{r-r_f}(T)|X(t)=x(t)}(x, x(t)) \, dx \tag{6.19}$$

$$= e^{-r(T-t)} E_\varphi \left(x(t) \, e^{(r-r_f-\sigma^2/2)(T-t)+Z\sigma\sqrt{T-t}} - K \right)^+ \tag{6.20}$$

$$= x(t) \, e^{-r_f(T-t)} \Phi\left(d_{r_f} \right) - Ke^{-r(T-t)} \Phi\left(d_{r_f} - \sigma\sqrt{T-t} \right) \tag{6.21}$$

where

$$d_{r_f} = \frac{(r - r_f + \sigma^2/2)(T - t) + \ln(x(t)/K)}{\sigma\sqrt{T-t}} \tag{6.22}$$

Use of put-call parity gives a put price as

$$c(t) + Ke^{-r(T-t)} = p(t) + x(t) \, e^{-r_f(T-t)}$$

6.3.4 Delta hedging

Now imagine a UK bank that sells a call option on a foreign currency that costs $x(t)$ at time t (for example $x(t)$ is the cost in pounds sterling of a block of 1000 euros). K, r, and σ represent the strike price, risk-free interest rate on pounds sterling, and the volatility of the block of currency respectively, while r_f is the risk-free interest rate earned by the foreign currency. By selling the call for R, say, the bank is now exposed to risk arising from the unknown price $X(T)$ at the exercise time, T. Its net position (discounted back to time t) on expiry will be $R - e^{-r(T-t)}(X(T) - K)^+$. Its expected net position will be $R - V(X(t), t)$, so of course taking into account the risk and transaction costs, it will certainly want to make R greater than the Black–Scholes price, $V(X(t), t)$. Theoretically, it can eliminate *all* the risk by continuous hedging, that is by holding a portfolio of not only -1 call options, but also Δ blocks of the currency. Here,

$$
\Delta = \frac{\partial V(x(t), t)}{\partial x(t)} = \frac{\partial c(t)}{\partial x(t)}
$$

$$
= \frac{\partial}{\partial x(t)} \left[e^{-r(T-t)} \int_{-\infty}^{\infty} (x - K)^+ f_{X_{r-r_f}(T)|X(t)=x(t)}(x, x(t)) \, dx \right]
$$

$$
= e^{-r(T-t)} \frac{\partial}{\partial x(t)} \int_{K}^{\infty} (x - K) f_{X_{r-r_f}(T)|X(t)=x(t)}(x, x(t)) \, dx
$$

$$
= e^{-r(T-t)} \frac{\partial}{\partial x(t)} \int_{-d_{r_f} + \sigma\sqrt{T-t}}^{\infty} \left(x(t) \, e^{\left(r - r_f - \sigma^2/2\right)(T-t) + z\sigma\sqrt{T-t}} - K \right) \phi(z) \, dz
$$

$$
= e^{-r_f(T-t)} \Phi\left(d_{r_f} \right)
$$

6.3.5 Discrete hedging

Of course, continuous hedging is not physically possible in practice, and in any case, the transaction costs associated with changing the position in the currency would preclude excessively frequent hedging. With *discrete* hedging at intervals of length h, say, two things will happen. Firstly, recalling that the actual price of the currency follows the stochastic differential equation

$$
\frac{dX}{X} = \mu \, dt + \sigma \, dB,
$$

unless the expected growth rate μ of the currency equals r, then the expected cost of writing and hedging the option will exceed $c(X(t), t)$, the Black–Scholes cost, since the discrete hedging policy is no longer optimal. Secondly, the actual cost of writing and hedging the option will no longer be known with certainty (and equal to $c(X(t), t)$), but will be a random variable. The larger its standard deviation, the greater is the bank's exposure to risk.

Simulation can be used to see these effects of discrete time hedging. For convenience, let $t = 0$. At time zero, the bank must borrow £$\Delta(0) X(0)$ to finance the purchase of euros for the initial hedge. Let $T = mh$. The call option that has been sold will be hedged at times $\{jh, j = 0, \ldots, m\}$. Let Δ_j, X_j, and C_j denote the delta, currency price, and cumulative amount borrowed respectively at time jh. Let μ denote the expected growth rate (assumed constant) of the price of the currency. During $[jh, (j+1)h)$ the amount borrowed has grown to £$C_j \, e^{rh}$ due to interest, while the amount of currency held has effectively grown from Δ_j to $\Delta_j \, e^{r_f h}$ due to the interest earned on that holding. Therefore, at time $(j+1)h$ the required holding in currency must be changed by $\Delta_{j+1} - \Delta_j e^{r_f h}$, at a cost of $X_{j+1} \left(\Delta_{j+1} - \Delta_j e^{r_f h} \right)$. Therefore, this gives the difference equations

$$C_{j+1} = C_j e^{rh} + X_{j+1} \left(\Delta_{j+1} - \Delta_j e^{r_f h} \right),$$

$$X_{j+1} = X_j e^{\left(\mu - \sigma^2/2 \right) h + \sigma B(h)}, \tag{6.23}$$

where

$$\Delta_j = e^{-r_f(m-j)h} \Phi \left(\frac{\left(r - r_f + \sigma^2/2 \right)(m-j)h + \ln\left(X_j/K \right)}{\sigma \sqrt{(m-j)h}} \right),$$

for $j = 0, \ldots, m-1$, subject to $C_0 = \Delta_0 X_0$. Of course, Δ_m will be either 1 or 0, according to whether $X_m > K$, or otherwise. This may be verified from Equation (6.22) where d_{r_f} is either ∞ or $-\infty$. Note the presence of the expected growth rate μ, rather than r, in Equation (6.23). The cost (discounted back to time zero) of writing and hedging the option is $\left(C_m - K 1_{X_m > K} \right) e^{-rT}$, and as $h \to 0$ this becomes $V(X(0), 0)$, which is independent of μ.

In Appendix 6.5 the Maple procedure 'hedge' evaluates this cost for a single realization where the option is hedged $nn + 1$ times. This procedure is called by another procedure 'effic', which samples 10 000 such paths to estimate the expectation and standard deviation of the cost of writing and hedging the option. The sample mean value (sample size = 10 000) is compared with the Black–Scholes price (result (6.21)) computed through the procedure 'bscurrency'. Table 6.1 shows that the mean and standard deviation increase as the frequency of hedging decreases. In this case it is supposed that the risk-free interest rate is 5 % per annum, that the euros earn interest continuously at 3 % per annum, and the expected growth rate of the euros (unknown in reality) is 15 % per annum. The worksheet can be used to verify that, if μ is changed from 0.15 to 0.05, then the expected cost reverts to the Black–Scholes cost (subject to sampling error) of £13.34 for all hedging frequencies. Of course, in this case, there is still variability in the actual hedging cost which reflects the risk of not hedging continuously.

Table 6.1 Cost of writing and discrete hedging on call option: $r = 0.05$ p.a., $r_f = 0.03$ p.a., $\mu = 0.15$ p.a., $t = 0$, $T = 0.5$ years, $\sigma = 0.1$ p.a., $X(0) = £680$, $K = £700$

Times hedged	Estimate of expected cost (£)	e.s.e. (£)	Sample standard deviation of cost (£)
∞	13.34	0	0
127	13.41	0.15	1.45
13	14.00	0.05	4.75
5	15.30	0.09	8.67
3	17.27	0.13	13.2
2	21.54	0.22	22.18

6.4 Asian options

6.4.1 Naive simulation

Result (6.15) showed that the price of a European call is the present value of the expected value of the payoff at time T in a risk-neutral world. This applies quite generally to other options as well, and this property will be used to price one of a well-known set of derivatives, known as Asian options.

An Asian option is one where the payoff depends upon the average price of the underlying asset during the option life. The payoff for an *average price call*, with strike price K is

$$(\overline{X}(T) - K)^+ \tag{6.24}$$

where $\overline{X}(T)$ is the *average* asset price during $[0, T]$. The *average price put* has payoff

$$(K - \overline{X}(T))^+, \tag{6.25}$$

an *average strike call* has payoff

$$(X(T) - \overline{X}(T))^+,$$

and an *average strike put* has payoff

$$(\overline{X}(T) - X(T))^+.$$

In all these cases the average is an *arithmetic mean* taken over the lifetime $[0, T]$ of the option. Therefore the price of the option at time t will depend upon the current asset price $X(t)$ and the quantity $I(t) = \int_0^t X(u) \, du$. To simplify matters we will take $t = 0$, so the price depends just on $X(0)$. (If we wish to price an option some time during its life (that is $T > t > 0$) it is a simple matter to convert this to the case $t = 0$. In practice, the average is a *discrete* time one, taken at times $h, 2h, \ldots, nh$ where $T = nh$. Let $X_j = X(jh)$.

We will concentrate on the *average price call*. Let c denote the price of this option (at time 0). Then, assuming a risk-neutral world,

$$X_j = X_{j-1} \exp[(r - \frac{1}{2}\sigma^2)h + \sigma\sqrt{h}Z_j]$$

for $j = 1, .., n$ where $\{Z_j\}$ are independent $N(0, 1)$ random variables. Therefore

$$c = \mathrm{e}^{-rT} E_{\mathbf{Z} \sim N(\mathbf{0}, \mathbf{I})} \left[\left(\frac{1}{n} \sum_{j=1}^{n} X_j \right) - K \right]^+ \tag{6.26}$$

This option is a *path-dependent* option. Such options usually have no closed-form solution as they involve evaluation of a definite multiple integral (of dimension n in this case). However, a simulation is easy to program and appears in Appendix 6.6.1 as the procedure 'asiannaive'. This simulates *npath* paths, each one producing an independent payoff. The estimator is

$$\widehat{c} = \frac{e^{-rT}}{npath} \sum_{i=1}^{npath} \left[\left(\frac{1}{n} \sum_{j=1}^{n} X_{ij} \right) - K \right]^+$$

where

$$X_{ij} = X_{i,j-1} \exp[(r - \frac{1}{2}\sigma^2)h + \sigma\sqrt{h}Z_{ij}]$$

for $i = 1, \ldots, npath$, $j = 1, \ldots, n$, and where $X_{i,0} = X(0)$ and $\{Z_{ij}\}$ are independent $N(0, 1)$ random variables. If P_i denotes the discounted payoff for the ith path and $\overline{P}$ the average of the $\{P_i\}$, then

$$\widehat{\text{Var}(\widehat{c})} = \frac{1}{npath(npath - 1)} \sum_{i=1}^{npath} (P_i - \overline{P})^2. \qquad (6.27)$$

In the specimen example considered in the worksheet, $npath = 25\,000$, and therefore Equation (6.27) can be taken to be the population variance, $\text{Var}(\widehat{c})$, with negligible error. This will be useful when we come to estimate a confidence interval for a variance reduction ratio in the next section. For the moment we will note that in view of the large sample size a 95 % confidence interval for c is

$$\left(\widehat{c} - 1.96 \ \widehat{\text{Var}(\widehat{c})}, \widehat{c} + 1.96 \ \widehat{\text{Var}(\widehat{c})} \right).$$

6.4.2 Importance and stratified version

An attempt will now be made to improve the efficiency of the previous procedure by incorporating importance and stratified sampling. The method used is motivated by a paper by Glasserman *et al.* (1999). There are some differences, however. The derivation of the importance sampling distribution is simplified, the connection between arithmetic and geometric averages is exploited, and the use of post stratification simplifies the implementation.

From Equation (6.26) let the discounted payoff be $P(\mathbf{Z})$ where

$$P(\mathbf{Z}) = e^{-rT} \left[\left(\frac{1}{n} \sum_{j=1}^{n} X_j \right) - K \right]^+$$

and where $\mathbf{Z}' = (Z_1, \ldots, Z_n)$. Then

$$c = E_{\mathbf{Z} \sim N(\mathbf{0}, \mathbf{I})} [P(\mathbf{Z})].$$

Now suppose that the importance sampling distribution is $N(\boldsymbol{\beta}, \mathbf{I})$ for some row vector $\boldsymbol{\beta}' = (\beta_1, \ldots, \beta_n)$. Then

$$c = E_{\mathbf{Z} \sim N(\boldsymbol{\beta}, \mathbf{I})} \left[\frac{P(\mathbf{Z}) \exp\{-\mathbf{Z}'\mathbf{Z}/2\}}{\exp\{-(\mathbf{Z} - \boldsymbol{\beta})'(\mathbf{Z} - \boldsymbol{\beta})/2\}} \right]$$

$$= E_{\mathbf{Z} \sim N(\boldsymbol{\beta}, \mathbf{I})} \left[P(\mathbf{Z}) \exp\left\{ \frac{\boldsymbol{\beta}'\boldsymbol{\beta}}{2} - \boldsymbol{\beta}'\mathbf{Z} \right\} \right]. \qquad (6.28)$$

Recall from result (5.13) that a bound on the variance of $P(\mathbf{Z}) \exp\{\boldsymbol{\beta}'\boldsymbol{\beta}/2 - \boldsymbol{\beta}'\mathbf{Z}\}$ is obtained by choosing the optimal envelope for a rejection algorithm that generates variates from a density proportional to $P(\mathbf{Z}) \exp\{-\mathbf{Z}'\mathbf{Z}/2\}$. Accordingly, we hope to find a good choice for $\boldsymbol{\beta}$ by finding z^* and $\boldsymbol{\beta}^*$ that solves

$$P(\mathbf{z}^*) \exp\left\{\frac{\boldsymbol{\beta}^{*\prime}\boldsymbol{\beta}^*}{2} - \boldsymbol{\beta}^{*\prime}\mathbf{z}^*\right\} = \min_{\boldsymbol{\beta}} \max_{P(z)>0} \left[P(z)\exp\left\{\frac{\boldsymbol{\beta}'\boldsymbol{\beta}}{2} - \boldsymbol{\beta}'z\right\}\right]$$

$$= \min_{\boldsymbol{\beta}} \max_{P(z)>0} \left[\exp\left\{\ln P(z) + \frac{\boldsymbol{\beta}'\boldsymbol{\beta}}{2} - \boldsymbol{\beta}'z\right\}\right].$$

There is a stationary point of this function at $\boldsymbol{\beta} = z$ and $\boldsymbol{\beta} = \nabla P(z)/P(z)$. It is conjectured that this is a saddle point and therefore that

$$\boldsymbol{\beta}^* = \frac{\nabla P(\boldsymbol{\beta}^*)}{P(\boldsymbol{\beta}^*)}. \tag{6.29}$$

It is possible to solve this numerically (Glasserman et al., 1999), but a simplification will be made by noting that empirical evidence suggests that for all $P(z) > 0$, it is often the case that $P(z)$ is approximately equal to the value it would take if the average over time was taken to be a *geometric* average, rather than an arithmetic one. Accordingly, a $\boldsymbol{\beta}^*$ is now found that solves $\boldsymbol{\beta}^* = \nabla P_g(\boldsymbol{\beta}^*)/P_g(\boldsymbol{\beta}^*)$ where $P_g(z)$, the discounted geometric average payoff, is given by

$$P_g(z) = e^{-rT}\left[\left(\prod_{j=1}^{n} X_0 \exp[(r - \frac{1}{2}\sigma^2)jh + \sigma\sqrt{h}\sum_{i=1}^{j} z_j]\right)^{1/n} - K\right]^+$$

$$= e^{-rT}\left[X_0 \exp\left[\left(r - \frac{1}{2}\sigma^2\right)\left(\frac{n+1}{2}\right)h + \frac{\sigma\sqrt{h}}{n}\sum_{i=1}^{n}(n-i+1)z_j\right] - K\right]^+ \tag{6.30}$$

where $X_0 = X(0)$. A solution is

$$\boldsymbol{\beta}_i^* = \left[\frac{P_g(\boldsymbol{\beta}^*) + Ke^{-rT}}{P_g(\boldsymbol{\beta}^*)}\right]\left(\frac{\sigma\sqrt{h}}{n}\right)(n-i+1) \tag{6.31}$$

for $i = 1, \ldots, n$. Therefore

$$\boldsymbol{\beta}_i^* = \lambda(n-i+1) \tag{6.32}$$

for some $\lambda > 0$. Substituting into Equation (6.30) it can be seen that $P_g(\boldsymbol{\beta}^*)$ is now a function of λ. When $P_g(\boldsymbol{\beta}^*) > 0$ call this function $e^{-rT}Q_g(\lambda)$. Then

$$\lambda = \frac{Q_g(\lambda) + K}{Q_g(\lambda)}\left(\frac{\sigma\sqrt{h}}{n}\right) \tag{6.33}$$

where

$$Q_g(\lambda) = X_0 \exp\left[\left(r - \frac{1}{2}\sigma^2\right)\left(\frac{n+1}{2}\right)h + \frac{\lambda\sigma\sqrt{h}}{n}\sum_{i=1}^{n}(n-i+1)^2\right] - K$$

$$= X_0 \exp\left[\left(r - \frac{1}{2}\sigma^2\right)\left(\frac{n+1}{2}\right)h + \frac{1}{6}\lambda\sigma\sqrt{h}(n+1)(2n+1)\right] - K.$$

Since the search is over those λ for which $Q_g(\lambda) > 0$, the search can be restricted to

$$\lambda > \left[\frac{\ln(K/X_0) - (r - \sigma^2/2)[(n+1)/2]h}{\sigma\sqrt{h}(n+1)(2n+1)/6} \right]^+. \tag{6.34}$$

Since $Q_g(\lambda)$ is increasing in λ, it is clear from Equation (6.33) that there is exactly one root. In summary, instead of sampling $Z_i \sim N(0,1)$ we will sample from $N(\lambda[n-i+1], 1)$ where λ is obtained from a one-dimensional search.

We now turn to a good stratification strategy. The price, c_g, of a *geometric* average price Asian call is given by

$$c_g = E_{\mathbf{Z} \sim N(\boldsymbol{\beta}^*, \mathbf{I})} \left[\exp\left\{ \ln P_g(\mathbf{Z}) + \frac{\boldsymbol{\beta}^{*'} \boldsymbol{\beta}^*}{2} - \boldsymbol{\beta}^{*'} \mathbf{Z} \right\} \right].$$

where $P_g(\mathbf{Z})$ is the discounted payoff. Observe from Equation (6.30) and (6.32) that $P_g(\mathbf{Z})$ is a function of $\boldsymbol{\beta}^{*'} \mathbf{Z}$ only. Consequently, the estimator

$$\exp\left\{ \ln P_g(\mathbf{Z}) + \frac{\boldsymbol{\beta}^{*'} \boldsymbol{\beta}}{2} - \boldsymbol{\beta}^{*'} \mathbf{Z} \right\}$$

is also a function of $\boldsymbol{\beta}^{*'} \mathbf{Z}$ only, and therefore a stratification variable based upon $\boldsymbol{\beta}^{*'} \mathbf{Z}$ will, with 'infinitely fine' stratification, give c_g with no error at all! Accordingly, a stratification variable is defined as

$$
\begin{aligned}
X &= \frac{\boldsymbol{\beta}^{*'} \mathbf{Z} - E_{\mathbf{Z} \sim N(\boldsymbol{\beta}^*, \mathbf{I})} (\boldsymbol{\beta}^{*'} \mathbf{Z})}{\sqrt{\boldsymbol{\beta}^{*'} \boldsymbol{\beta}^*}} \\
&= \frac{\boldsymbol{\beta}^{*'} \mathbf{Z} - \boldsymbol{\beta}^{*'} \boldsymbol{\beta}^*}{\sqrt{\boldsymbol{\beta}^{*'} \boldsymbol{\beta}^*}} \\
&= \frac{\sum_{i=1}^{n} \lambda(n-i+1) Z_i - \lambda^2 \sum_{i=1}^{n} (n-i+1)^2}{\sqrt{\lambda^2 \sum_{i=1}^{n} (n-i+1)^2}} \\
&= \frac{\sum_{i=1}^{n} (n-i+1) Z_i - \lambda n(n+1)(2n+1)/6}{\sqrt{n(n+1)(2n+1)/6}}.
\end{aligned}
$$

Clearly, $X \sim N(0,1)$. In fact, simulation would not be used for a geometric average price Asian call option, since a closed-form expression for c_g is available (see Problem 6). However, given the similarity between arithmetic and geometric averages, it is hoped that the optimal stratification variable for the geometric will yield a good stratification for the arithmetic. Employing the standard stratification algorithm with m strata, it is then necessary, for each path generated in stratum j to sample $X \sim N(0,1)$ subject to $X \in [\Phi^{-1}[(j-1)/m], \Phi^{-1}(j/m)]$. One way is to set $X = \Phi^{-1}[(j-1+U)/m]$ where $U \sim N(0,1)$. There are accurate and efficient approximations to the inverse cumulative normal; see, for example, the Beasley–Springer (1977) approximation as modified by Moro (1995). The Maple one is particularly slow. This fact motivates another approach, whereby $\Phi^{-1}(j/m)$ is evaluated just once for each $j = 1, \ldots, m-1$ at the beginning of a simulation. Then for a path in the jth stratum, generate X using rejection with a uniform envelope if $j = 2, \ldots, m - 1$, and with an envelope proportional to $x \exp(-x^2/2)$ for $j = 1$ and m (see Problem 7). The remaining problem is to sample from the conditional distribution of $\mathbf{Z} \sim N(\boldsymbol{\beta}^*, \mathbf{I})$ given that $X = x$. This is a standard problem concerning a multivariate normal distribution. One way that is efficient and requires only $O(n)$ multiplications is suggested by Glasserman *et al.* (1999), who set $\boldsymbol{v} = \boldsymbol{\beta}^*/\sqrt{\boldsymbol{\beta}^{*'} \boldsymbol{\beta}^*}$, samples $\mathbf{W} \sim N(\mathbf{0}, \mathbf{I})$, and then put

$$\mathbf{Z} = \boldsymbol{\beta}^* + \mathbf{W} + (x - \boldsymbol{v}'\mathbf{W}) \boldsymbol{v}.$$

Alternatively, we can, as described in Chapter 5, avoid sampling from the conditional distribution by employing *post stratification*. This leads to a somewhat shorter procedure, which gives comparable precision and efficiency providing the expected number in each stratum is of the order of 20 or above (Cochran, 1977, p. 134). A Maple procedure that does this, 'asianimppoststrat', is shown in Appendix 6.6.2. Now numerical results for the procedures 'asiannaive' and 'asianimppoststrat' will be compared in Table 6.2. The parameter values are $x_0 = 50$, $r = 0.05$, $T = 1$, and $n = 16$.

Note that an approximate 95% confidence interval is given for the variance reduction ratio. This is obtained as follows. For naive sampling there are $npath_1 (= 25\,000)$ independent payoffs, and so without too much error we may take $\mathrm{Var}(\widehat{c}_1) = \widehat{\mathrm{Var}(\widehat{c}_1)}$, that is the point estimate of variance of $\widehat{c}_1$. For the importance plus post stratified estimator, $\widehat{c}_2$, there are p (=number of replications= 100) independent estimates of the price. Each one is made up of an average of m (=number of strata = 100) stratum sample means, where the expected number of payoffs in each sample mean is $npath_2/m = 2500/100 = 25$. This latter number is probably sufficient to justify a near-normal distribution for the stratum sample means, and therefore of the individual replication estimates. Therefore if $\widehat{\mathrm{Var}(\widehat{c}_2)}$ is a point estimate of $\mathrm{Var}(\widehat{c}_2)$, an approximate 95 % confidence interval for $[\mathrm{Var}(\widehat{c}_2)]^{-1}$ is $\left[\widehat{\mathrm{Var}(\widehat{c}_2)}\right]^{-1} \left(\chi^2_{p-1,0.025}/(p-1), \chi^2_{p-1,0.975}/(p-1)\right)$. The sample sizes for

Table 6.2   A comparison of naive Monte Carlo (asiannaive) with combined importance and post stratification (asianimppoststrat) for an Asian average price call

		asiannaive[a]		asianimppoststrat[b]		
σ	K	$\widehat{c}_1$	$\sqrt{\widehat{\mathrm{Var}(\widehat{c}_1)}}$	$\widehat{c}_2$	$\sqrt{\widehat{\mathrm{Var}(\widehat{c}_2)}}$	v.r.r.[c]
0.3	55	2.2149	0.0300	2.2116	0.000313	(681,1192)
0.3	50	4.1666	0.0399	4.1708	0.000374	(843,1476)
0.3	45	7.1452	0.0486	7.1521	0.000483	(750,1313)
0.1	55	0.2012	0.00462	0.2024	0.0000235	(2864,5014)
0.1	50	1.9178	0.0140	1.9195	0.0000657	(3365,5890)
0.1	45	6.0482	0.0186	6.0553	0.000191	(703,1230)

[a] 25 000 paths.
[b] 100 replications, each consisting of 2500 paths over 100 equiprobable strata.
[c] An approximate 95 % confidence interval for the variance reduction ratio.

the two simulations are $npath_1$ and $p \times npath_2$ respectively. Therefore, an approximate 95 % confidence interval for the variance reduction ratio is

$$\frac{npath_1 \times \widehat{\mathrm{Var}}\,(\widehat{c_1})}{p \times npath_2 \times \widehat{\mathrm{Var}}\,(\widehat{c_2})} \left(\frac{\chi^2_{p-1,0.025}}{p-1}, \frac{\chi^2_{p-1,0.975}}{p-1} \right).$$

The last column of Table 6.2 shows that the combination of importance sampling and post stratification is highly effective. Similar results and variance reductions are achievable with a standard stratified sampling algorithm in which exactly one path is generated in each stratum. The wide confidence intervals in Table 6.2 are to be expected when estimating variances with a small number of replications.

6.5 Basket options

Consider a basket (or portfolio) consisting of n assets. The basket contains a quantity q_i of asset i where $i = 1, \ldots, n$. Let μ_i, σ_i, and $X_i(t)$ denote the mean growth rate, volatility, and price of one unit of the ith asset at time t respectively, where

$$\frac{dX_i}{X_i} = \mu_i\,dt + \sigma_i\sqrt{dt}\,dB_i,$$

and where $\{B_i(t)\}$ are standard Brownian motions that are not necessarily independent. Suppose the current time is s and that $X_i(s) = x_i(s)$ and $X_j(s) = x_j(s)$. Choose any $t \geq s$. Then

$$\ln X_i(t) = \ln x_i(s) + \left(r - \frac{1}{2}\sigma_i^2 \right)(t - s) + \sigma_i\sqrt{t - s}\,Z_i$$

and

$$\ln X_j(t) = \ln x_j(s) + \left(r - \frac{1}{2}\sigma_j^2 \right)(t - s) + \sigma_j\sqrt{t - s}\,Z_j. \tag{6.35}$$

where $\mathbf{Z}' = (Z_1, \ldots, Z_n) \sim N(0, \boldsymbol{\rho})$ and $\ln X_i(t) - \ln x_i(s)$ and $\ln X_j(t) - \ln x_j(s)$ are the *returns* in $[s, t]$ from assets i and j respectively in a risk-neutral world. Therefore, the variance of return on the assets are $\sigma_i^2\,(t - s)$ and $\sigma_j^2\,(t - s)$ respectively, and the covariance between the returns is $\sigma_i\sigma_j\,(t - s)\,\mathrm{cov}\,(Z_i, Z_j) = \rho_{ij}\sigma_i\sigma_j\,(t - s)$. It follows that ρ_{ij} is the correlation between the returns for all s and $t \geq s$. The value of the basket at time t (the spot price) is $\sum_{i=1}^{n} q_i X_i(t)$ and the price of a European call option at time t, with strike price K and exercise time T, is the discounted expected payoff in a risk-neutral world, that is

$$c = e^{-r(T-t)} E_{\mathbf{Z} \sim N(0, \boldsymbol{\rho})} \left[\sum_{i=1}^{n} q_i x_i(t) e^{\left(r - \sigma_i^2/2 \right)(T-t) + \sigma_i\sqrt{T-t}\,Z_i} - K \right]^+. \tag{6.36}$$

The matrix $\boldsymbol{\rho}$, being positive definite, can be decomposed as

$$\boldsymbol{\rho} = \boldsymbol{b}\boldsymbol{b}'$$

where b is the unique lower triangular matrix. This is known as a *Cholesky decomposition*. If $Y \sim N(0, I)$ then the density of Y is proportional to $\exp\left(-\frac{1}{2}y'y\right)$. Let $Z = bY$ and let J denote the Jacobian of the transformation from y to z. Then J is independent of z. The density of Z is proportional to $\exp\left(-\frac{1}{2}\left[b^{-1}z\right]'b^{-1}z\right)|J|$ which is proportional to $\exp\left(-\frac{1}{2}z'\left[bb'\right]^{-1}z\right) = \exp\left(-\frac{1}{2}z'\rho^{-1}z\right)$. Therefore

$$Z = bY \sim N(0, \rho).$$

For example, suppose

$$\rho = \begin{pmatrix} 1.0 & 0.7 & 0.5 & 0.3 \\ 0.7 & 1.0 & 0.6 & 0.2 \\ 0.5 & 0.6 & 1.0 & 0.4 \\ 0.3 & 0.2 & 0.4 & 1.0 \end{pmatrix} \tag{6.37}$$

Then Maple computes b as follows:

```
with(LinearAlgebra):
> rho := Matrix([[1,0.7,0.5,0.3],[0.7,1,0.6,0.2],[0.5,0.6,1,0.4],[0.3,0.2,0.4,1]]):
> b := LUDecomposition(A, method='Cholesky');
  [1]
  [0.699999999999999954 , 0.714142842854284976 , 0. , 0.]
  [0.500000000000000000 , 0.350070021007002463 , 0.792118034381339431 , 0.]
  [0.299999999999999988 , -0.0140028008402800531 , 0.321797951467419185 ,
   0.897914249803398512]
```

and so

$$\begin{pmatrix} Z_1 \\ Z_2 \\ Z_3 \\ Z_4 \end{pmatrix} = \begin{pmatrix} 1 & & & \\ 0.7 & 0.714143 & & \\ 0.5 & 0.350070 & 0.792118 & \\ 0.3 & -0.014003 & 0.321798 & 0.897914 \end{pmatrix} \begin{pmatrix} Y_1 \\ Y_2 \\ Y_3 \\ Y_4 \end{pmatrix}$$

In Appendix 6.7.1 the procedure 'basket' uses a naive Monte Carlo to estimate c.

In order to improve the estimate, importance sampling will now be employed with post stratification. Referring to Equation (6.36), define a set of weights using

$$w_i = \frac{q_i x_i(t) e^{(r-\sigma_i^2/2)(T-t)}}{\sum_{i=1}^n q_i x_i(t) e^{(r-\sigma_i^2/2)(T-t)}}$$

for $i = 1, \ldots, n$, and put $x_0 = \sum_{i=1}^n q_i x_i(t) e^{(r-\sigma_i^2/2)(T-t)}$. Then

$$c = e^{-r(T-t)} x_0 E_{Z \sim N(0,\rho)} \left[\sum_{i=1}^n w_i e^{\sigma_i \sqrt{T-t} Z_i} - \frac{K}{x_0} \right]^+$$

$$= e^{-r(T-t)} x_0 E_{Z \sim N(0,I)} \left[\sum_{i=1}^n w_i e^{\sigma_i \sqrt{T-t}(bZ)_i} - \frac{K}{x_0} \right]^+$$

Now, instead of sampling from $N(0, \mathbf{I})$, an importance sampling distribution $N(\boldsymbol{\beta}, \mathbf{I})$ is used. Thus

$$c = \mathrm{e}^{-r(T-t)} x_0 E_{\mathbf{Z} \sim N(\boldsymbol{\beta}, \mathbf{I})} \left\{ \left[\sum_{i=1}^{n} w_i \mathrm{e}^{\sigma_i \sqrt{T-t}(b\mathbf{Z})_i} - \frac{K}{x_0} \right]^+ \mathrm{e}^{\frac{1}{2}\boldsymbol{\beta}'\boldsymbol{\beta} - \boldsymbol{\beta}'\mathbf{Z}} \right\}. \tag{6.38}$$

A good $\boldsymbol{\beta}$ and stratification variable is now chosen. Observe that if $\sum_{i=1}^{n} w_i \sigma_i \sqrt{T-t}(b\mathbf{Z})_i$ is small, then

$$\sum_{i=1}^{n} w_i \mathrm{e}^{\sigma_i \sqrt{T-t}(b\mathbf{Z})_i} \approx \exp\left[\sum_{i=1}^{n} w_i \sigma_i \sqrt{T-t}(b\mathbf{Z})_i \right],$$

since $\sum_{i=1}^{n} w_i = 1$. Therefore, the aim is to find a good $\boldsymbol{\beta}$ and stratification variable for an option with price c_g, where

$$c_g = \mathrm{e}^{-r(T-t)} x_0 E_{\mathbf{Z} \sim N(\boldsymbol{\beta}, \mathbf{I})} \left\{ \left[\exp\left(\sum_{i=1}^{n} w_i \sigma_i \sqrt{T-t}(b\mathbf{Z})_i \right) - \frac{K}{x_0} \right]^+ \mathrm{e}^{\frac{1}{2}\boldsymbol{\beta}'\boldsymbol{\beta} - \boldsymbol{\beta}'\mathbf{Z}} \right\}$$

Now define $c_i = w_i \sigma_i$. Then

$$c_g = \mathrm{e}^{-r(T-t)} x_0 E_{\mathbf{Z} \sim N(\boldsymbol{\beta}, \mathbf{I})} \left\{ \left[\exp\left(\sqrt{T-t}\, c'b\mathbf{Z} \right) - \frac{K}{x_0} \right]^+ \mathrm{e}^{\frac{1}{2}\boldsymbol{\beta}'\boldsymbol{\beta} - \boldsymbol{\beta}'\mathbf{Z}} \right\} \tag{6.39}$$

Following the same approach as for the Asian option, a good choice of $\boldsymbol{\beta}$ is $\boldsymbol{\beta}^*$ where

$$\boldsymbol{\beta}^* = \nabla \ln\left[\exp\left(\sqrt{T-t}\, c'b\boldsymbol{\beta}^* \right) - \frac{K}{x_0} \right]$$

Therefore,

$$\boldsymbol{\beta}^* = b'c \frac{\sqrt{T-t}\exp\left(\sqrt{T-t}\, c'b\boldsymbol{\beta}^* \right)}{\exp\left(\sqrt{T-t}\, c'b\boldsymbol{\beta}^* \right) - K/x_0}$$

where $\exp\left(\sum_{i=1}^{n} \sqrt{T-t}\, c'b\boldsymbol{\beta}^* \right) - K/x_0 > 0$. Therefore

$$\boldsymbol{\beta}^* = \lambda b'c \tag{6.40}$$

where

$$\lambda = \frac{\sqrt{T-t}\exp\left(\lambda\sqrt{T-t}\, c'\boldsymbol{\rho}c \right)}{\exp\left(\lambda\sqrt{T-t}\, c'\boldsymbol{\rho}c \right) - K/x_0} \tag{6.41}$$

and where $\lambda > \ln(K/x_0)/(\sqrt{T-t}\, c'\boldsymbol{\rho}c)$. With this choice of λ, and therefore $\boldsymbol{\beta}^*$, we obtain from Equation (6.39)

$$c_g = \mathrm{e}^{-r(T-t)} x_0 E_{\mathbf{Z} \sim N(\boldsymbol{\beta}, \mathbf{I})} \left\{ \left[\exp\left(\lambda^{-1}\sqrt{T-t}\boldsymbol{\beta}^{*\prime}\mathbf{Z} \right) - \frac{K}{x_0} \right]^+ \mathrm{e}^{\frac{1}{2}\boldsymbol{\beta}^{*\prime}\boldsymbol{\beta}^* - \boldsymbol{\beta}^{*\prime}\mathbf{Z}} \right\}.$$

Table 6.3 Results for basket option, using naive Monte Carlo (basket) and importance sampling with post stratification (basketimppostratv2)

		basket[a]		basketimppostratv2[b]		
σ	K	$\hat{c}$	$\sqrt{\widehat{\text{Var}}(\hat{c})}$	$\hat{c}$	$\sqrt{\widehat{\text{Var}}(\hat{c})}$	v.r.r.[c]
σ_1	600	84.03	0.881	85.18	0.0645	(96,306)
σ_1	660	47.21	0.707	48.03	0.0492	(107,338)
σ_1	720	23.49	0.514	24.04	0.0282	(171,544)
σ_2	600	71.68	0.390	72.25	0.0390	(51,164)
σ_2	660	22.74	0.283	23.09	0.0096	(447,1420)
σ_2	720	2.74	0.100	2.87	0.00297	(585,1869)

[a] 10 000 paths.
[b] 25 replications, each consisting of 400 paths over 20 equiprobable strata.
[c] Approximate 95 % confidence interval for the variance reduction ratio.

Since this is the expectation of a function of $\boldsymbol{\beta}^{*\prime}\mathbf{Z}$ only, the ideal stratification variable for the option with price c_g is

$$X = \frac{\boldsymbol{\beta}^{*\prime}\mathbf{Z} - \boldsymbol{\beta}^{*\prime}\boldsymbol{\beta}^*}{\sqrt{\boldsymbol{\beta}^{*\prime}\boldsymbol{\beta}^*}} \sim N(0, 1). \qquad (6.42)$$

From Equation (6.38), for the original option with price c the estimator

$$e^{-r(T-t)}x_0 \left[\sum_{i=1}^{n} w_i e^{\sigma_i \sqrt{T-i}(b\mathbf{Z})_i} - \frac{K}{x_0} \right]^{+} e^{\frac{1}{2}\boldsymbol{\beta}^{*\prime}\boldsymbol{\beta}^* - \boldsymbol{\beta}^{*\prime}\mathbf{Z}} \qquad (6.43)$$

is used, where $\mathbf{Z} \sim N(\boldsymbol{\beta}^*, \mathbf{I})$, $\boldsymbol{\beta}^*$ is determined from Equations (6.40) and (6.41), and Equation (6.42) defines the stratification variable.

The procedure 'basketimppoststratv2' in Appendix 6.7.2 implements this using post stratified sampling. Table 6.3 compares results using this and the naive method for a call option on an underlying basket of four assets. The data are $r = 0.04, x' = (5, 2.5, 4, 3), q' = (20, 80, 60, 40), T = 0.5, t = 0$, and ρ as given in Equation (6.37). Two sets of cases were considered, one with $\sigma = \sigma_1 = (0.3, 0.2, 0.3, 0.4)'$, the other with $\sigma = \sigma_2 = (0.05, 0.1, 0.15, 0.05)'$. The *spot price* is $q'x = 660$.

6.6 Stochastic volatility

Although the Black–Scholes model is remarkably good, one of its shortcomings is that it assumes a constant volatility. What happens if the parameter σ is replaced by a known function of time $\sigma(t)$? Then

$$\frac{dX}{X} = \mu\,dt + \sigma(t)\,dB_1(t),$$

so using Itô's lemma

$$d(\ln X) = \frac{dX}{X} - \frac{\sigma^2(t) X^2}{2X^2} dt$$

$$= \mu dt + \sigma(t) dB(t) - \frac{1}{2}\sigma^2(t) dt$$

$$= \left[\mu - \frac{1}{2}\sigma^2(t)\right] dt + \sigma(t) dB_1(t). \tag{6.44}$$

Now define an average squared volatility, $V(t) = (1/t)\int_0^t \sigma^2(u)/2 \, du$. Given that $X(0) = x_0$, Equation (6.44) can be integrated to give

$$X(t) = x_0 \exp\left\{\left[\mu - \frac{1}{2}V(t)\right]t + \int_0^t \sigma(u) \, dB_1(u)\right\}$$

$$= x_0 \exp\left\{\left[\mu - \frac{1}{2}V(t)\right]t + \sqrt{V(t)} \, B_1(t)\right\}.$$

Using the principle that the price at time zero of a European call with exercise time T and strike K is the discounted (present) value of the payoff in a risk-neutral world, the price is given by

$$c = e^{-rT} E_{Z \sim N(0,1)} \left[x(0) \, e^{[r - V(T)/2]T + \sqrt{TV(T)}Z} - K\right]^+. \tag{6.45}$$

Therefore, the usual Black–Scholes equation may be used by replacing the constant volatility with the average squared volatility.

A more realistic model is one that models the variable $\sigma(t)$ as a function of a stochastic process $Y(t)$. An example is given in Figure 6.1. For example, Fouque and Tullie (2002) suggested using an Ornstein–Uhlenbeck process (see, for example, Cox and Miller, 1965, pp. 225–9),

$$dY = \alpha(m - Y) \, dt + \beta \, dB_2(t) \tag{6.46}$$

where the correlation between the two standard Brownian motions $\{B_1(t)\}$ and $\{B_2(t)\}$ is ρ, and where $\alpha, \beta > 0$. A possible choice ensuring that $\sigma(t) > 0$ is

$$\sigma(t) = e^{Y(t)}.$$

A rationale for Equation (6.46) is that the further Y strays from m the larger the drift towards m. For this reason, $\{Y(t)\}$ is an example of a *mean reverting random walk*. To solve Equation (6.46), Itô's lemma is used to give

$$d[(m - Y) e^{\alpha t}] = -e^{\alpha t} dY + \alpha e^{\alpha t}(m - Y) \, dt$$

$$= -e^{\alpha t}[\alpha(m - Y) \, dt + \beta \, dB_2(t)] + \alpha e^{\alpha t}(m - Y) \, dt$$

$$= -e^{\alpha t}\beta \, dB_2(t)$$

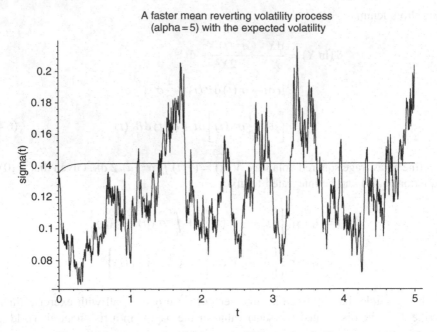

Figure 6.1 An exponential Ornstein–Uhlenbeck volatility process

Integrating between s and $t\,(>s)$ gives

$$[m - Y(t)]\mathrm{e}^{\alpha t} - [m - Y(s)]\mathrm{e}^{\alpha s} = -\beta \int_s^t \mathrm{e}^{\alpha u}\, dB_2(u)$$

$$= -\beta \sqrt{\frac{\int_s^t \mathrm{e}^{2\alpha u}\, du}{t - s}}\, \{B_2(t) - B_2(s)\}$$

$$= -\sqrt{\frac{\beta^2 (\mathrm{e}^{2\alpha t} - \mathrm{e}^{2\alpha s})}{2\alpha (t - s)}}\, \{B_2(t) - B_2(s)\}.$$

Now define $\nu^2 = \beta^2/2\alpha$. Then

$$[m - Y(t)]\mathrm{e}^{\alpha t} - [m - Y(s)]\mathrm{e}^{\alpha s} = -\nu \sqrt{\frac{\mathrm{e}^{2\alpha t} - \mathrm{e}^{2\alpha s}}{t - s}}\, \{B_2(t) - B_2(s)\} \qquad (6.47)$$

or

$$Y(t) = \mathrm{e}^{-\alpha(t-s)} Y(s) + \left(1 - \mathrm{e}^{-\alpha(t-s)}\right) m + \nu \sqrt{1 - \mathrm{e}^{-2\alpha(t-s)}}\, Z_{(s,t)}$$

where $\{Z_{(s,t)}\}$ are independent $N(0,1)$ random variables for disjoint intervals $\{(s,t)\}$. Putting $s = 0$ and letting $t \to \infty$, it is apparent that the stationary distribution of the process is $N(m, \nu^2)$, so if $\sigma(t) = \mathrm{e}^{Y(t)}$, then for large t, $E[\sigma(t)] \sim \exp(m + \nu^2/2)$. To simulate an Ornstein–Uhlenbeck (OU) process in $[0, T]$ put $T = nh$ and $Y_j = Y(jh)$. Then

$$Y_j = \mathrm{e}^{-\alpha h} Y_{j-1} + \left(1 - \mathrm{e}^{-\alpha h}\right) m + \nu \sqrt{1 - \mathrm{e}^{-2\alpha h}}\, Z_j$$

where Z_j are independent $N(0, 1)$ random variables for $j = 1, \ldots, n$. Note that there is no Euler approximation here. The generated discrete time process is an exact copy at times $0, h, \ldots, nh$ of a randomly generated continuous time OU process. The procedure 'meanreverting' in Appendix 6.8 simulates the volatility process $\{e^{Y(t)}\}$. Additional plots in the Appendix show the effect of changing α. As α increases the process reverts to the mean more quickly.

Now we turn to the pricing of a European call option subject to a stochastic volatility. The dynamics of this are

$$\frac{dX}{X} = \mu \, dt + \sigma(Y) \, dB_1, \tag{6.48}$$

$$dY = \alpha(m - Y) \, dt + \beta \left\{ \rho \, dB_1 + \sqrt{1 - \rho^2} \, dB_2 \right\}, \tag{6.49}$$

$$\sigma(Y) = \exp(Y), \tag{6.50}$$

where B_1 and B_2 are independent standard Brownian motions. Note the correlation of ρ between the instantaneous return on the asset and dY that drives the volatility $\sigma(Y)$. There are now two sources of randomness and perfect hedging would be impossible unless there were another traded asset that is driven by B_2. As it stands, in order to price a derivative of X and Y, theory shows that the drift in Equations (6.48) and (6.49) should be reduced by the corresponding *market price of risk* multiplied by the volatility of X and Y respectively. The resulting drift is called the *risk-neutral drift*. Call the two market prices λ_X and λ_Y. A market price of risk can be thought of in the following way. For the process X, say, there are an infinite variety of derivatives. Suppose $d^{(i)}$ and $s^{(i)}$ are the instantaneous drift and volatility respectively of the ith one. To compensate for the risk, an investor demands that $d^{(i)} = r + \lambda_X s^{(i)}$ for all i, where λ_X is a function of the process $\{X\}$ only, and not of any derivative of it. Remembering that a derivative is a tradeable asset, we notice that one derivative of X is the asset itself, so $\mu = r + \lambda_X \sigma(Y)$. Therefore the risk-neutral drift for Equation (6.48) is $\mu - \lambda_X \sigma(Y) = r$, which is consistent with what has been used previously. In the case of Y, volatility is not a tradeable asset so we cannot reason similarly; and can only say that the risk-neutral drift for Equation (6.49) is $\alpha(m - Y) - \lambda_Y \beta$. It turns out that

$$\lambda_Y = \rho \lambda_X + \sqrt{1 - \rho^2} \, \gamma(Y)$$

$$= \rho \left(\frac{\mu - r}{\sigma(Y)} \right) + \sqrt{1 - \rho^2} \, \gamma(Y)$$

(Hobson, 1998) where $\rho \lambda_X$ and $\sqrt{1 - \rho^2} \gamma(Y)$ are the components arising from randomness in B_1 and B_2. The fact that both λ_X and $\gamma(Y)$ are unknown is unfortunate and accounts for the fact that there is no unique pricing formula for stochastic volatility. Given some view on what λ_Y should be, a derivative is priced by solving

$$\frac{dX}{X} = r \, dt + \sigma(Y) \, dB_1,$$

$$dY = [\alpha(m - Y) - \lambda_Y \beta] \, dt + \beta \left\{ \rho \, dB_1 + \sqrt{1 - \rho^2} dB_2 \right\}, \tag{6.51}$$

$$\sigma(Y) = \exp(Y). \tag{6.52}$$

The call option price is the present value of the expected payoff in a risk-neutral world. It is now a simple matter to modify the procedure 'meanreverting' to find the payoff, $[X(T) - K]^+$, for a realized path

$$\{(X(t), Y(t)), 0 \leq t \leq T\}$$

(see Problem 10).

If the X and Y processes are *independent* then the valuation is much simplified. Let c denote the call price at time zero for such an option expiring at time T. Then

$$c = e^{-rT} E_{B_2, B_1} [X(T) - K]^+$$

where X is sampled in a risk-neutral world as described above. Therefore

$$c = e^{-rT} E_{B_2} \left\{ E_{B_1 | B_2} \left([X(T) - K]^+ \right) \right\}.$$

Since B_1 is independent of B_2, it follows that $E_{B_1 | B_2} \left([X(T) - K]^+ \right)$ is simply the Black–Scholes price for a call option, with average squared volatility

$$V_{B_2}(t) = \frac{1}{t} \int_0^t \frac{1}{2} \sigma_{B_2}^2(u) \, du$$

where $\{\sigma_{B_2}^2(u)\}$ is a realization of the volatility path. Therefore, an unbiased estimate of c is obtained by sampling such a volatility path using Equations (6.51) and (6.52) with $\rho = 0$. This is an example of conditional Monte Carlo. If $T = nh$, there are usually $2n$ variables in the integration. However, with independence, $\rho = 0$. This design integrates out n of the variables analytically. The remaining n variables are integrated using Monte Carlo.

6.7 Problems

1. Show that the delta for a European put, at time t, on an asset earning interest continuously at rate r_f, is $-e^{-r_f(T-t)} \Phi\left(-d_{r_f}\right)$ where d_{r_f} is as given in Equation (6.22).

2. Use the procedure 'bscurrency' in Appendix 6.5 to price European put options on a block of 1000 shares offering no dividends, where the current price is 345 pence per share, the volatility is 30 % per annum, the risk-free interest rate is 4.5 % per annum, and the strike prices are (a) 330, (b) 345, and (c) 360 pence respectively. The options expire in 3 months time. If you have just sold these puts to a client and you wish to hedge the risk in each case, how many shares should you 'short' (i.e. borrow and sell) initially in each case?

3. A bank offers investors a bond with a life of 4 years on the following terms. At maturity the bond is guaranteed to return £1. In addition if the FTSE index at maturity is higher than it was when the bond was purchased, interest on £1 equal to one-half of

the % rise in the index is added. However, this interest is capped at £0.30. The risk-free interest rate is 4 % per annum and the volatility of the FTSE is 0.2 per annum. The aim is to find a fair (arbitrage-free) price for the bond $V(x(t), t)$ for $0 \leq t \leq 4$ where $x(t)$ is the index value at time t, using the principle that the price of any derivative of the FTSE is the present value of the expected payoff at $t = 4$ years, in a risk neutral world.

(a) Deduce a definite integral whose value equals $V(x(t), t)$. The integrand should contain the standard normal density $\phi(z)$.

(b) Since integration is over one variable only, Monte Carlo is not justified providing numerical integration is convenient. Therefore, use numerical integration with Maple to find $V(x(0), 0)$.

(c) After 2 years the index is standing at $1.8x(0)$. What is the value of the bond now?

4. The holder of a call option has the *right* to buy a share at time T for price K. However, the holder of a *forward* contract has the *obligation* to do so. The derivation of the price of such a derivative is easier than that for a call option:

(a) Consider a portfolio A consisting at time zero of one forward contract on the share and an amount of cash Ke^{-rT}. Consider another portfolio B comprising one share. Show that the two portfolios always have equal values in $[0, T]$. Hence show that if a forward contract is made at time zero, its value at time t when the price of the share is $x(t)$ is given by

$$V(x(t), t) = x(t) - Ke^{-r(T-t)}.$$

(b) Show that $V(X(t), t)$ satisfies the Black–Scholes differential equation (6.14).

(c) What is the hedging strategy for a forward contract that results in a riskless portfolio?

(d) What happens if $K = x(0)e^{rT}$?

5. A bank has sold a European call option to a customer, with exercise time T from now, for a (divisible) share. The price of one share (which does not yield a dividend) is determined by

$$\frac{dX}{X} = \mu\, dt + \sigma\, dB.$$

The risk-free interest rate is r and the volatility is σ. The following policy is used by the bank to hedge its exposure to risk (recall that it will have a payoff of $[X(T) - K]^+$ at time T): at time $t = 0$ it borrows $\Delta(0)X(0)$ to purchase $\Delta(0)$ shares, while at time $t = T$ it purchases an additional $[\Delta(1) - \Delta(0)]$ shares. Therefore, it is employing an extreme form of discrete hedging, changing its position in the shares only at the beginning and end of the option's life. At these times the bank has decided it will use the deltas calculated for continuous hedging. Let C denote the total cost of writing the option and hedging it. Show that

$$E(C) = c + X(0) e^{(\mu-r)T} \left[\Phi(d_\mu) - \Phi(d) \right]$$
$$+ Ke^{-rT} \left[\Phi\left(d - \sigma\sqrt{T}\right) - \Phi\left(d_\mu - \sigma\sqrt{T}\right) \right]$$

where c is the Black–Scholes price of the option at time zero,

$$d_\mu = \frac{(\mu + \sigma^2/2) T + \ln(x(0)/K)}{\sigma\sqrt{T}},$$

and

$$d = \frac{(r + \sigma^2/2) T + \ln(x(0)/K)}{\sigma\sqrt{T}}.$$

Plot $E(C) - c$ when $r = 0.05$, $\sigma = 0.1$, $X(0) = 680$, $K = 700$, and $T = 0.5$ for $\mu \in [-0.3, 0.3]$.

6. Consider an average price Asian call with expiry time T. The average is a *geometric* one sampled at discrete times $h, 2h, \ldots, nh = T$. Let c_g denote the price of the option at time zero. Show that

$$c_g = e^{-rT} E_{Z \sim N(0,1)} \left[X_0 \exp\left[\left(r - \frac{1}{2}\sigma^2 \right) \left(\frac{n+1}{2} \right) h + \frac{\sigma\sqrt{a}hZ}{n} \right] - K \right]^+$$

where $a = n(n+1)(2n+1)/6$. By taking the limit as $n \to \infty$, show that the price for the continuously sampled geometric average is the same as the price of a European call where the volatility is $\sigma/\sqrt{3}$ and where the asset earns a continuous income at rate $r/2 + \sigma^2/12$.

7. When stratifying a standard normal variate, as, for example, in Section 6.4.2, an algorithm for sampling from $N(0, 1)$ subject to $X \in \left[\Phi^{-1}((j-1)/m), \Phi^{-1}(j/m) \right]$ for $j = 1, \ldots, m$ is required. Write a Maple procedure for this. Use a uniform envelope for $j = 2, \ldots, m-1$ and one proportional to $x \exp(-x^2/2)$ for $j = 1$ and m. Derive expressions for the probability of acceptance for each of the m strata. Such a procedure can be used for sampling from $N(0, 1)$ by sampling from stratum j, $j = 1, \ldots, m$, with probability $1/m$. What is the overall probability of acceptance in that case? For an alternative method for sampling from intervals $j = 1$ and m (the tails of a normal) see Dagpunar (1988b).

8. Using the result in Problem 6, obtain the prices of continuous time geometric average price Asian call options. Use the parameter values given in Table 6.2. Obtain results for the corresponding arithmetic average price Asian call options when the number of sampling points in the average are 50, 200, and 500 respectively. Compare the last of these with the continuous time geometric average price Asian call option prices.

9. Let c denote the price at $t = 0$ of a European arithmetic average price call with expiry time T. The risk-free interest rate is r, the strike price is K, and the asset price at time t is $X(t)$. The average is computed at times $h, 2h, \ldots, nh$ where $nh = T$.

   (a) Make the substitutions $x_j = x(0)\exp[r(jh - T)]$ and $\sigma_j = \sigma\sqrt{jh/T}$ for $j = 1, \ldots, n$. Hence show that c is also the price of a basket option on n assets, where there are $1/n$ units of asset j which has a price of x_j at time zero, $j = 1, \ldots, n$, and where the correlation between the returns on assets j and m is $\sqrt{j/m}$ for $1 \le j \le m \le n$.

   (b) Refer to the results in Table 6.2. Verify any of these by estimating the price of the equivalent basket option.

10. (a) Modify the procedure 'meanreverting' in Appendix 6.8 so that it prices a European call option on an asset subject to stochastic volatility, assuming $\lambda_Y = 0$.

    (b) Suggest how the precision may be improved by variance reduction techniques.

11. Use the modified procedure referred to in Problem 10(a) to evaluate options when the Brownian motions driving the asset price and volatility processes are independent. Then repeat using conditional Monte Carlo. Estimate the variance reduction achieved.

7

Discrete event simulation

A *discrete event system* will be defined as one in which the time variable is discrete and the state variables may be discrete or continuous. Correspondingly, a *continuous event system* is one in which the time variable is continuous and all state variables are continuous. This leaves one remaining type of system, one where the time variable is continuous and the state variables are discrete. In that case, the process can be *embedded* at those points in time at which a state change occurs. Thus such a system can be reduced to a discrete event system.

Of course a system may be a mixture of discrete and continuous events. The essential feature of a pure discrete event system is that the state remains unchanged between consecutive discrete time points (events). In simulating such a process, it is necessary only to advance time from one event to the next without worrying about intermediate times. A continuous event system can always be approximated by a discrete event system through an appropriate discretization of the time variable. In fact, if the (stochastic) differential equations cannot be solved to give a closed form solution, this is a sensible way to proceed.

In this chapter we will show how to simulate some standard discrete event stochastic processes and then move on to examples of nonstandard processes. We do not deal with the simulation of large scale systems such as complex manufacturing processes where it is an advantage to use one of the dedicated simulation languages/packages. There are a large number of these available, many of them incorporating visual interactive components. Examples of these are Simscript II.5, Witness, Simul8, Microsaint, and Extend. From a historical viewpoint the book by Tocher (1963) is interesting. Other books emphasizing practical aspects of building discrete event simulation models include those by Banks *et al.* (2005), Fishman (1978), Law and Kelton (2000), and Pidd (1998).

7.1 Poisson process

Consider the following situation. You are waiting for a taxi on a street corner and
have been told that these pass by in a 'completely random' fashion. More precisely,
in a small time interval of duration h, the chance that a taxi arrives is approximately
proportional to h, and does *not* depend upon the previous history. Let $N(t)$ denote the
number of taxis passing during the interval $[0, t]$. Suppose there is a positive constant
λ, the *arrival rate*, such that for any small time interval $(t, t+h]$, the probability of
a single arrival is approximately λh and the probability of no arrival is approximately
$1 - \lambda h$. These probabilities are assumed to be *independent* of occurrences in $[0, t]$.
Specifically,

$$P[N(t+h) = n+1 | N(t) = n] = \lambda h + o(h)$$

and

$$P[N(t+h) = n | N(t) = n] = 1 - \lambda h + o(h).$$

Note that the two conditions ensure that

$$P[N(t+h) > n+1 | N(t) = n] = o(h).$$

Such a process is a *Poisson process*. It has the following important properties:

 (i) *Independent increments*. The numbers arriving in nonoverlapping intervals are
     independently distributed. This is the *Markov* or 'loss of memory' property.

 (ii) The chance of two or more arrivals in a small time interval may be neglected.

(iii) *Stationary increments*. The distribution of the number of arrivals in an interval
      depends only on the duration of the interval.

   The probability distribution of $N(t)$ is now derived. Define $p_n(t) = P(N(t) = n)$.
Conditioning on the number of arrivals in $[0, t]$ gives

$$p_n(t+h) = p_{n-1}(t)[\lambda h + o(h)] + p_n(t)[1 - \lambda h + o(h)] + o(h). \tag{7.1}$$

Note that the first term on the right-hand side of Equation (7.1) is the probability that
there are $n-1$ arrivals in $[0, t]$ and one arrival in $(t, t+h]$. The second term is the
probability that there are n arrivals in $[0, t]$ and no arrivals in $(t, t+h]$. There is no need
to consider any other possibilities as the Poisson axioms allow the occurrence of two or
more arrivals to be neglected in a small time interval. Equation (7.1) is valid for $n \geq 1$
and also for $n = 0$ if the convention is adopted that $p_{-1}(t) = 0$ $\forall t$. Now rewrite as

$$\frac{p_n(t+h) - p_n(t)}{h} = \lambda p_{n-1}(t) - \lambda p_n(t) + \frac{o(h)}{h}$$

and take the limit as $h \to 0$. Then

$$p'_n(t) = \lambda p_{n-1}(t) - \lambda p_n(t). \tag{7.2}$$

To solve this, multiply through by $e^{\lambda t}$ and rewrite as

$$\frac{d}{dt}\left[e^{\lambda t} p_n(t)\right] = \lambda \left[e^{\lambda t} p_{n-1}(t)\right]. \tag{7.3}$$

Solving this for $n = 0$ gives $d/dt\left[e^{\lambda t} p_0(t)\right] = 0$, which gives $e^{\lambda t} p_0(t) = A$, say. However, $p_0(0) = 1$, so $e^{\lambda t} p_0(t) = 1$. Therefore, $d/dt\left[e^{\lambda t} p_1(t)\right] = \lambda$, which gives $e^{\lambda t} p_1(t) - \lambda t = B$, say. However, $p_n(0) = 0$ for $n > 0$, so $e^{\lambda t} p_1(t) = \lambda t$. At this stage it is guessed that the solution is $e^{\lambda t} p_n(t) = (\lambda t)^n/n!$ $\forall n \geq 0$. Suppose it is true for $n = k$. Then from Equation (7.3), $d/dt\left[e^{\lambda t} p_{k+1}(t)\right] = \lambda (\lambda t)^k/k!$, so, on integrating, $e^{\lambda t} p_{k+1}(t) - (\lambda t)^{k+1}/(k+1)! = C$, say. However, $p_{k+1}(0) = 0$, so $e^{\lambda t} p_{k+1}(t) = (\lambda t)^{k+1}/(k+1)!$. Therefore, since it is true for $n = 0$, it must, by the principle of induction, be true for $\forall n \geq 0$. Therefore

$$p_n(t) = \frac{e^{-\lambda t}(\lambda t)^n}{n!} \tag{7.4}$$

and so $N(t)$ follows a Poisson distribution with mean and variance both equal to λt.

We now determine the distribution of the time between consecutive events in a Poisson process. Let T_n denote the time between the $(n-1)th$ and nth arrivals for $n = 1, 2, \ldots$. Then

$$P(T_n > t \mid T_{n-1} = t_{n-1}, \ldots, T_1 = t_1) = P\left(0 \text{ events in } \left(\sum_{i=1}^{n-1} t_i, t + \sum_{i=1}^{n-1} t_i\right)\right)$$

$$= P(0 \text{ events in } (0, t))$$

by the stationarity property. Since this is independent of $t_{n-1}, \ldots, t_1$, it follows that the interarrival times are independent. Now from Equation (7.4),

$$P(0 \text{ events in } (0, t)) = e^{-\lambda t}$$

so

$$P(T_n > t) = e^{-\lambda t}.$$

Therefore, the interarrival times are i.i.d. with density

$$f(t) = -\frac{d}{dt}\left(e^{-\lambda t}\right)$$

$$= \lambda e^{-\lambda t}. \tag{7.5}$$

The mean and standard deviation of the interarrival times are both λ^{-1}.

Note that a Poisson process $\{N(t), t \geq 0\}$ is a discrete state stochastic process in continuous time. By embedding the process at arrival times we can construct a discrete state, discrete time process, $\{N(t), t = T_{(1)}, T_{(2)}, \ldots\}$, where $T_{(n)}$ is the time of the nth arrival. All the information in any realization of the original process is captured in the corresponding realization of $\{N(t), t = T_{(1)}, T_{(2)}, \ldots\}$. Using the definition at the start of this chapter, the latter is a discrete event system. It may be simulated using result (7.5), recalling that $\text{Exp}(\lambda) := -(1/\lambda)\ln R$, where $R \sim U(0, 1)$. Thus

$$T_{(n)} = T_{(n-1)} - \frac{1}{\lambda}\ln R_n$$

for $n = 1, 2, \ldots$, where $T_{(0)} = 0$.

This immediately leads to a method for generating a Poisson variate from the distribution

$$P(X = n) = \frac{e^{-m}m^n}{n!}, \quad (n = 0, 1, \ldots).$$

Here X is also the number of events in $[0, 1]$ in a Poisson process of rate m. So $X = n$ if and only if $T_{(n)} \le 1$ and $T_{(n+1)} > 1$. This occurs if and only if $\sum_{i=1}^{n}(-(1/\lambda)\ln R_i) \le 1$ and $\sum_{i=1}^{n+1}(-(1/\lambda)\ln R_i) > 1$, that is if $\prod_{i=1}^{n+1} R_i < e^{-\lambda} \le \prod_{i=1}^{n} R_i$. Since $\prod_{i=1}^{n} R_i$ is decreasing in n,

$$X = \min\left(n : \prod_{i=1}^{n+1} R_i < e^{-\lambda}\right).$$

The expected number of uniform variates required in such a method is $E(X+1) = \lambda + 1$. This method becomes quite expensive in computer time if m is large. Other methods are discussed in Chapter 4.

Another way of simulating a Poisson process in $[0, T]$ is first to sample $N(T)$ using an efficient Poisson variate generator and then to sample the arrival times conditional on the Poisson variate. Using this method, suppose that $N(t) = n$. It turns out that the conditional joint density of the arrival times, $T_{(1)}, \ldots, T_{(n)}$, is the same as the joint density of the uniform order statistics from the $U(0, T)$ density, as now shown. For any $x \in (0, T]$ and $i \in [1, \ldots, n]$,

$$P\left(T_{(i)} \le x | N(T) = n\right) = \frac{\sum_{j=i}^{n} P(N(x) = j \text{ and } N(T-x) = n-j)}{P(N(T) = n)}$$

$$= \frac{\sum_{j=i}^{n}\left[(\lambda x)^j e^{-\lambda x}/j!\right]\left[\lambda(T-x)\right]^{n-j}e^{-\lambda(T-x)}/(n-j)!}{(\lambda T)^n e^{-\lambda t}/n!}$$

$$= \sum_{j=i}^{n}\binom{n}{j}\left(\frac{x}{T}\right)^j\left(1 - \frac{x}{T}\right)^{n-j} \tag{7.6}$$

Now let $X_{(1)}, \ldots, X_{(n)}$ denote the n uniform order statistics from $U[0, T]$. The number of these falling in the interval $[0, x]$ is a binom($n, x/T$) random variable. So

$$P\left(X_{(i)} \le x\right) = \sum_{j=i}^{n}\binom{n}{j}\left(\frac{x}{T}\right)^j\left(1 - \frac{x}{T}\right)^{n-j}. \tag{7.7}$$

which is the same distribution as Equation (7.6), which proves the result. Using this property, the following Maple code and results show five random realizations of a Poisson process in [0,3.5], each of rate 2. Note the use of the Maple 'sort' command:

```
> randomize(75321):
  lambda := 2;T := 3.5;m := lambda*T;
  for u from 1 to 5 do;
    n := stats[random,poisson[m]](1):
    for j from 1 to n do:
      t[j] := evalf(T*rand()/10^12):
    end do:
    d := [seq(t[j], j = 1..n)]:
    arrival_times:= sort(d):
    print(arrival_times);
  end do:
```

lambda := 2
T := 3.5
m := 7.0
[0.4683920028, 0.8584469999, 1.324848195, 1.564463956, 2.342103589, 2.753604757, 3.161013255, 3.355203918]
[0.3105425825, 0.6851910142, 1.025506152, 1.036301499, 1.247404803, 1.370810129, 2.376811957, 2.377386193, 2.564390192, 3.436339776]
[0.8816330302, 0.9995187699, 1.733006037, 1.926557959, 1.926642493, 2.803064014]
[1.596872439, 2.036243709, 2.042999552, 2.341445360, 2.513656874, 2.987985832, 3.185727007, 3.370120432]
[0.9566889486, 1.358244739, 2.998496576]

Another way to obtain the uniform order statistics avoids sorting. Let $U_{(1)}, \ldots, U_{(n)}$ be the n-order statistics from $U(0, 1)$. Then

$$P\left(U_{(1)} > u_1\right) = (1 - u_1)^n$$

and

$$P\left(U_{(j+1)} > u_{j+1} \mid U_{(j)} = u_j\right) = \left(\frac{1 - u_{j+1}}{1 - u_j}\right)^{n-j}$$

for $j = 1, \ldots, n-1$. Let $R_1, \ldots, R_n$ be n uniform random numbers in $U(0, 1)$. Inverting the complementary cumulative distribution functions gives

$$1 - U_{(1)} = R_1^{1/n}$$

and

$$1 - U_{(j+1)} = \{1 - U_{(j)}\} R_{j+1}^{1/(n-j)}$$

for $j = 1, \ldots, n-1$. Whether or not this is quicker than the sort method depends upon the nature of the sorting algorithm. The Maple one appears to have a sort time of $O(n \ln n)$, and for any reasonable λT, the sort method seems to be faster.

7.2   Time-dependent Poisson process

This is an important generalization of the (simple) Poisson process considered above. If there is *time-dependence* then λ is replaced by a known non-negative function $\lambda(t)$. This gives a *heterogeneous* or *nonhomogeneous* Poisson process. Replacing λ by $\lambda(t)$ in Equation (7.1) gives

$$p'_n(t) = \lambda(t) p_{n-1}(t) - \lambda(t) p_n(t) \quad (n = 0, 1, \dots).\qquad(7.8)$$

The solution to this is obtained by replacing λt in Equation (7.4) by

$$\Lambda(t) = \int_0^t \lambda(u)\ du.$$

Noting that $\Lambda'(t) = \lambda(t)$ it can be verified that

$$p_n(t) = \frac{e^{-\lambda t}[\Lambda(t)]^n}{n!}$$

satisfies Equation (7.8) and the initial condition that $p_0(0) = 1$. A number of methods of simulating $\{T_{(i)}, i = 1, 2, \dots\}$ are discussed in Lewis and Shedler (1976, 1979a, 1979b).

One such method uses a *time scale transformation*. Define $\tau = \Lambda(t)$. Since Λ is increasing in t, the order of arrivals is preserved in the transformed time units and the *ith* such arrival occurs at time $\tau_{(i)} = \Lambda\left(T_{(i)}\right)$. Let $M(\tau)$ denote the number of arrivals in $[0, \tau]$ in the transformed time units and let $q_n(\tau) = P(M(\tau) = n)$. Then $p_n(t) = q_n(\tau)$, so $p'_n(t) = q'_n(\tau)\,d\tau/dt = \lambda(t)\,q'_n(\tau)$. Substituting these into Equation (7.8) gives

$$q'_n(\tau) = q_{n-1}(\tau) - q_n(\tau).$$

It follows that the process $\{M(\tau)\}$ is a simple (or homogeneous) Poisson process of rate 1. Therefore, $\tau_{(n)} = \tau_{(n-1)} - \ln R_n$ where $\{R_n\}$ are uniform random numbers. Arrival times are obtained in the original process using $T_{(i)} = \Lambda^{-1}\left(\tau_{(i)}\right)$. If $\{N(t), 0 \le t \le t_0\}$ is to be simulated, then the simulation is stopped at arrival number $\max_{i=0,1,\dots}\left[i : \tau_{(i)} \le \Lambda(t_0)\right]$. An exponential polynomial intensity, $\lambda(t) = \exp\left(\sum_{i=0}^k \lambda_i t^i\right)$, often proves to be a useful model, since the parameters $\{\lambda_i\}$ may be fitted to data, without the need to specify constraints to ensure $\lambda(t) \ge 0$. The efficiency of the method depends upon how easily Λ may be inverted.

Another method involves *thinning*. In order to simulate in the interval $[0, t_0]$, an attempt is made to find λ such that $\lambda \ge \lambda(t)$ for all $t \in [0, t_0]$. A realization comprising prospective events is then generated from a simple Poisson process with rate λ. Suppose this realization is $\{S_{(i)}, i = 1, \dots, N(t_0)\}$. Then the process is thinned so that the prospective event at $S_{(i)}$ is accepted with probability $\lambda\left(S_{(i)}\right)/\lambda$. The thinned process comprising accepted events is then a realization from the desired time-dependent Poisson process. This is because, in the thinned process, the probability of one event in $(t, t+h]$ is $\{\lambda h + o(h)\}\lambda(t)/\lambda = \lambda(t)h + o(h)$, while the probability of no such event is $\{1 - \lambda h + o(h)\} + \{\lambda h + o(h)\}\{1 - \lambda(t)/\lambda\} = 1 - \lambda(t)h + o(h)$. These probabilities are independent of the history of the thinned process in $[0, t]$. The *ith* event is accepted if and only if $R_i < \lambda \left(S_{(i)} \right) / \lambda$ where $R_i \sim U(0, 1)$. Clearly, the efficiency of the process is determined by the expected proportion of prospective events that are accepted, and this is $\Lambda(t_0) / \lambda t_0$. The method is most suited when $\lambda(t)$ varies little. Otherwise, some improvement can be made by splitting $[0, t_0]$ into disjoint subintervals, each subinterval having its own λ bound.

It was noted from results (7.6) and (7.7) that the conditional joint density of the arrival times $T_{(1)}, \ldots, T_{(n)}$ in $[0, T]$ in a homogeneous Poisson process is the joint density of n uniform order statistics. In the case of a heterogeneous Poisson process, a similar result holds. Here, the joint density is that of the n-order statistics from the density

$$f(t) = \frac{\lambda(t)}{\Lambda(t_0)}$$

where $0 \le t \le t_0$. This has the distribution function

$$F(t) = \frac{\Lambda(t)}{\Lambda(t_0)}.$$

Frequently, analytical inversion of F will not be possible. One possibility then is to use envelope rejection. If $\lambda(t)$ varies considerably, a nonuniform envelope will be needed for an efficient implementation. The general algorithm is

Sample $N(t) \sim \text{Poisson}\left(\Lambda(t_0)\right)$
Sample $T_1, \ldots, T_{N(t)}$ independently from $f(t) = \lambda(t) / \Lambda(t_0)$
Sort into ascending order and deliver $T_{(1)}, \ldots, T_{(N(t))}$

As mentioned previously, Maple's sort procedure is relatively efficient.

7.3 Poisson processes in the plane

Consider a two-dimensional domain, D. Let $C \subseteq D$ and $E \subseteq D$ where $C \cap E = \phi$ and ϕ is the empty set. Let $N(C)$ and $N(E)$ denote the number of randomly occuring points in C and E respectively. Suppose there exists a positive constant λ, such that, for all such C and E, $N(C)$ and $N(E)$ are independent Poisson random variables with means $\lambda \int_C dx\, dy$ and $\lambda \int_E dx\, dy$ respectively. Then the point process $\{N(H) : H \subseteq D\}$ is defined to be a two-dimensional Poisson process. We can think of λ as being the density of points, that is the expected number of points per unit area.

Suppose D is a circle of radius a. Let $R_{(1)}, \ldots, R_{(N(D))}$ denote the ranked distances from the centre of the circle in the two-dimensional Poisson process, $\{N(H) : H \subseteq D\}$, with rate λ. Let $A_{(i)} = \pi \left(R_{(i)}^2 - R_{(i-1)}^2 \right)$ for $i = 1, \ldots, N(D)$ with $R_{(0)} = 0$. Then

$$P\left(A_{(i)} > x \mid R_{(1)} = r_1, \ldots, R_{(i-1)} = r_{i-1} \right) = P(0 \text{ points in a region of area } x)$$

$$= e^{-\lambda x}.$$

This is independent of $r_1, \ldots, r_{i-1}$. Therefore, $\left\{ \pi \left(R_{(i)}^2 - R_{(i-1)}^2 \right) \right\}$ are independently distributed as negative exponentials with mean λ^{-1}. Therefore, while $R_{(i)} \leq a$, the following may be set:

$$\pi \left(R_{(i)}^2 - R_{(i-1)}^2 \right) = -\frac{1}{\lambda} \ln U_i$$

where $\{U_i\}$ are uniform random numbers. Since the process is homogeneous, the angular component is independent of the radial one and $\Theta_i \sim U(0, 2\pi)$. Therefore,

$$\Theta_i = 2\pi V_i$$

where $\{V_i\}$ are uniform random numbers.

Suppose now that $D = \{(x, y) : 0 \leq x \leq x_0, h(x) \leq y \leq g(x)\}$ where x_0, $h(x)$, and $g(x)$ are all given, and $g(x) \geq h(x)$ for all $x \in [0, x_0]$. Without loss of generality we can take $h(x) = 0$ and $g(x) \geq 0$. Observe that the projection of the process on to the x axis is a one-dimensional heterogeneous process of rate $\lambda(x) = \lambda g(x)$. Accordingly, use any method from Section 7.2 to generate the $N(D)$ abscissae for the points. For example, if it is not convenient to invert Λ, then use the algorithm at the end of section 7.2, where

$$f(x) = \frac{\lambda(x)}{\Lambda(x_0)}$$

$$= \frac{g(x)}{\int_0^{x_0} g(u)\, du},$$

together with rejection with an envelope that is similar in shape to $g(x)$. Suppose the ith such abscissa is X_i (it is not necessary to generate these in increasing order). Then the homogeneity of the two-dimensional process means that $Y_i \sim U(0, g(X_i))$.

7.4   Markov chains

7.4.1   Discrete-time Markov chains

Let X_n be the state at the nth step in a homogeneous discrete-time, discrete-state Markov chain with state space S and probability transition matrix $\mathbf{P}$. Given that $X_{n-1} = i$ and given a random number R_n, inversion of the cumulative distribution function may be used to obtain

$$X_n = \min_{j \in S} \left\{ j : R_n < \sum_{k \in S, k \leq j} p_{i,k} \right\}.$$

This method can be inefficient if there is a large proportion of transitions where the transition is from a state to itself. In such cases a better method is to identify those steps that lead to a *change* of state. In the terminology of discrete event systems these steps are *events*. Let the state immediately after the kth event be $X^{(k)}$ and let $M^{(k)}$ denote the number of steps between the $(k-1)$th and kth events. Then $\{M^{(k)}\}$ are independent geometric random variables and may be generated, given that $X^{(k-1)} = i$, using inversion of the c.d.f. to yield

$$M^{(k)} = \left\lfloor 1 + \frac{\ln R^{(k)}}{\ln p_{ii}} \right\rfloor$$

where $R^{(k)} \sim U(0,1)$. Now observe that for $j \neq i$,

$$P\left(X^{(k)} = j \mid X^{(k-1)} = i\right) = \frac{p_{ij}}{1 - p_{ii}}.$$

Therefore, given $U^{(k)} \sim U(0,1)$,

$$X^{(k)} = \min_{j \in S} \left\{ j : U^{(k)}(1 - p_{ii}) < \sum_{m \in S, m \neq i, m \leq j} p_{i,m} \right\}.$$

7.4.2 Continuous-time Markov chains

Consider a discrete state, continuous-time Markov chain with state space S and stationary infinitesimal generator $\mathbf{Q}$. Then

$$P[X(t+h) = j \mid X(t) = i] = q_{ij}h + o(h)$$

for $j \neq i$ where $\sum_{j \in S} q_{ij} = 0$ for all $i \in S$ and

$$P[X(t+h) = i \mid X(t) = i] = 1 + q_{ii}h + o(h).$$

The simulation method used here is analogous to that described for discrete-time Markov chains. Let $T_{(k)}$ be the time of the kth event (state change) and let $X\left(T_{(k-1)}\right)$ be the state immediately after the $(k-1)$th event. Given $X\left(T_{(k-1)}\right) = i$ and $T_{(k-1)} = t_{(k-1)}$, it is found that $T_{(k)} - t_{(k-1)}$ is exponentially distributed with mean $(-q_{ii})^{-1}$ (the 'leaving rate' for state i is $-q_{ii}$). Therefore,

$$T_{(k)} - t_{(k-1)} = \frac{\ln R_k}{q_{ii}}$$

where $R_k \sim U(0,1)$. Given $X\left(T_{(k-1)}\right) = i$, the conditional probability that the next state is $j \, (\neq i)$ is $-q_{ij}/q_{ii}$ (this is the ijth element of the probability transition matrix for a chain embedded at state changes only). Therefore, given another random number $U_k \sim U(0,1)$,

$$X\left(T_{(k)}\right) = \min_{j \in S} \left\{ j : -q_{ii}U_k < \sum_{m \in S, m \neq i, m \leq j} q_{im} \right\}. \qquad (7.9)$$

A birth–death process is one in which $q_{ii} = -(\lambda_i + \mu_i)$, $q_{i,i+1} = \lambda_i$, and $q_{i,i-1} = \mu_i$ for all $i \in S$. This means that it is impossible to move to a nonadjacent state. Therefore, in performing the inversion implicit in Equation (7.9), for a state space $S = \{0, 1, 2, \ldots\}$, say, $X\left(T_{(k)}\right) = i-1, i+1$ with probabilities $\mu_i/(\lambda_i + \mu_i)$ and $\lambda_i/(\lambda_i + \mu_i)$ respectively.

7.5 Regenerative analysis

First the definition of a *renewal reward* process will be recalled. Let τ_i denote the time between the $(i-1)$th and ith events in a renewal process. We can think of τ_i as being the duration of the ith 'cycle'. Since it is a renewal process, $\{\tau_i\}$ are identically and independently distributed. Let $R(t)$ denote the cumulative 'reward' earned by the process in $[0, t]$ and let R_i denote the reward earned during the ith cycle. Suppose $\{R_i\}$ are identically and independently distributed. We allow R_i and τ_i to be statistically dependent. The presence of rewards allows such a process to be called a renewal reward process. Frequently, in simulation, we wish to estimate θ where θ is the long-run reward per unit time, where this exists. Thus

$$\theta = \lim_{t \to \infty} \frac{R(t)}{t}.$$

For example, we may wish to estimate the long-run occupancy of beds in a hospital ward, and here $R(t) = \int_0^t O(u)\, du$ would be identified, where $O(u)$ is the bed occupancy at time u. The *renewal reward theorem* states that

$$\lim_{t \to \infty} \frac{R(t)}{t} = \frac{E(R_i)}{E(\tau_i)}.$$

Therefore one method of estimating θ is to use

$$\widehat{\theta} = \frac{\overline{R}}{\overline{\tau}}$$

where $\overline{R}$ and $\overline{\tau}$ are the sample mean cycle rewards and durations respectively, over a *fixed* number, n, of cycles. To determine the variance of $\widehat{\theta}$ consider the random variable $\overline{R} - \theta\overline{\tau}$. This has expectation zero and variance

$$\sigma^2 = \frac{1}{n} \left[\sigma_{R_i}^2 + \theta^2 \sigma_{\tau_i}^2 - 2\theta\, \mathrm{Cov}\,(R_i, \tau_i) \right].$$

Let S_R^2, S_τ^2, and $S_{R,\tau}$ denote the sample, variances of $\{R_i\}$, $\{\tau_i\}$, and covariance between them respectively. Let

$$S^2 = S_R^2 + \widehat{\theta}^2 S_\tau^2 - 2\widehat{\theta} S_{R,\tau} \tag{7.10}$$

Then as $n \to \infty$, $S^2/n \to \sigma^2$, and so

$$\frac{\overline{R} - \theta\overline{\tau}}{S/\sqrt{n}} \to N(0, 1).$$

Therefore,

$$\frac{\widehat{\theta} - \theta}{S/(\overline{\tau}\sqrt{n})} \to N(0, 1). \tag{7.11}$$

It follows that a confidence interval can be found when n is large. However, it should be noted that since $\widehat{\theta}$ is a ratio estimator, it is biased. The bias is $O(1/n)$. A number of estimators that reduce this to $O(1/n^2)$ have been proposed. We will use

$$\widetilde{\theta} = \widehat{\theta}\left[1 + \frac{1}{n}\left(\frac{S_{R,\tau}}{\overline{R}\overline{\tau}} - \frac{S_{\tau,\tau}}{\overline{\tau}^2}\right)\right] \tag{7.12}$$

due to Tin (1965). Now

$$\frac{\widetilde{\theta} - \theta}{S/(\overline{\tau}\sqrt{n})} \sim N(0, 1) \tag{7.13}$$

as $n \to \infty$, where $\widetilde{\theta}$ has replaced $\widehat{\theta}$ in Equation (7.10). For large n it is safe to use (7.13) as an improvement on (7.11).

The *regenerative* method for analysing simulation output data seeks to exploit the theory described in the previous paragraph by attempting to find renewal or regeneration points. It is an elegant mode of analysis developed in a series of papers by P. Heidelberger, D. Iglehart, S. Lavenberg, and G. Shedler. For more details on the method refer to Shedler (1993).

As an example of the method consider an M/G/1 queueing system where the continuous time system is embedded at *departure* times. Let the state of the system be the current number of customers in the system (including any customer being served) immediately after a departure. Assume that the mean service duration is less than the mean interarrival time, so that if the system is in state i it is certain to return to that state in a finite period of time, for all i. Assume that the reward per unit time is a function of the current state only. Regeneration points can be defined as being those times at which the system enters state i (following a departure) for any predetermined i. To understand why this is so let E_i denote the event 'system enters state i' (following a departure). Let τ_j denote the time between the $(j-1)$th and jth occurences of E_i and let $T_{(m-1)} = \sum_{j=1}^{m-1}\tau_j$. Given $\tau_1, \ldots, \tau_{m-1}$, it is noted that τ_m is independent of $\tau_1, \ldots, \tau_{m-1}$. This is so because the arrival process is Poisson, since service durations are independent and since at time $T_{(m-1)}$ there is no partially completed service. Therefore $\{\tau_j\}$ are independently distributed and obviously identically distributed since the return is always to state i. It can similarly be shown that $\{R_j\}$ are i.i.d., which completes the requirements for a renewal reward process.

For a G/G/1 system, that is a single server system with i.i.d distributed interarrival times from an arbitrary distribution and service durations also i.i.d. from an arbitrary distribution, the only regeneration points are at those instants when the system leaves the empty and idle state. The event E_i (following a departure) is *not* (unless arrivals are Poisson) a regeneration point when $i > 1$ since the distribution of time until the next arrival depends upon the state of the system prior to E_i.

One of the advantages of regenerative analysis is that there is no need to allow for a 'burn-in' period at the beginning of any realization in order to reach a point in the sample record at which it is believed the behaviour is stationary (assuming, of course, that it is the stationary behaviour that is to be investigated). Another advantage is that only one realization is required. If regenerative analysis is not used, the usual approach (bearing in mind that observations within a sample realization are often highly dependent) is to perform several independent realizations (replications) and to use the independent responses from these to make an inference about parameter values. A major disadvantage of the regenerative approach is that for many systems the cycle length (time between successive regeneration points) is very large. For example, in a G/G/1 system, if the arrival rate is only slightly less than the service rate, stationarity is assured, but the system returns very infrequently to the empty and idle state.

7.6　Simulating a G/G/1 queueing system using the three-phase method

There is a method of writing programs for discrete event systems known as the *three-phase method*. It has gathered favour since the 1970s in the UK, while the *process* and *event* based methods are perhaps more prominent in the US. The three-phase method is not the most efficient and can involve some redundancy in the program logic. It does, however, have the advantage of being an easy way of programming and of visualizing the structure of a discrete event system, and so it will be used here.

First some terms are introduced that are fairly standard in discrete event simulation. An *entity* is a component part of a system. Entities can be permanent (for example a machine or a bus) or temporary (for example a customer who will eventually leave the system). An *attribute* describes some aspect of an entity. It may be static or dynamic. For example, an aircraft will have a number of seats (static) but the number of passengers occupying them (dynamic) may vary during the simulation. A particular attribute that an entity may have is a time attribute. This may represent, for example, the time at which something is next known to happen to the entity. An *event* is an instant of time at which one or more state changes takes place. At an event an entity may finish one *activity* and start another. For example, a customer may finish the activity 'queuing' and start 'service'. Note that an event is instantaneous, while an activity generally has a duration. A *list* is an array used for storing entities and their attributes. An example is a queue of customers waiting for some facility to become available.

The three-phase method involves steps A, B, and C at each event. A represents the time advance from one discrete event to the next. It is performed by keeping a list of scheduled or *bound* state changes. B stands for execution of *bound* state changes. C stands for execution of *conditional* state changes. The simulation program keeps a list of bound (scheduled state changes). One way to do this is to keep a set of time attributes for the entities. In phase A, a timing routine scans the list to find the smallest time, and then advances the variable representing the present simulation time (*clock*, say) to this smallest time. In phase B those state changes that are bound to occur at this new time are executed. They are bound to occur in the sense that their occurrence is not conditional upon the state of the system but merely upon the passage of the correct amount of time. An example of a bound event is an arrival. Once an arrival has occurred at an event, the time of the next arrival can be scheduled by generating an interarrival time and adding it to the present value of *clock*. One of the entities in such a simulation could be an 'arrival generator', and the time attribute for it is just the scheduled or bound time of the next arrival. Once all the bound state changes have been made at an event, the simulation program checks in phase C for *every* possible conditional state change that could occur at this event. An example of a conditional state change is 'start service for a customer'. It is conditional since its execution depends upon the state of the queue being nonempty and on at least one server being idle. To ease construction of the program *every* conditional state change is checked, even though logic might indicate that a particular one is impossible given that a certain bound state change has occurred at this event. This redundancy is deemed worthwhile to save programming effort. Once every conditional state change has been checked and if necessary executed, a further pass is made through phase C. This is because, as a result of state changes executed during the first pass, it may then be possible for further conditional state changes being executed at this event. At the current event this is done repeatedly until a pass through phase C results in no additional state changes being executed at this event. At that stage the work done at this event is completed and control is then passed back to phase A.

Turning now to the simulation of a G/G/1 queue, an arrival generator, a server, and individual customers can be identified as entities. The time attribute for the server will be the time at which the next departure is scheduled. If the server is currently busy this will have been calculated by adding a service duration to the time at which this service started. If not, then the time attribute will be set to ∞ until the server next moves to the busy state. Suppose the number of customers in the system (line length) at time t is $L(t)$ and that $L(t)$ has a limit distribution as $t \to \infty$. The latter will exist if the expected interarrival time is greater than the expected service duration. We wish to estimate μ_L, the long-run average number of customers in the system, and μ_W, the expected waiting time for a customer under steady state behaviour. Now,

$$\mu_L = \lim_{t \to \infty} \frac{R(t)}{t}$$

where $R(t) = \int_0^t L(u)\, du$ is the cumulative 'reward' in $[0, t]$. Also, $\mu_W = \mu_A \mu_L$ where μ_A is the mean interarrival time. This is is an application of Little's formula (see, for example, Tijms, 2003, pp. 50–3). Little's formula means that it is possible to dispense with a list holding joining times for each customer. However, such a list would be needed if the *individual* customer waiting times in the line or queue were required. The bound events are (i) customer arrival and (ii) customer departure. There is only one conditional event, which is (iii) customer starts service. A regenerative analysis will be performed of the simulation output data (which will consist of the times at which events take place and the line length immediately after each event) and the regeneration points are recognized, which are those times at which the system enters the busy state from the empty and idle state.

An algorithm for simulating *one* cycle of the regenerative process is:

$\tau := 0$
$R := 0$
$L := 1$
sample A and D
$t_A := \tau + A,\ t_D := \tau + D$
do
$\tau_{\text{prev}} := \tau$
$\tau := \min(t_A, t_D)$
$\delta := \tau - \tau_{\text{prev}}$

$R := R + \delta \times L$
if $\tau = t_A$ and $L = 0$ then break end if
if $\tau = t_A$ then $L := L+1$, sample A, $t_A := \tau + A$, end if
if $\tau = t_D$ then $L := L-1$, $t_D := \infty$, end if
if $L > 0$ and $t_D = \infty$ then sample D, $t_D := \tau + D$, end if
end do
Store R, τ.

where

τ = partial length of the current cycle
τ_{prev} = partial length of the current cycle at a previous event
t_A = time of next arrival
t_D = time of next departure
L = number of customers in the system
A = interarrival time
D = service duration
$R = \int_0^\tau L(u)\, du$
δ = time since the last event

The data collected are used to estimate μ_L and μ_W, but it is a simple matter to modify the above to estimate other quantities such as the long-run proportion of time that the line length exceeds a specified threshold. In this system simultaneous 'starts of service' are impossible, so it is unnecessary to make repeated passes through phase C at an event. The procedure 'gg1' in Appendix 7.1 is written to estimate μ_L when the interarrival times are i.i.d. with complementary cumulative density $\overline{F}(x) = \exp[(-\lambda x)^\gamma]$ on support $[0, \infty)$, and service durations are i.i.d. with complementary cumulative distribution $\overline{G}(x) = \exp\left[(-\mu x)^\beta\right]$. These are Weibull distributions. If $\gamma = 1$ then the arrival process is Poisson. Other distributions can be used by making suitable modifications to 'gg1'. For example, if arrivals are based upon appointments, as in a doctor's surgery, then the interarrival time could be modelled as a constant plus a random variable representing the lateness/earliness of customer arrival times. In the procedure, the algorithm shown above is nested within a loop that executes n regenerative cycles. The n cycle rewards $\{R\}$ and durations $\{\tau\}$ are stored in arrays. These data are then used to compute a confidence interval for μ_L using (7.13). In Appendix 7.1, some example runs are shown for $n = 10\,000$, $\lambda = 1$, for the case of M/M/1. This refers to a single server system with Poisson arrivals and exponential service durations. Let $\rho = \lambda/\mu$. This is the *traffic intensity*. For this queue, standard theory shows that a stationary (steady state) distribution exists if and only if $0 < \rho < 1$ and then $\mu_L = \rho/(1-\rho)$ and the expected cycle length, $\mu_\tau = [(1-\rho)\lambda]^{-1}$. The results are summarized in Table 7.1, where the superscripts p and a refer to the primary and antithetic realizations respectively.

Note how the expected length of a regenerative cycle increases as the traffic intensity approaches 1 from below. This means that for fixed n, the simulation run length increases as ρ increases, demonstrating that as $\rho \to 1$ it is even harder to obtain a precise estimate for μ_L than the confidence intervals might suggest. The reason for this is that as $\rho \to 1$ the stochastic process $\{L(t), t \geq 0\}$ is highly *autocorrelated*, so a simulation run of given length (measured in simulated time) yields less information than it would for smaller values of ρ. The situation can be improved by replicating the runs using antithetic random

Table 7.1 Regenerative analysis of the M/M/1 queue

| ρ | $\widehat{\mu_\tau^{(p)}}$ | μ_τ | $\widehat{\mu_L^{(p)}}$ | $\widehat{\mu_L^{(a)}}$ | $\dfrac{\widehat{\mu_L^{(p)}} + \widehat{\mu_L^{(a)}}}{2}$ | μ_L | 95% confidence interval | |
							μ_L (primary)	μ_L (antithetic)
$\dfrac{1}{2}$	2.0038	2	0.9994	1.0068	1.0031	1	(0.950, 1.048)	(0.960, 1.054)
$\dfrac{2}{3}$	3.0150	3	2.0302	1.9524	1.9913	2	(1.903, 2.157)	(1.839, 2.066)
$\dfrac{10}{11}$	11.449	11	10.3273	10.1597	10.2435	10	(9.087, 11.568)	(8.897, 11.422)

numbers. For best effect a given random number and its antithetic counterpart should be used for sampling the primary variate and its exact antithetic variate counterpart. Therefore two random number streams are used, one for interarrival times and another for service durations. This is achieved by using the Maple 'randomize' function with an argument equal to 'seeda' or 'seedb', these being the previous $U[0, 10^{12} - 11]$ seeds used for interarrival and service durations respectively. The fact that Weibull variates are generated by the inversion method ensures that there is a one-to-one correspondence between uniform random numbers and variates. This would not be the case for distributions where a rejection method was employed. The table does not show a single confidence interval for μ_L constructed from both the primary and antithetic point estimates $\widehat{\mu_L^{(p)}}$ and $\widehat{\mu_L^{(a)}}$. A little reflection should convince the reader that the estimation of the variance of $(\widehat{\mu_L^{(p)}} + \widehat{\mu_L^{(a)}})/2$ from a single primary and antithetic realization is not straightforward. However, in two of the three cases the point estimates for the average appear to indicate that the antithetic method is worthwhile with respect to the cost of implementation, which is almost zero extra programming cost and a doubling of processing time.

For more adventurous variance reduction schemes, the reader should refer to the control variate methods suggested by Lavenberg *et al.* (1979).

7.7   Simulating a hospital ward

Now a simulation of a hospital ward comprising n beds will be designed. Suppose arrivals are Poisson rate λ. This might be appropriate for a ward dealing solely with emergency admissions. Patient occupancy durations are assumed to be Weibull distributed with the complementary cumulative distribution function $\overline{G}(x) = \exp\left[(-\mu x)^\beta\right]$. If an arriving patient finds all beds occupied, the following protocol applies. If the time until the next 'scheduled' departure is less than a specified threshold, α, then that patient leaves now (early departure) and the arriving patient is admitted to the momentarily vacant bed. Otherwise, the arriving patient cannot be admitted and is referred elsewhere. Using the 'three-phase' terminology, the bound events are (i) an arrival and (ii) a normal (scheduled) departure. The conditional events are (iii) normal admission of patient, (iv) early departure of patient followed by admission of patient, and (v) referral elsewhere of arriving patient. The following variables are used:

$simtim$ = duration of simulation

n = number of beds in unit

$t0[j]$ = ranked (ascending) scheduled departure times at *start* of simulation, with $t0[j]$ = infinity indicating bed is currently unoccupied, $j = 1, \ldots, n$

α = threshold early departure parameter

$seeda$ = seed for random number stream generating arrivals

$seedb$ = seed for random number stream generating (nonreduced) patient length of stay

$clock$ = present time

$clockprev$ = time of previous event

δ = time since last event

$t[j]$ = ranked (ascending) scheduled departure times at *current* time, with $t[j]$ = infinity indicating bed is currently unoccupied, $j = 1, \ldots, n$

a = time of next arrival

$nocc$ = number of occupied beds

$q = 1, 0$ according to whether or not there is a patient momentarily requiring admission

na = cumulative number of arrivals

$nout1$ = cumulative number of normal departures

$nout2$ = cumulative number of early departures

na = cumulative number of admissions

$nrefer$ = cumulative number of referrals

$cum[j]$ = cumulative time for which occupancy is j beds, $j = 1, \ldots, n$

The simulation will be used to determine the proportion of patients referred and the long-run utilization of beds for given arrival and service parameters and given α. We wish to visualize the stationary (steady state) distribution of bed occupancy. A stationary distribution will exist for arbitrary stationary arrival and service processes because the admission and discharge protocols ensure that there is never a queue. For small wards (n small) there will occasionally be times when all the beds are unoccupied and so regenerative analysis based upon this empty and idle state would be possible. However, for larger values of n, even though return to the empty and idle state is certain, it may be so infrequent as to never occur within even a lengthy simulation. It is worth noting that if a ward were to be modelled subject to, say, Poisson closures due to infection, then these would provide convenient regeneration points. This is not included in this case so simulation is for a given period of time rather than for a specified number of regeneration cycles. Appendix 7.2.1 shows a Maple procedure 'hospital_ward' with the main sections of code dealing with bound and conditional state changes duly flagged. Note the sorting of scheduled departure times after each state change, and the collection and updating of essential information needed to produce summary statistics at the end of the simulation. Maple's 'sort' function is particularly useful in discrete event simulation.

Any opportunity to verify the correctness of the logic in a simulation program should be taken. If service durations are exponentially distributed and no early departures are allowed ($\alpha = 0$) then this model reduces to an M/M/n/n queueing system, for which the stationary distribution of number of occupied beds is

$$p_j = \frac{(1/j!)\,(\lambda/\mu)^j}{\sum_{i=0}^{n}(1/i!)\,(\lambda/\mu)^i}$$

for $j = 0, \ldots, n$. A 10 bedded example in Appendix 7.2.2 compares these probabilities with the sample proportions of time spent in state j ($j = 0, \ldots, n$), when the simulation is run for 100 000 days. The agreement is close enough to verify the correctness of the program for this special case without any formal test of hypothesis.

In Appendix 7.2.3 the assumption of exponentially distributed lengths of stay is relaxed. Here, $n=20$ $\lambda = 2.5$ per day, $\mu = 0.15$ per day, and $\beta = 2.5$, giving a mean and standard deviation of length of stay of 5.92 and 2.53 days respectively. Early departures are not possible ($\alpha = 0$).

Appendix 7.2.4 shows some point estimates for a 20 bedded ward when $\lambda = 2.5$ per day and the mean and standard deviation of a patient's length of stay are 5.92 days and 2.53 days respectively. Early departures are possible up to one day ($\alpha = 1$) prior to scheduled departure times. If interval estimates are required then independent replications should be made in the usual way. The initial state of the ward is determined by the scheduled departure times for beds occupied at time zero, and this forms part of the input data. This might reflect the true state of the ward now, in which case performing several replications of the simulation could show the variability in bed occupancy in 10 days time, say. This could be used for short-term planning purposes. Alternatively, we may be interested in steady state (stationary) behaviour only. In that case it would certainly be inappropriate to specify an initial state in which all beds were unoccupied. That would incur a burn-in period that is longer than necessary. It is better to guess an initial state that might be thought typical under steady state behaviour. For these data with a lengthy 10 000 day simulation, the long-run utilization is approximately 0.7. With a ward of 20 beds this corresponds to a typical occupancy of 14 beds, so the initial state chosen (18 beds occupied) is not atypical and therefore the burn-in can effectively be ignored for estimating long-run measures of performance. Finally, Appendix 7.2.5 shows the effect of altering α (the number of days in advance of scheduled departure that a patient may be discharged) upon the utilization and the proportion of patients referred.

7.8 Problems

1. Classify the following stochastic processes, first according to discrete/continuous state and discrete/continuous time, then as discrete/continuous event systems:

   (a) $dX = \mu X\,dt + \sigma X\,dB$, $t \geq 0$;

   (b) $X_j = X_{j-1}\exp\left[(\mu - \sigma^2/2)\,h + \sigma B(h)\right]$, $j = 1, 2, \ldots$, where h is a positive known constant;

   (c) $\{X(t), t \geq 0\}$ where $X(t)$ denotes the cumulative amount of alcohol consumed by an individual in $(0, t)$.

2. Consider a heterogeneous Poisson process with intensity $\lambda(t) = \exp(\alpha + \beta t)$. Write Maple procedures for simulating events in $[0, t_0]$ using

   (a) thinning;

   (b) time scale transformation.

3. Write a Maple procedure to generate events according to a two-dimensional Poisson process of rate λ per unit area, over the ellipse

$$\frac{x^2}{a^2} + \frac{y^2}{b^2} = 1.$$

4. Customers arrive at a bank requiring one of three services: private banking, business banking, and travel and securities. These form three independent Poisson processes of rate 40, 20, and 10 per hour respectively.

   (a) Write a procedure to simulate the arrival times and type of customer for the first n arrivals. (*Hint*: the sum of independent Poisson processes is a Poisson process.)

   (b) Use the procedure 'ipois1' in Appendix 4.5 and the Maple 'sort' function to simulate the times of arrival and type of customer in a 2 hour period.

5. Let $T_{(i)}$ denote the time of the ith event in a nonhomogeneous Poisson process of intensity $\lambda(t)$. Let $\Lambda(t) = \int_0^t \lambda(u) \, du$. Show that

$$P\left[T_{(i)} \geq t \mid T_{(i-1)} = s\right] = \exp[\Lambda(s) - \Lambda(t)]$$

   for $i = 1, 2, \ldots$ and for all $t > s$. Use this to show that $\{T_{(i)}\}$ may be generated by solving

$$\Lambda(T_{(i)}) = \Lambda(T_{(i-1)}) - \ln R_i$$

   where $\{R_i\}$ are $U(0, 1)$ random numbers. Hence show that this method will give exactly the same realization as time scale transformation.

6. Consider a two-dimensional Poisson process of rate λ over the circle $D = \{(r, \theta) : r \in [0, r_0], \Theta \in [0, 2\pi]\}$. Let $N(u)$ denote the number of events for which $r < u$. Show that $\{N(r) : r \in [0, r_0]\}$ is a nonhomogeneous Poisson process in one dimension that has rate $\lambda(r) = 2\pi r \lambda$. Hence show that the following algorithm generates events over D.

Sample $N \sim \text{Poisson}\left(\lambda \pi r_0^2\right)$
For $i = 1, \ldots, N$ do
Generate $U, V, W \sim U(0, 1)$ random numbers
$R_i := r_0 \max(U, V)$
$\Theta_i := 2\pi W$
End do;

7. At the beginning of each day a piece of equipment is inspected and classified as one of the working conditions $i = 1, \ldots, 5$. State 1 corresponds to 'new' and state 5 to 'equipment needs replacing'. State i is always better than state $i + 1$. The transition matrix is

$$P = \begin{pmatrix} 0.8 & 0.15 & 0.03 & 0.02 & 0 \\ 0 & 0.7 & 0.2 & 0.1 & 0 \\ 0 & 0 & 0.6 & 0.35 & 0.05 \\ 0 & 0 & 0 & 0.9 & 0.1 \\ 0 & 0 & 0 & 0 & 1 \end{pmatrix}$$

Starting in state 1, simulate the times at which the machine changes state until the Markov chain is absorbed into state 5. Perform 1000 replications of this and find a 95 % confidence interval for the mean lifetime of the equipment. Compare this with the theoretical result.

8. Suppose there are k infected and n noninfected individuals in a population. During an interval $(t, t + \delta t)$ the probability that a specified noninfected individual will contact any infected is $k\beta\,\delta t + o(\delta t)$, independent of the previous history of the system. Therefore the probability that one contact takes place in the entire population is $nk\beta\,\delta t + o(\delta t)$. There is a probability p that such contact results in infection. Therefore, the probability that the number infected increases by one is $nk\beta p\,\delta t + o(\delta t)$. The probability that the number infected increases by more than one is $o(\delta t)$. Once infected the time to death for the individual is negative exponential, mean $1/\mu_1$, and this is independent of corresponding times for other individuals. During the interval $(t, t + \delta t)$, for each noninfected individual there is an independent probability, $\mu_2\,\delta t + o(\delta t)$, that the individual will die.

   Let $N(t)$, $K(t)$, and $D(t)$ denote the number of noninfected, infected, and deceased individuals at time t. The *duration* of an epidemic is $\min\{t : K(t) = 0\}$. The *size* of an epidemic is the total number of individuals who were infected over the duration. Simulate realizations of an epidemic for chosen values of λ, μ_1, μ_2 and initial conditions to demonstrate the features of such a system.

9. The price of a share follows a geometric Brownian motion $\{X(t), t \geq 0\}$ with expected growth rate μ and volatility σ. In addition, there is a Poisson process, $\{N(t) : t \geq 0\}$, of rate λ that produces 'jumps' in the price. Suppose the jumps occur at times $t_1, t_2, \ldots$. Then

   $$X(t_j^-) = X(t_{j-1}^+) \exp\left\{ (\mu - \sigma^2/2)(t_j - t_{j-1}) + \sigma\left[B(t_j) - B(t_{j-1}) \right] \right\}$$

   for $j = 1, 2, \ldots$, where $t_j^+ (t_j^-)$ is the time immediately after (before) the jth jump, $t_0 = 0$, $X(t_j^+) = Y_j X(t_j^-)$, and where $\{Y_j, j = 1, 2, \ldots\}$ are identically and independently distributed positive random variables. The sets $\{Y_j\}$ and $\{B(t_j)\}$ and the process $\{N(t) : t \geq 0\}$ are mutually independent.

   (a) Show that

   $$X(t) = X(0) \exp\left[(\mu - \sigma^2/2)t + \sigma B(t) \right] \prod_{j=0}^{N(t)} Y_j$$

   where $Y_0 = 1$ and $B(t)$ is a standard Brownian motion.

(b) A plausible distribution for $\{Y_j\}$ is lognormal. Recall that Y is lognormal when $Y = \exp(W)$ and $W \sim N(\alpha, \beta^2)$. Now, $E(Y) = \exp(\alpha + \beta^2/2)$ and $\sigma(Y) = E(Y)\sqrt{\exp(\beta^2) - 1}$. Find α and β when $E(Y) = 1.2$ and $\text{Var}(Y) = 0.1$. Use Maple to simulate realizations of $X(1)$ when $\mu = 0.08$, $\sigma = 0.3$, $\lambda = 0.5$ per annum and $X(0) = 100$ pence.

(c) By conditioning on $N(t)$ or otherwise, show that

$$E\{X(t)\} = X(0)\exp\{\mu t + \lambda t [E(Y) - 1]\}$$

Estimate $E\{X(1)\}$ from your simulation in (b) and compare with the theoretical answer.

10. Consider an M/M/1 queueing system with arrival rate λ and service rate $\mu > \lambda$. Determine all possible sets of regeneration points (events). Do you have any preference for the set you would choose for estimating the expected line length in the steady state?

11. Modify the procedure 'gg1' in Appendix 7.1 so that arrivals are still Poisson but service durations are constant. Compare the results with those in Table 7.1 and comment.

12. A hospital has five ambulances for emergencies. The catchment area may be approximated by a circle of radius 5 miles, the hospital being at the centre. The physical distribution of accidents is a two-dimensional Poisson process, with a rate μ per hour. If an ambulance is not available the patient must wait until one becomes free. An ambulance always takes a straight-line route to the scene of the emergency and returns to the hospital. Assume that ambulances travel at a constant speed of v miles per hour and that one ambulance only is required for each emergency.

(a) Show that the return travel time (x hours) may be sampled by setting $x = 10\sqrt{U}/v$ where $U \sim U(0, 1)$.

(b) Simulate the system to print out for each patient the time between the occurrence of the emergency and arrival at hospital.

13. A maintenance department responsible for m machines having identical characteristics consists of n repairmen and p testers. Machines break down after a random running time with known distribution. If no repairman is free it enters the repair queue. Once a repair is completed it is ready for testing. If no tester is free it is put in a testing queue. The test includes rectifying any faults found and the testing time has a known distribution. Simulate the system with the aim of estimating the long-run average proportion of time that machines are working. Under what conditions would regenerative analysis be possible?

14. A drive-in burger bar has two windows, each manned by an assistant, and each with a separate drive-in lane. The drive-in lanes are adjacent. Customer arrivals are Poisson, with rate 0.4 per minute. The service duration follows a known distribution which is not negative exponential and has a mean which is less than 5 minutes. Arriving customers choose the shortest lane. The last customer in a longer queue will change queues ('queue hop') once the difference in queue lengths becomes 2.

(a) Simulate this system in order to estimate, under steady state conditions, the mean time a customer has to wait between arriving and leaving.

(b) If customers cannot queue hop, estimate the difference in this time for your selected parameter values.

(c) When is a regenerative analysis feasible?

15. A job shop has two machines, A and B. Arriving items are processed first by machine A and then by machine B. The queue discipline is first come, first served. Each machine processes one job at a time and q_A and q_B are the queue lengths at A and B. Simulate this process for arbitrary interarrival and service time distributions, subject to them leading to steady state behaviour in the long run. Use this to produce a histogram of distribution of time between arriving at and leaving the job shop.

16. Patients present themselves at ward A (six beds) according to a Poisson process of rate 1.5 per day. Other patients arrive at a ward B (three beds) according to a Poisson process of rate 0.5 per day. The lengths of stay in each ward are gamma distributed with means of 3.4 and 3.8 days respectively and standard deviations of 3.5 and 1.6 days respectively. At the end of their stay patients are transferred, if room is available, to a convalescent home C (14 beds), where the length of stay is gamma with mean 5 days and standard deviation 4 days. If room is not available a patient stays in their original ward until a bed becomes available in C. This means that the A or B beds are sometimes blocked for incoming patients and these patients have to be directed to other units. The objective is to find the proportion of patients to whom this happens. Assume that the strategy for admitting patients to the home C is on the basis of longest wait since a scheduled departure gets first priority.

17. Records arrive in a buffer according to a Poisson process of rate λ per second. The buffer can hold a maximum of M records. Any arrival when the buffer is full is lost. The buffer is emptied at times $0, \tau_1, \tau_1 + \tau_2, \ldots$, where $\{\tau_i\}$ are independently distributed with density $f_\tau(x), x > 0$.

(a) Show that the long-run average number of records lost per unit time is

$$\frac{E_f\left(\sum_{n=M+1}^{\infty} (n-M)(\lambda\tau)^n e^{-\lambda\tau}/n!\right)}{E_f(\tau)}.$$

(b) Show that the long-run proportion of time, p_i, that there are i records in the buffer is

$$\frac{(1/\lambda)E_f\left(\sum_{n=i+1}^{\infty} (\lambda\tau)^n\, e^{-\lambda\tau}/n!\right)}{E_f(\tau)}$$

for $i = 0, \ldots, M-1$.

(c) Hence derive an expression for the long-run average over time of the number of records in the buffer.

(d) Show that if $\{\tau_i\}$ are exponentially distributed with $E_f(\tau) = \mu^{-1}$, then the answers to previous parts of this question are:

(a) $\lambda \left(\dfrac{\lambda}{\lambda + \mu} \right)^M$,

(b) $\dfrac{\mu}{\lambda} \left(\dfrac{\lambda}{\lambda + \mu} \right)^{i+1}$, $i = 0, \ldots, M - 1$,

(c) $M - \dfrac{\mu}{\lambda} \sum_{i=0}^{M-1} (M - i) \left(\dfrac{\lambda}{\lambda + \mu} \right)^{i+1}$.

(e) Write a Maple simulation procedure to estimate at least one of the quantities in (a) to (c) for arbitrary f (not exponential), λ, and M.

8

Markov chain Monte Carlo

Markov chain Monte Carlo (MCMC) refers to a class of methods for sampling random vectors $X^{(0)}, X^{(1)}, \ldots$ (generally, they are not independent) from a multivariate distribution f. Note that in this chapter the convention will be dropped of using bold face type to indicate that a quantity is a vector or matrix. Vectors will be assumed to be column vectors. MCMC works by simulating a Markov chain in discrete time. An appropriate (it is not unique) Markov chain is chosen so that it has a limit (and therefore unique stationary) distribution which is f. MCMC is therefore useful for Monte Carlo based integration. In particular, since the early 1990s it has transformed the practice of Bayesian statistics. Only the bare bones of MCMC are described in this chapter. For a more in-depth treatment there are many excellent books including those by Gamerman (1997), Gilks *et al.* (1996), Morgan (2000, chapter 7), and Robert (2001, chapter 6), Robert and Casella (2004). It will be seen that one popular MCMC approach is *Gibbs sampling*. For this type of sampling, a purpose-built package, under the name of BUGS (Bayesian inference using Gibbs sampling), is available from http://www.mrc-bsu.cam.ac/bugs/welcome.shtml. This software includes 'WINBUGS', which has a graphical user interface. In addition to the sampling, it performs Bayesian inference and also includes facilities for assessing when the sample output has converged in distribution to f.

8.1 Bayesian statistics

Suppose we wish to perform inference on a vector of model parameters, $\theta' = (\theta_1, \ldots, \theta_d) \in S \subseteq R^d$. Let $P(\theta)$ denote the *prior distribution* of θ, that is the distribution of θ before observing the values of certain random variables in the model. Suppose we now observe those values and let D denote these data. Let $P(D|\theta)$ denote the likelihood of the data and let $P(D) = \int_S P(D|\theta) P(\theta) \, d\theta$. Using Bayes' theorem, the conditional probability distribution of θ given the observed data is

$$P(\theta|D) = \frac{P(D|\theta) P(\theta)}{P(D)}. \tag{8.1}$$

Simulation and Monte Carlo: With applications in finance and MCMC   J. S. Dagpunar
© 2007 John Wiley & Sons, Ltd

$P(\theta|D)$ is known as the *posterior distribution*. It is often useful to write this as

$$P(\theta|D) \propto L(\theta|D) P(\theta)$$

where $L(\theta|D)$ (or simply $L(\theta)$ for short) is the *likelihood function*. Typically,

$$\mu_h = E_{P(\theta|D)}[h(\theta)] = \int_S h(\theta) P(\theta|D) \, d\theta \tag{8.2}$$

where $h(\theta)$ is a known function of θ. For example, the posterior marginal expectation of θ_i, $i = 1, \ldots, d$, is

$$E_{P(\theta|D)}(\theta_i) = \int_S \theta_i P(\theta|D) \, d\theta_1 \cdots d\theta_d.$$

The expectation (8.2) can be estimated by simulating $\{\theta^{(i)}, i = 1, \ldots, n\}$ from $P(\theta|D)$. Then an unbiased estimator of μ_h is

$$\widehat{\mu_h} = \frac{1}{n} \sum_{i=1}^{n} h\left(\theta^{(i)}\right). \tag{8.3}$$

Note that $\theta^{(1)}, \ldots, \theta^{(n)}$ are identically distributed but do not have to be independent. In Bayesian analysis $P(\theta|D)$ is usually multivariate and it is often very difficult to sample independent values from it. One possibility is to sample from the posterior marginal distribution of θ_1, then to sample from the posterior conditional distribution of θ_2 given θ_1, and so on. The difficulty with this approach is that to obtain the posterior marginal density of θ_1 requires the evaluation of a $(d-1)$-dimensional integral, and this is unlikely to be known theoretically. That is why the requirement for the random variables $\theta^{(1)}, \ldots, \theta^{(n)}$ to be independent is relaxed. There is usually a price to be paid for this lack of independence since

$$\text{Var}_{P(\theta|D)}(\widehat{\mu_h}) = \frac{1}{n}\text{Var}_{P(\theta|D)}[h(\theta)] + \frac{1}{n^2} \sum_{j=1}^{n} \sum_{i=1, i \neq j}^{n} \text{Cov}_{P(\theta|D)}\left[h\left(\theta^{(i)}\right), h\left(\theta^{(j)}\right)\right].$$

Certainly, if all the covariances are positive, the variance of the estimator might be orders of magnitude larger than what it would be if the sample comprised independent observations. On the other hand, the variance would be smaller than the independence case if the net covariance could be arranged to be negative, but the MCMC methods developed so far tend to fall into the first category. One further comment concerns the denominator $P(D)$ in Equation (8.1). It might be thought that in order to sample from $P(\theta|D)$ it is necessary to evaluate $P(D)$ which involves a d-dimensional integration. However, $P(D)$ is just a multiplicative constant in the distribution and methods will now be used that require the sampling distribution to be known only up to a multiplicative constant.

8.2   Markov chains and the Metropolis–Hastings (MH) algorithm

The ideas can be traced back to Metropolis *et al.* (1953). This was then generalized by Hastings (1970). In the previous section it was seen that the central issue in using Monte Carlo for Bayesian inference is to sample vectors from a multivariate probability density $P(\theta|D)$. Now think of $f(x)$, with support $S \subseteq R^d$, as replacing $P(\theta|D)$. Here, $x' = (x_1, \ldots, x_d)$. A sequence of random vectors $X^{(0)}, X^{(1)}, \ldots$ is generated from a homogeneous discrete time Markov chain on state space S, with *transition kernel* $P(y|x)$. This means that

$$P(y|x) = P\left(X^{(t)} \le y | X^{(t-1)} = x, \ldots, X^{(0)} = x^{(0)}\right)$$
$$= P\left(X^{(t)} \le y | X^{(t-1)} = x\right)$$

and

$$P\left(X^{(t)} \le y | X^{(t-1)} = x\right) = P\left(X^{(1)} \le y | X^{(0)} = x\right)$$

for all $x, y \in S$ and for $t = 1, 2, \ldots$. Note that $X \le y$ denotes that $X_i \le y_i$ for $i = 1, \ldots, d$. The associated conditional probability density is $p(y|x)$ where

$$p(y|x) = \frac{\partial P(y|x)}{\partial y}.$$

The chain is chosen to be *ergodic*. Consequently, it has a unique *stationary* probability density f. It follows that after a suitable *burn-in* time (m steps, say, of the chain), the random vectors $X^{(m+1)}, X^{(m+2)}, \ldots$ have a marginal distribution which is approximately f. Of course they are not independent. We can now estimate μ_h where

$$\mu_h = E_f[h(x)] = \int_S h(x) f(x) \, dx \tag{8.4}$$

by

$$\widehat{\mu_h} = \frac{1}{n} \sum_{t=m+1}^{m+n} h\left(X^{(t)}\right) \tag{8.5}$$

where n is large enough to make the variance of $\widehat{\mu_h}$ small enough.

The basic Metropolis–Hastings algorithm works as follows. Given $X^{(t)} = x$, the state at the next step, $X^{(t+1)}$, is y with probability $\alpha(x, y)$, or remains unchanged at x with probability $1 - \alpha(x, y)$. Here, y is a *candidate point* from a *proposal density* $q(y|x)$. The acceptance probability is

$$\alpha(x, y) = \min\left[1, \frac{f(y) q(x|y)}{f(x) q(y|x)}\right]. \tag{8.6}$$

Subject to certain regularity conditions $q(.|.)$ can take any form (providing the resulting Markov chain is ergodic), which is a mixed blessing in that it affords great flexibility in design. It follows that the sequence $X^{(0)}, X^{(1)}, \ldots$ is a homogeneous Markov chain with

$$p(y|x) = \alpha(x, y) q(y|x)$$

for all $x, y \in S$, with $x \neq y$. Note that the conditional probability of remaining in state x at a step in this chain is a *mass* of probability equal to

$$\int_S [1 - \alpha(x, y)] q(y|x) \, dy.$$

Suppose $\alpha(x, y) < 1$. Then according to Equation (8.6), $\alpha(y, x) = 1$. Similarly, if $\alpha(x, y) = 1$ then $\alpha(y, x) < 1$. It follows from Equation (8.6) that for all $x \neq y$

$$\alpha(x, y) f(x) q(y|x) = \alpha(y, x) f(y) q(x|y).$$

This shows that the chain is time reversible in equilibrium with

$$f(x) p(y|x) = f(y) p(x|y)$$

for all $x, y \in S$. Summing over y gives

$$f(x) = \int_S f(y) p(x|y) \, dy,$$

showing that f is indeed a stationary distribution of the Markov chain. Providing the chain is ergodic, then the stationary distribution of this chain is unique and is also its limit distribution. This means that after a suitable burn-in time, m, the marginal distribution of each $X^{(t)}, t > m$, is almost f, and the estimator (8.5) can be used.

To estimate μ_h, the Markov chain is replicated K times, with widely dispersed starting values. Let $X_i^{(t)}$ denote the tth *equilibrium* observation (i.e. the tth observation following burn-in) on the *i*th replication. Let

$$\widehat{\mu}_h^{(i)} = \frac{1}{n} \sum_{t=1}^{n} h\left(X_i^{(t)}\right)$$

and

$$\widehat{\mu}_h = \frac{1}{K} \sum_{i=1}^{K} \widehat{\mu}_h^{(i)}.$$

Then $\widehat{\mu}_h$ is unbiased and its estimated standard error is

$$\text{e.s.e.}\left(\widehat{\mu}_h\right) = \sqrt{\frac{1}{K} \sum_{i=1}^{K} \frac{\left(\widehat{\mu}_h^{(i)} - \widehat{\mu}_h\right)^2}{K - 1}}.$$

There remains the question of how to assess when a realization has burned in. This can be a difficult issue, particularly with high-dimensional state spaces. One possibility is to plot a (several) component(s) of the sequence $\{X^{(t)}\}$. Another is to plot some function of $X^{(t)}$ for $t = 0, 1, 2, \ldots$. For example, it might be appropriate to plot $\left\{ h\left(X_i^{(t)}\right), t = 1, 2, \ldots \right\}$. Whatever choice is made, repeat for each of the K independent replications. Given that the initial state for each of these chains is different, equilibrium is perhaps indicated when t is of a size that makes all K plots similar, in the sense that they fluctuate about a common central value and explore the same region of the state space. A further issue is how many equilibrium observations, n, there should be in each realization. If the chain has strong positive dependence then the realization will move slowly through the states (*slow mixing*) and n will need to be large in order that the entire state space is explored within a realization. A final and positive observation relates to the calculation of $\alpha(x, y)$ in Equation (8.6). Since f appears in both the numerator and denominator of the right-hand side it need be known only up to an arbitrary multiplicative constant. Therefore it is unnecessary to calculate $P(D)$ in Equation (8.1).

The original Metropolis (Metropolis *et al.*, 1953) algorithm took $q(y|x) = q(x|y)$. Therefore,

$$\alpha(x, y) = \min\left(1, \frac{f(y)}{f(x)}\right).$$

A suitable choice for q might be

$$q(y|x) \propto \exp\left[-(y-x)' \Sigma^{-1} (y-x)\right]; \tag{8.7}$$

that is, given x, $Y \sim N(x, \Sigma)$. How should Σ, which controls the average step length, be chosen? Large step lengths potentially encourage good mixing and exploration of the state space, but will frequently be rejected, particularly if the current point x is near the mode of a unimodal density f. Small step lengths are usually accepted but give slow mixing, long burn-in times, and poor exploration of the state space. Clearly, a compromise value for Σ is called for.

Hastings (1970) suggested a *random walk sampler*; that is, given that the current point is x, the candidate point is $Y = x + W$ where W has density g. Therefore

$$q(y|x) = g(y-x).$$

This appears to be the most popular sampler at present. If g is an even function then such a sampler is also a Metropolis sampler. The sampler (8.7) is a random walk algorithm with

$$Y = x + \Sigma^{1/2} Z$$

where $\Sigma^{1/2} \Sigma^{1/2'} = \Sigma$ and Z is a column of i.i.d. standard normal random variables.

An *independence sampler* takes $q(y|x) = q(y)$, so the distribution of the candidate point is independent of the current point. Therefore,

$$\alpha(x, y) = \min\left[1, \frac{f(y) q(x)}{f(x) q(y)}\right].$$

In this case, a good strategy is to choose q to be similar to f. This results in an acceptance probability close to 1, with successive variates nearly independent, which of course is good from the point of view of reducing the variance of an estimator. In a Bayesian context q might be chosen to be the prior distribution of the parameters. This is a good choice if the posterior differs little from the prior.

Let us return to the random walk sampler. To illustrate the effect of various step lengths refer to the procedure 'mcmc' in Appendix 8.1. This samples values from $f(x) \propto \exp(-x^2/2)$ using

$$Y = x + W$$

where $W \sim (a, -a)$. This is also a Metropolis sampler since the density of W is symmetric about zero. The acceptance probability is

$$\alpha(x, y) = \min\left[1, \frac{f(y)}{f(x)}\right]$$
$$= \min\left(1, e^{-(y^2-x^2)/2}\right)$$

To illustrate the phenomenon of burn-in initialization with $X^{(0)} = -2$ will take place, which is a relatively rare state in the equilibrium distribution of $N(0, 1)$. Figure 8.1(a) $(a = 0.5)$ shows that after 200 iterations the sampler has not yet reached equilibrium status. With $a = 3$ in Figure 8.1(b) it is possible to assess that equilibrium has been achieved after somewhere between 50 and 100 iterations. Figures 8.1(c) to (e) are for an initial value of $X^{(0)} = 0$ (no burn-in is required, as knowledge has been used about

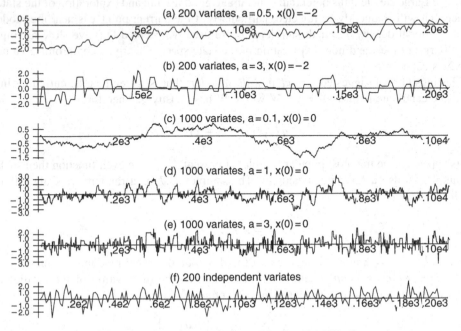

Figure 8.1 The x value against iteration number for $N(0, 1)$ samplers

the most likely state under equilibrium conditions) for $N(0, 1)$ over 1000 iterations with $a = 0.1, 1$, and 3 respectively. Note how, with $a = 0.1$, the chain is very slow mixing and that after as many as 1000 iterations it has still not explored the tails of the normal distribution. In Figure 8.1(d) ($a = 1$), the support of $N(0, 1)$ is explored far better and the mixing of states is generally better. In Figure 8.1(e) ($a = 3$) there is rapid mixing, frequent rejections, and perhaps evidence that the extreme tails are not as well represented as in Figure 8.1(d). Figure 8.1(f) is of 200 independent $N(0, 1)$ variates. In effect, this shows an ideal mixing of states and should be compared in this respect with Figures 8.1(a) and (b).

8.3 Reliability inference using an independence sampler

The Weibull distribution is frequently used to model the time to failure of equipment or components. Suppose the times to failure, $\{X_i\}$, of identically manufactured components are identically and independently distributed with the survivor function

$$\bar{F}(x) = P(X > x) \tag{8.8}$$

$$= \exp\left[-\left(\frac{x}{\beta}\right)^\alpha\right]$$

where $x, \alpha, \beta > 0$. It follows that the probability density function is

$$f(x) = \alpha\beta^{-\alpha}x^{\alpha-1}\exp\left[-\left(\frac{x}{\beta}\right)^\alpha\right].$$

The *failure rate* at age $x, r(x)$, given that the component is still working at age x, is defined to be the conditional probability density of failure at age x, given that the component has survived to age x. Therefore,

$$r(x) = \frac{f(x)}{\bar{F}(x)} = \alpha\beta^{-\alpha}x^{\alpha-1}. \tag{8.9}$$

For some components the failure rate is independent of age ($\alpha = 1$) but for many the failure rate is increasing with age ($\alpha > 1$) due to wear and tear or other effects of ageing.

Consider a set of components where no data on failure times is available. Engineers believed that the failure rate is increasing with age ($\alpha > 1$), with the worst case scenario being a linear dependence ($\alpha = 2$). Moreover, the most likely value of α was felt to be approximately 1.5, with the prior probability decreasing in a similar manner for values on either side of 1.5. Therefore, a suitable choice for the marginal prior of α might be

$$g(\alpha) = \begin{cases} 4(\alpha - 1), & 1 < \alpha < 1.5, \\ 4(2 - \alpha), & 1.5 < \alpha \le 2. \end{cases}$$

This is a symmetric triangular density on support $(1, 2)$. To sample from such a density, take $R_1, R_2 \sim U(0, 1)$ and put

$$\alpha = 1 + \frac{1}{2}(R_1 + R_2). \tag{8.10}$$

It is also thought that the expected lifetime lies somewhere between 2000 and 3000 hours, depending upon the α and β values. Accordingly, given that

$$E(X \mid \alpha, \beta) = \int_0^\infty \bar{F}(x)\,dx$$

$$= \beta \Gamma\left(\frac{1}{\alpha}+1\right),$$

a choice might be made for the conditional prior of β given α, the $U(2000/[\Gamma(1/\alpha+1)], 3000/[\Gamma(1/\alpha+1)])$ density. Once α has been sampled, β is sampled using

$$\beta = \frac{1000\,(2+R_3)}{\Gamma(1/\alpha+1)} \tag{8.11}$$

where $R_3 \sim U(0, 1)$. Note that the joint prior is

$$g(\alpha, \beta) = \begin{cases} \dfrac{4}{1000}(\alpha-1)\,\Gamma\left(\dfrac{1}{\alpha}+1\right), & 1 < \alpha < 1.5, \\[2mm] \dfrac{4}{1000}(2-\alpha)\,\Gamma\left(\dfrac{1}{\alpha}+1\right), & 1.5 < \alpha \le 2, \end{cases} \tag{8.12}$$

where $2000/[\Gamma(1/\alpha+1)] < \beta < 3000/[\Gamma(1/\alpha+1)]$.

In order to implement a maintenance policy for such components, it was required to know the probabilities that a component will survive to ages 1000, 2000, and 3000 hours respectively. With no failure time data available the predictive survivor function with respect to the joint prior density is

$$P_{\text{prior}}(X > x) = E_{g(\alpha,\beta)}\left\{\exp\left[-\left(\frac{x}{\beta}\right)^\alpha\right]\right\}$$

$$= \int_1^2 d\alpha \int_{2000/[\Gamma(1/\alpha+1)]}^{3000/[\Gamma(1/\alpha+1)]} \exp\left[-\left(\frac{x}{\beta}\right)^\alpha\right] g(\alpha, \beta)\,d\beta.$$

Now suppose that there is a random sample of failure times $x_1, \ldots, x_n$. Table 8.1 shows these data where $n = 43$. It is known that the posterior density of α and β is

$$\pi(\alpha, \beta) \propto L(\alpha, \beta)\,g(\alpha, \beta) \tag{8.13}$$

where L is the likelihood function. The *posterior predictive* survivor function is

$$P_{\text{post}}(X > x) = E_{\pi(\alpha,\beta)}\left\{\exp\left[-\left(\frac{x}{\beta}\right)^\alpha\right]\right\}$$

$$= \int_1^2 d\alpha \int_{2000/[\Gamma(1/\alpha+1)]}^{3000/[\Gamma(1/\alpha+1)]} \exp\left[-\left(\frac{x}{\beta}\right)^\alpha\right] \pi(\alpha, \beta)\,d\beta. \tag{8.14}$$

Table 8.1   Failure times (hours) for 43 components

293	1902	1272	2987	469	3185	1711	8277	356	822	2303
317	1066	1181	923	7756	2656	879	1232	697	3368	486
6767	484	438	1860	113	6062	590	1633	2425	367	712
953	1989	768	600	3041	1814	141	10511	7796	1462	

In order to find a point estimate of Equation (8.14) for specified x, we will sample k values $\{(\alpha_i, \beta_i)\}$ from $\pi(\alpha, \beta)$ using MCMC, where the proposal density is simply the joint prior, $g(\alpha, \beta)$. This is therefore an example of an independence sampler. The k values will be sampled when the chain is in a (near) equilibrium condition. The estimate is then

$$\widehat{P}_{post}(X > x) = \frac{1}{k} \sum_{i=1}^{k} \exp\left[-\left(\frac{x}{\beta_i}\right)^{\alpha_i}\right].$$

In Appendix 8.2 the procedure 'fail' is used to estimate this for $x = 1000$, 2000, and 3000 hours. Given that the current point is (α, β) and the candidate point sampled using Equations (8.10) and (8.11) is (α_c, β_c), the acceptance probability is

$$\min\left[1, \frac{\pi(\alpha_c, \beta_c)\, g(\alpha, \beta)}{\pi(\alpha, \beta)\, g(\alpha_c, \beta_c)}\right] = \min\left[1, \frac{L(\alpha_c, \beta_c)}{L(\alpha, \beta)}\right]$$

where

$$L(\alpha, \beta) = \prod_{i=1}^{n} \alpha x_i^{\alpha-1} \beta^{-\alpha} \exp\left[-\left(\frac{x_i}{\beta}\right)^{\alpha}\right]$$

$$= \alpha^n \beta^{-n\alpha} \exp\left[\sum_{i=1}^{n} -\left(\frac{x_i}{\beta}\right)^{\alpha}\right](x_1 \ldots x_n)^{\alpha-1}.$$

The posterior estimates for a component surviving 1000, 2000, and 3000 hours are 0.70, 0.45, and 0.29 respectively. It is interesting to note that the maximum likelihood estimates (constrained so that $\alpha \geq 1$) are $\widehat{\alpha}_{ml} = 1$ and $\widehat{\beta}_{ml} = 2201$. This represents a component with a constant failure rate (exponential life). However, the prior density indicated the belief that the failure rate is increasing and indeed this must be the case with the *Bayes estimates* (i.e. the posterior marginal expectations of α and β). These are $\widehat{\alpha}_{bayes} = 1.131$ and $\widehat{\beta}_{bayes} = 2470$.

8.4   Single component Metropolis–Hastings and Gibbs sampling

Single component Metropolis–Hastings in general, and Gibbs sampling in particular, are forms of the Metropolis–Hastings algorithm, in which just one component of the vector x is updated at a time. It is assumed as before that we wish to sample variates x from a density f. Let $x' \equiv (x_1, \ldots, x_d)$ denote a vector state in a Markov chain which has a limit density f. Let x_{-i} denote this vector with the ith component removed. Let $[y, x_{-i}]$ denote the original vector with the ith component replaced by y. Given that the current state is $(x_1, \ldots, x_d)$, which is the same as $[x_i, x_{-i}]$, single component Metropolis–Hastings samples y from a *univariate* proposal density

$$q(y_i \mid [x_i, x_{-i}]).$$

This samples a prospective value y_i for the ith component (conditional on the current point) and generates a candidate point $[y_i, x_{-i}]$. This is accepted with probability

$$\alpha = \min\left[1, \frac{f([y_i, x_{-i}]) \, q(x_i \mid [y_i, x_{-i}])}{f([x_i, x_{-i}]) \, q(y_i \mid [x_i, x_{-i}])}\right].$$

However, $f([x_i, x_{-i}]) = f(x_{-i}) f(x_i \mid x_{-i})$ and $f([y_i, x_{-i}]) = f(x_{-i}) f(y_i \mid x_{-i})$. Therefore,

$$\alpha = \min\left[1, \frac{f(y_i \mid x_{-i}) \, q(x_i \mid [y_i, x_{-i}])}{f(x_i \mid x_{-i}) \, q(y_i \mid [x_i, x_{-i}])}\right].$$

The essential feature of this approach is that either we remain at the same point or move to an 'adjacent' point that differs only in respect of one component of the vector state. This means that univariate sampling is being performed.

Now suppose the proposal density is chosen to be

$$q(y_i \mid [x_i, x_{-i}]) = f(y_i \mid x_{-i}).$$

Then the acceptance probability becomes one. This is the Gibbs sampler. Note that we sample (with respect to the density f) a value for the ith component conditional upon the current values of all the other components. Such a conditional density, $f(y_i \mid x_{-i})$, is known as a *full conditional*. As only one component changes, the point is updated in small steps, which is perhaps a disadvantage. The main advantage of this type of algorithm compared with the more general Metropolis–Hastings one is that it is expected to be much simpler to sample from d univariate densities than from a single d variate density. In some forms of this algorithm the component i is chosen at random. However, most implementations sample the components 1 through to d sequentially, and this constitutes one iteration of the algorithm shown below:

$t := 0$

1. $X_1^{(t+1)} \sim f\left(x_1 \mid x_2^{(t)}, x_3^{(t)}, \ldots, x_d^{(t)}\right)$

   $X_2^{(t+1)} \sim f\left(x_2 \mid x_1^{(t+1)}, x_3^{(t)}, \ldots, x_d^{(t)}\right)$

   $\vdots$

   $X_d^{(t+1)} \sim f\left(x_d \mid x_1^{(t+1)}, x_2^{(t+1)}, \ldots, x_{d-1}^{(t+1)}\right)$

   $t := t + 1$

   goto 1

Sampling is from univariate *full conditional distributions*. For example, at some stage there is a need to sample from

$$f(x_3 \mid x_1, x_2, x_4, \ldots, x_d).$$

However, this is *proportional* to the joint density

$$f(x_1, x_2, x_3, x_4, \ldots, x_d)$$

where $x_1, x_2, x_4, \ldots, x_d$ are known. The method is therefore particularly efficient if there are univariate generation methods that require the univariate density to be known up to a multiplicative constant only. Note, however, that the full conditionals are changing not only between the different components sampled within an iteration but also between the same component sampled in different iterations (since the parameter values, being the values of the remaining components, have also changed). This means that the univariate sampling methods adopted will need to have a small set-up time. Therefore, a method such as adaptive rejection (Section 3.4) may be particularly suitable.

Given that the method involves sampling from full conditionals, finally check that this is likely to be much simpler than a direct simulation in which X_1 is sampled from the marginal density of $f(x_1)$, X_2 from the conditional density $f(x_2 \mid x_1), \ldots,$ and X_d from the conditional density $f(x_d \mid x_1, \ldots, x_{d-1})$. To show that this is so, note that in order to obtain $f(x_1)$ it would first be necessary to integrate out the other $d-1$ variables, which is likely to be very expensive, computationally.

As an illustration of the method, suppose

$$f(x_1, x_2, x_3) = \exp\left[-(x_1 + x_2 + x_3) - \theta_{12}x_1x_2 - \theta_{23}x_2x_3 - \theta_{31}x_3x_1\right] \qquad (8.15)$$

for $x_i \geq 0$ for all i, where $\{\theta_{ij}\}$ are known positive constants, as discussed by Robert and Casella (2004, p. 372) and Besag (1974). Then

$$f(x_1 \mid x_2, x_3) = \frac{f(x_1, x_2, x_3)}{f(x_2, x_3)}$$

$$\propto f(x_1, x_2, x_3)$$

$$\propto \exp\left(-x_1 - \theta_{12}x_1x_2 - \theta_{31}x_3x_1\right).$$

Therefore the full conditional of X_1 is

$$X_1 \mid x_2, x_3 \sim \text{Exp}\left(1 + \theta_{12}x_2 + \theta_{31}x_3\right)$$

or a negative exponential with expectation $(1 + \theta_{12}x_2 + \theta_{31}x_3)^{-1}$. The other full conditionals are derived similarly.

8.4.1   Estimating multiple failure rates

Gaver and O'Muircheartaigh (1987) estimated the failure rates for 10 different pumps in a power plant. One of their models had the following form. Let X_i denote the number of failures observed in $[0, t_i]$ for the ith pump, $i = 1, \ldots, 10$, where the $\{t_i\}$ are known. It is assumed that $X_i \mid \lambda_i \sim \text{Poisson}(\lambda_i t_i)$ where $\{\lambda_i \mid \alpha, \beta\}$ are independently distributed as $g_{\lambda_i \mid \beta}(\lambda \mid \beta) \sim \text{gamma}(\alpha, \beta)$ and β is a realization from $g_\beta(\beta) \sim \text{gamma}(\gamma, \delta)$. The *hyperparameter* values, α, γ, and δ are assumed to be known. The first four columns of Table 8.2 show the sample data comprising x_i, t_i, and the raw failure rate, $r_i = x_i / t_i$.

The aim is to obtain the posterior distribution of the ten failure rates, $\{\lambda_i\}$. The likelihood is

$$L(\{\lambda_i\}) = \prod_{i=1}^{10} \frac{e^{-\lambda_i t_i}(\lambda_i t_i)^{x_i}}{x_i!}$$

$$\propto \prod_{i=1}^{10} e^{-\lambda_i t_i}\lambda_i^{x_i} \qquad (8.16)$$

Table 8.2 Pump failures. (Data, excluding last column are from Gaver and O'Muircheartaigh, 1987)

Pump	x_i	$t_i \times 10^{-3}$ (hours)	$r_i \times 10^3 (\text{hours}^{-1})$	Bayes estimate $= \widehat{\lambda}_i \times 10^3 (\text{hours}^{-1})$
1	5	94.320	0.053	0.0581
2	1	15.720	0.064	0.0920
3	5	62.860	0.08	0.0867
4	14	125.760	0.111	0.114
5	3	5.240	0.573	0.566
6	19	31.440	0.604	0.602
7	1	1.048	0.954	0.764
8	1	1.048	0.954	0.764
9	4	2.096	1.91	1.470
10	22	10.480	2.099	1.958

and the prior distribution is

$$g_{\{\lambda_i\},\beta}(\{\lambda_i\},\beta) = g_\beta(\beta) \prod_{i=1}^{10} g_{\lambda_i|\beta}(\lambda_i \mid \beta)$$

$$= \frac{e^{-\delta\beta}\beta^{\gamma-1}\delta^\gamma}{\Gamma(\gamma)} \prod_{i=1}^{10} \frac{e^{-\beta\lambda_i}\beta^\alpha \lambda_i^{\alpha-1}}{\Gamma(\alpha)}$$

$$\propto e^{-\delta\beta}\beta^{\gamma-1} \prod_{i=1}^{10} e^{-\beta\lambda_i}\beta^\alpha \lambda_i^{\alpha-1}. \qquad (8.17)$$

The posterior joint density is

$$\pi_{\{\lambda_i\},\beta}(\{\lambda_i\},\beta) \propto L(\{\lambda_i\}) g_{\{\lambda_i\},\beta}(\{\lambda_i\},\beta)$$

The posterior full conditional of β is

$$\pi(\beta \mid \{\lambda_i\}) \propto e^{-\delta\beta}\beta^{\gamma-1} \prod_{i=1}^{10} e^{-\beta\lambda_i}\beta^\alpha$$

$$= \beta^{10\alpha+\gamma-1} e^{-\beta\left(\delta+\sum_{i=1}^{10}\lambda_i\right)}.$$

which is a gamma $\left(10\alpha + \gamma, \delta + \sum_{i=1}^{10}\lambda_i\right)$ density. The posterior full conditional of λ_j is

$$\pi_{\lambda_j|\{\lambda_i, i \neq j\},\beta}(\lambda_j \mid \{\lambda_i, i \neq j\},\beta) \propto e^{-\beta\lambda_j}\lambda_j^{\alpha-1}e^{-\lambda_j t_j}\lambda_j^{x_j}$$

which is a gamma $(\alpha + x_j, \beta + t_j)$ density. Note that this is independent of λ_i for $i \neq j$. Gibbs sampling is therefore particularly easy for this example, since the full conditionals are standard distributions. The Bayes estimate of the jth failure rate is

$$\widehat{\lambda}_j = E_{\pi_\beta}\left(E_{\pi_{\lambda_j|\beta}}[\lambda_j]\right)$$

$$= E_{\pi_\beta}\left(\frac{\alpha + x_j}{\beta + t_j}\right) \qquad (8.18)$$

where π_β is the posterior marginal density of β.

There remains the problem of selecting values for the hyperparameters α, γ, and δ. When $\gamma > 1$, the prior marginal expectation of λ_j is

$$
\begin{aligned}
E_{g_{\{\lambda_i\},\beta}}(\lambda_j) &= E_{g_\beta}\left(E_{g_{\lambda_j|\beta}}[\lambda_j]\right) \\
&= E_{g_\beta}\left(\frac{\alpha}{\beta}\right) \\
&= \int_0^\infty \frac{\alpha}{\beta} \frac{e^{-\delta\beta}\beta^{\gamma-1}\delta^\gamma}{\Gamma(\gamma)} d\beta \\
&= \frac{\alpha\delta\Gamma(\gamma-1)}{\Gamma(\gamma)} \\
&= \frac{\alpha\delta}{\gamma-1}
\end{aligned}
\tag{8.19}
$$

for $j = 1, \ldots, 10$. Similarly, when $\gamma > 2$,

$$
\begin{aligned}
\mathrm{Var}_{g_{\{\lambda_i\},\beta}}(\lambda_j) &= E_{g_\beta}\left(\mathrm{Var}_{g_{\lambda_j|\beta}}[\lambda_j]\right) \\
&\quad + \mathrm{Var}_{g_\beta}\left(E_{g_{\lambda_j|\beta}}[\lambda_j]\right) \\
&= E_{g_\beta}\left(\frac{\alpha}{\beta^2}\right) + \mathrm{Var}_{g_\beta}\left(\frac{\alpha}{\beta}\right) \\
&= E_{g_\beta}\left(\frac{\alpha}{\beta^2}\right) + E_{g_\beta}\left(\frac{\alpha^2}{\beta^2}\right) - \left[E_{g_\beta}\left(\frac{\alpha}{\beta}\right)\right]^2 \\
&= \alpha(1+\alpha)\int_0^\infty \frac{e^{-\delta\beta}\beta^{\gamma-1}\delta^\gamma}{\beta^2\Gamma(\gamma)} d\beta - \left(\frac{\alpha\delta}{\gamma-1}\right)^2 \\
&= \frac{\alpha(1+\alpha)\delta^2}{(\gamma-1)(\gamma-2)} - \left(\frac{\alpha\delta}{\gamma-1}\right)^2 \\
&= \frac{\alpha\delta^2(\alpha+\gamma-1)}{(\gamma-1)^2(\gamma-2)}.
\end{aligned}
\tag{8.20}
$$

$$
\tag{8.21}
$$

In the original study by Gaver and O'Muircheartaigh (1987) and also in several follow-up analyses of this data set, including those by Robert and Casella (2004, pp. 385–7) and Gelfand and Smith (1990), an *empirical* Bayes approach is used to fit the hyperparameters, α, γ, and δ (apparently set arbitrarily to 1). In empirical Bayes the data are used to estimate the hyperparameters and therefore the prior distribution. Here a true Bayesian approach is adopted. It is supposed that a subjective assessment of the hyperparameters, α, γ, and δ, is based upon the belief that the prior marginal expectation and standard deviation of any

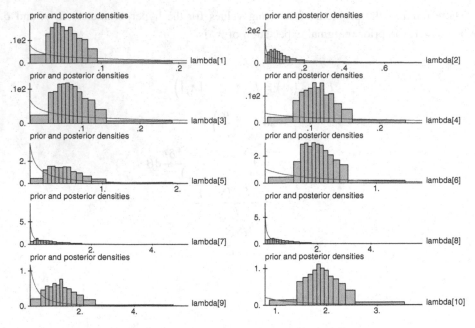

Figure 8.2   Posterior and prior densities for 10 failure rates

of the failure rates are (per thousand hours) $\frac{1}{2}$ and 2 respectively. Using Equations (8.19) and (8.21) two of the three parameters can be fitted. A third hyperparameter value is fitted by the prior belief that, for any pump, the marginal probability that the failure rate exceeds 5 per 1000 hours is 0.01. This fitting results in

$$\alpha = 0.54, \quad \gamma = 2.20, \quad \delta = 1.11. \tag{8.22}$$

The plots in Figure 8.2 show the results of Gibbs sampling over 2000 iterations following a burn-in time of 200 iterations (see the procedure 'pump' in Appendix 8.3). The histograms are representations of the posterior distributions of the 10 failure rates (per thousand hours). Superimposed on each plot is the (common) prior marginal density of the failure rate for any pump. Note that the latter is unbounded as $\lambda_j \to 0$ since $\alpha < 1$. The (estimate of the) Bayes estimate of λ_j is the sample mean of the simulated values, $\{(\alpha+x_j)/(\beta^{(i)}+t_j), i = 1, \ldots, 2000\}$.

Such a simulation could also be used, for example, to determine the posterior survival probability for pump j say, $E_{\pi_{\{\lambda_i\},\beta}(\{\lambda_i\},\beta)}\left(e^{-\lambda_j y}\right)$. Then

$$E_{\pi_{\{\lambda_i\},\beta}}\left(e^{-\lambda_j y}\right) = E_{\pi_\beta}\left[E_{\pi_{\lambda_j}|\beta}\left(e^{-\lambda_j y}\right)\right]$$

$$= E_{\pi_\beta}\left[\int_0^\infty \frac{e^{-\lambda_j y}e^{-\lambda_j(\beta+t_j)}\lambda_j^{x_j+\alpha-1}(\beta+t_j)^{x_j+\alpha}\,\mathrm{d}\lambda_j}{\Gamma(x_j+\alpha)}\right]$$

$$= E_{\pi_\beta}\left[\left(\frac{\beta+t_j}{\beta+t_j+y}\right)^{x_j+\alpha}\right]$$

Therefore, an estimate is the sample mean of $\{[(\beta^{(i)}+t_j)/(\beta^{(i)}+t_j+y)]^{x_j+\alpha}$, $i=1,\ldots,2000\}$.

8.4.2 Capture–recapture

The aim is to estimate the unknown population size N of a collection of animals, as, for example, given in Castledine (1981) and George and Robert (1992). In the first episode, for every animal in the population, let p denote the common probability of capture. Each captured animal is marked and then returned to the population. In the second episode, assume that N and p are as before. Let $n_{1,0}$, $n_{0,1}$, and $n_{1,1}$ denote the number of animals caught on the first episode only, the second episode only, and both episodes respectively. Let

$$n = n_{1,0} + n_{0,1} + 2n_{1,1}$$

which represents the combined count of animals captured on the two episodes. Let

$$n' = n_{1,0} + n_{0,1} + n_{1,1}$$

which is the number of *separate* animals caught in the two episodes (some animals caught and marked in the first episode may be caught again in the second one).

Using a multinomial distribution the likelihood of these data is

$$L(N,p) = \frac{N!}{n_{1,1}!n_{1,0}!n_{0,1}!(N-n')!}[p(1-p)]^{n_{1,0}}[(1-p)p]^{n_{0,1}}[p^2]^{n_{1,1}}[(1-p)^2]^{N-n'}$$

$$= \binom{N}{n_{1,1}n_{1,0}n_{0,1}} p^n(1-p)^{2N-2n'+n_{1,0}+n_{0,1}}$$

$$= \binom{N}{n_{1,1}n_{1,0}n_{0,1}} p^n(1-p)^{2N-n}.$$

Suppose there is little idea of the value of p. Then a suitable prior distribution is $U(0,1)$. Suppose that the prior distribution for N is Poisson (λ) where $\lambda = 250$. The aim is to find the posterior distribution of N given n and n'. In particular we wish to find the *Bayes* estimate of N, that is the posterior expectation of N.

The joint posterior distribution of N and p is

$$\pi(N,p) \propto \frac{p^n(1-p)^{2N-n}\lambda^N}{(N-n')!}.$$

Now put $N' = N - n'$. This is the number of animals in the population that were not captured in either episode. Then the posterior distribution of N' and p is

$$\pi_{N',p}(N',p) \propto \frac{p^n(1-p)^{2N'+2n'-n}\lambda^{N'+n'}}{N'!}. \tag{8.23}$$

Then the full conditionals of p and N' are

$$p|N' \sim \text{beta}\left(n+1, 2N'+n_{1,0}+n_{0,1}+1\right)$$

and

$$N' \mid p \sim \text{Poisson}\left(\lambda \left[1 - p\right]^2\right).$$

Using Gibbs sampling m equilibrium values $\left(N'^{(1)}, p^{(1)}\right), \ldots, \left(N'^{(m)}, p^{(m)}\right)$ are obtained. Now $E_{\pi(N'\mid p)}\left(N' \mid p\right) = \lambda \left[1 - p\right]^2$, so a Bayes estimate of N' is $E_{\pi(p)}\left(\lambda \left[1 - p\right]^2\right)$. The MCMC estimate of this expectation is

$$\frac{\lambda}{m} \sum_{i=1}^{m} \left[1 - p^{(i)}\right]^2$$

(See the procedure 'gibbscapture' in Appendix 8.4.)

Gibbs sampling works out quite well for this example. However, the opportunity will also be taken to see whether the use of Monte Carlo can be avoided altogether. By summing over N' in Equation (8.23) the posterior marginal density of p can be obtained as

$$\pi_p(p) \propto p^n (1 - p)^{2n' - n} e^{\lambda [1 - p]^2}.$$

Using the data values $\lambda = 250$, $n = 160$, and $n' = 130$, as in Appendix 8.4, the Bayes estimate of N' is

$$E_{\pi_p}\left[E_{\pi_{N'\mid p}}\left(N'\mid p\right)\right] = E_{\pi_p}\left[\lambda \left[1 - p\right]^2\right]$$

$$= \frac{\lambda \int_0^1 p^n (1 - p)^{2n' - n + 2} e^{\lambda [1 - p]^2} dp}{\int_0^1 p^n (1 - p)^{2n' - n} e^{\lambda [1 - p]^2} dp}$$

$$= 111.30 \tag{8.24}$$

using numerical integration in Maple. This can be compared with the estimate of 113.20 obtained using Gibbs sampling with a burn-in of 200 iterations followed by a further 500 iterations. This is shown in Appendix 8.4.

8.4.3 Minimal repair

This example concerns the failure of several components. However, unlike the example in Section 8.4.1 it cannot be assumed that failure for any component follows a Poisson process. This is because the failure rate here is believed to increase with the age of a component.

The ith of m components is new at time 0 and put on test for a fixed time $T^{(i)}$. The density of time to first failure is specified as $f(x \mid \alpha_i, \lambda_i)$ where (α_i, λ_i), $i = 1, \ldots, m$, are identically and independently distributed as $g(\alpha, \lambda \mid \mu, b)$, g being a prior density. The parameter b is itself distributed as a *hyperprior* density $w(b \mid \gamma, \delta)$. The *hyperparameters* μ, γ, and δ are assumed to have known values.

Failures of the m components are independent, given $\{(\alpha_i, \lambda_i)\}$. Let $\overline{F}_i(x) = \int_x^{\infty} f(u \mid \alpha_i, \lambda_i) \, du$. On failure during $\left[0, T^{(i)}\right]$ the ith component is instantaneously *minimally repaired*. This means the following. Suppose the jth failure is at age $x_{(j)}^{(i)} < T^{(i)}$.

Then, conditional upon $x_{(1)}^{(i)}, \ldots, x_{(j)}^{(i)}$, the density of age, $X_{(j+1)}^{(i)}$, at the next failure, is $f_i(x)/\overline{F}_i\left(x_{(j)}^{(i)}\right)$ for $T^{(i)} > x > x_{(j)}^{(i)}$. There is a probability $\overline{F}_i(T^{(i)})/\overline{F}_i\left(x_{(j)}^{(i)}\right)$ that the next failure lies outside the observation period $[0, T^{(i)}]$. Physically, minimal repair means that immediately after the jth repair the component behaves exactly as though it were a component of age $x_{(j)}^{(i)}$ that had not failed since new. In other words, the component is 'as bad as old'. Suppose that n_i such failures and repairs of component i are observed in $[0, T^{(i)}]$ at times $x_{(1)}^{(i)} < \cdots < x_{(n_i)}^{(i)} < T^{(i)}$. It follows that the joint distribution/density of $n_i, \left\{x_{(j)}^{(i)}, j = 1, \ldots, n_i\right\}$ is

$$h\left(n_i, \left\{x_{(j)}^{(i)}, j = 1, \ldots, n_i\right\} | \alpha_i, \lambda_i\right) = \frac{\overline{F}_i(T^{(i)})}{\overline{F}_i\left(x_{(n_i)}^{(i)}\right)} \prod_{j=1}^{n_i} \frac{f_i\left(x_{(j)}^{(i)}\right)}{\overline{F}_i\left(x_{(j-1)}^{(i)}\right)}$$

where $x_{(0)}^{(i)} = 0$. So a summary of the model is

$$N_i, \left\{X_{(j)}^{(i)}, j = 1, \ldots, N_i\right\} \overset{\text{independent}}{\sim} h\left(n_i, \left\{x_{(j)}^{(i)}, j = 1, \ldots, n_i\right\} | \alpha_i, \lambda_i\right), \quad i = 1, \ldots, m,$$

$$(\alpha_i, \lambda_i) \overset{\text{i.i.d.}}{\sim} g(\alpha, \lambda | \mu, b), \quad i = 1, \ldots, m,$$

$$b \sim w(b | \gamma, \delta).$$

The likelihood function for these data for component i alone is

$$L^{(i)} = h\left(n_i, \left\{x_{(j)}^{(i)}, j = 1, \ldots, n_i\right\} | \alpha_i, \lambda_i\right).$$

Now suppose that

$$\overline{F}_i(x) = e^{-(\lambda_i x)^{\alpha_i}},$$

a Weibull distribution. Then

$$L^{(i)} = e^{-\left(\lambda_i T^{(i)}\right)^{\alpha_i}} \prod_{j=1}^{n_i} \alpha_i \left(x_{(j)}^{(i)}\right)^{\alpha_i - 1} \lambda_i^{\alpha_i}$$

$$= e^{-\left(\lambda_i T^{(i)}\right)^{\alpha_i}} \alpha_i^{n_i} \lambda_i^{n_i \alpha_i} \left(\prod_{j=1}^{n_i} x_{(j)}^{(i)}\right)^{\alpha_i - 1}.$$

Therefore, the full likelihood is

$$L\left(\{\alpha_i\}, \{\lambda_i\}\right) = e^{-\sum_{i=1}^m \left(\lambda_i T^{(i)}\right)^{\alpha_i}} \prod_{i=1}^m \alpha_i^{n_i} \lambda_i^{n_i \alpha_i} \left(\prod_{j=1}^{n_i} x_{(j)}^{(i)}\right)^{\alpha_i - 1}.$$

The prior distribution of $\{\alpha_i\}, \{\lambda_i\}$ is taken to be

$$g\left(\{\alpha_i\}, \{\lambda_i\}\right) = \prod_{i=1}^m g_\alpha(\alpha_i) \left[\frac{\mu(b-1)}{\Gamma(1/\alpha_i + 1)}\right]^b \frac{\lambda_i^{b-1}}{\Gamma(b)} \exp\left[-\frac{\mu(b-1)\lambda_i}{\Gamma(1/\alpha_i + 1)}\right] \quad (8.25)$$

where $b - 2$ has a gamma (γ, δ) density on support $[0, \infty)$. The condition $b > 2$ arises as the prior variance of X is taken to be finite. Note that Equation (8.25) indicates that the prior conditional density of λ_i given $\{\alpha_k\}$, $\{\lambda_k, k \neq i\}$ and b is gamma $(b, [\mu (b-1)]/[\Gamma (1/\alpha_i + 1)])$, $i = 1, \dots, m$, and that $g_\alpha (\alpha_i)$ is the prior marginal density of α_i. The hyperparameters μ and δ are both positive, while $\gamma > 1$ is necessary if the prior variance of X is taken to be finite. The only restriction on $g(\alpha)$ is that the support excludes $(-\infty, 1]$, as the prior belief is that components have an increasing failure rate. Suppose X_i is the time to first failure for the ith component. Then the standard result is that

$$E (X_i|\alpha_i, \lambda_i) = \int_0^\infty \overline{F}_i (x) \, dx$$

$$= \lambda_i^{-1} \Gamma \left(\frac{1}{\alpha_i} + 1 \right)$$

and

$$E \left(X_i^2|\alpha_i, \lambda_i \right) = \int_0^\infty 2x \, \overline{F}_i (x) \, dx$$

$$= \lambda_i^{-2} \Gamma \left(\frac{2}{\alpha_i} + 1 \right).$$

Therefore, the prior expectation of X_i given α_i and b is

$$E (X_i|\alpha_i, b) = \Gamma \left(\frac{1}{\alpha} + 1 \right) \int_0^\infty \frac{\lambda_i^{b-2}}{\Gamma (b)} \exp \left[- \frac{\mu (b-1) \lambda_i}{\Gamma (1/\alpha_i + 1)} \right] \left[\frac{\mu (b-1)}{\Gamma (1/\alpha_i + 1)} \right]^b \, d\lambda_i$$

$$= \Gamma \left(\frac{1}{\alpha_i} + 1 \right) \frac{\Gamma (b-1)}{\Gamma (b)} \frac{\mu (b-1)}{\Gamma \left(\frac{1}{\alpha_i} + 1 \right)}$$

$$= \mu$$

which is independent of α_i and b. Therefore,

$$E (X_i) = \mu. \tag{8.26}$$

The prior expectation of X_i^2 given α_i and b is

$$E \left(X_i^2|\alpha_i, b \right) = \Gamma \left(\frac{2}{\alpha_i} + 1 \right) \int_0^\infty \frac{\lambda_i^{b-3}}{\Gamma (b)} \exp \left[- \frac{\mu (b-1) \lambda_i}{\Gamma (1/\alpha_i + 1)} \right] \left[\frac{\mu (b-1)}{\Gamma (1/\alpha_i + 1)} \right]^b \, d\lambda_i$$

$$= \Gamma \left(\frac{2}{\alpha_i} + 1 \right) \left[\frac{\mu (b-1)}{\Gamma (1/\alpha_i + 1)} \right]^2 \frac{\Gamma (b-2)}{\Gamma (b)}$$

$$= \frac{\mu^2 \Gamma (2/\alpha_i + 1) (b-1)}{\Gamma^2 (1/\alpha_i + 1) (b-2)}.$$

Integrating out b,

$$E \left(X_i^2|\alpha_i \right) = \frac{\mu^2 \Gamma (2/\alpha_i + 1)}{\Gamma^2 (1/\alpha_i + 1)} E_{w(b|\gamma, \delta)} \left(\frac{b-1}{b-2} \right)$$

$$= \frac{\mu^2 \Gamma(2/\alpha_i+1)}{\Gamma^2(1/\alpha_i+1)} \left(1 + \frac{\delta}{\gamma-1}\right)$$

and therefore

$$E(X_i^2) = \mu^2 \left(1 + \frac{\delta}{\gamma-1}\right) E_{g_\alpha(\alpha_i)} \left[\frac{\Gamma(2/\alpha_i+1)}{\Gamma^2(1/\alpha_i+1)}\right],$$

and so the prior variance of X_i is

$$\mathrm{Var}(X_i) = \mu^2 \left\{\left(1 + \frac{\delta}{\gamma-1}\right) E_{g_\alpha(\alpha_i)} \left[\frac{\Gamma(2/\alpha_i+1)}{\Gamma^2(1/\alpha_i+1)}\right] - 1\right\}. \tag{8.27}$$

Results (8.26) and (8.27) suggest a simple way of setting parameters. Set μ to the prior expected time to first failure of any of the m components, and obtain $\delta/(\gamma-1)$ using the prior knowledge on its variance. To keep the fitting process simple, a subjective assessment of the expected value, γ/δ, of $b-2$ allows the final hyperparameter to be fitted.

The aim here is to sample from the posterior distribution of $\{(\alpha_i, \lambda_i), i = 1, \ldots, m\}$ and b, given the observed failure data $\left\{\left\{x_{(1)}^{(i)}, \ldots, x_{(n_i)}^{(i)}\right\}, i = 1, \ldots, m\right\}$. The posterior density is

$$\pi(\{\alpha_i\}, \{\lambda_i\}, b) \propto \frac{\delta^\gamma (b-2)^{\gamma-1} e^{-\delta(b-2)}}{\Gamma(\gamma)}$$

$$\times \prod_{i=1}^{m} \left\{\alpha_i^{n_i} \lambda_i^{n_i\alpha_i} \left(\prod_{j=1}^{n_i} x_{(j)}^{(i)}\right)^{\alpha_i-1} g(\alpha_i) \left[\frac{\mu(b-1)}{\Gamma(1/\alpha_i+1)}\right]^b \right.$$

$$\times \left. \frac{\lambda_i^{b-1}}{\Gamma(b)} \exp\left[-\frac{\mu(b-1)\lambda_i}{\Gamma(1/\alpha_i+1)} - \left(\lambda_i T^{(i)}\right)^{\alpha_i}\right]\right\}.$$

The full conditionals are

$$\pi(\alpha_i | \{\alpha_k, k \neq i\}, \{\lambda_k\}, b) \propto \exp\left[-\frac{\mu(b-1)\lambda_i}{\Gamma(1/\alpha_i+1)} - \left(\lambda_i T^{(i)}\right)^{\alpha_i}\right]$$

$$\times \alpha_i^{n_i} \lambda_i^{n_i\alpha_i} \left(\prod_{j=1}^{n_i} x_{(j)}^{(i)}\right)^{\alpha_i-1} g(\alpha_i) \left[\Gamma\left(\frac{1}{\alpha_i}+1\right)\right]^{-b}, \tag{8.28}$$

$$\pi(\lambda_i | \{\alpha_k\}, \{\lambda_k, k \neq i\}, b) \propto \exp\left[-\frac{\mu(b-1)\lambda_i}{\Gamma(1/\alpha_i+1)} - \left(\lambda_i T^{(i)}\right)^{\alpha_i}\right] \lambda_i^{n_i\alpha_i+b-1} \tag{8.29}$$

and

$$\pi(b | \{\alpha_k\}, \{\lambda_k\}) \propto \frac{(b-2)^{\gamma-1} [\{\mu(b-1)\}^m \prod_{i=1}^{m} \lambda_i/\Gamma(1/\alpha_i+1)]^b}{\Gamma^m(b)}$$

$$\times \exp\left[-\delta(b-2) - \mu(b-1) \sum_{i=1}^{m} \frac{\lambda_i}{\Gamma(1/\alpha_i+1)}\right] \tag{8.30}$$

for $i = 1, \ldots, m$. Note that (8.29) is log concave, so is suitable for adaptive rejection, while (8.28) and (8.30) are not obviously so. Fortunately, Gilks *et al.* (1995) have shown how to relax the log concavity condition by introducing a Metropolis step into adaptive rejection. This method is known as adaptive rejection Metropolis sampling (ARMS).

8.5   Other aspects of Gibbs sampling

8.5.1   Slice sampling

Suppose we wish to sample from the univariate density

$$f(x) \propto f_1(x) \cdots f_k(x)$$

where $f_1, \ldots, f_k$ are non-negative functions. Define g to be a joint density of random variables $U_1, \ldots, U_k, X$, where

$$g(u_1, \ldots, u_k, x) = B \tag{8.31}$$

is a constant on support

$$S = \{(u_1, \ldots, u_k, x) : 0 \le u_i \le f_i(x) \, \forall i, \, x \in \text{support}(f)\} \tag{8.32}$$

Integrating out $u_1, \ldots, u_k$, it can be seen that the marginal density of X is

$$g_X(x) = B \int_0^{f_k(x)} du_k \cdots \int_0^{f_1(x)} du_1$$
$$= B f_1(x) \cdots f_k(x).$$

It follows that $g_X(x) = f(x)$. Successive values of X can now be sampled (not independently) using Gibbs sampling on the set of random variables $\{U_1, \ldots, U_k, X\}$. From Equation (8.31) and (8.32) the full conditional of U_1, say, is

$$g_{U_1|}(u_1 | u_2, \ldots, u_k, x) \propto B$$

on support $[0, f_1(x)]$. The full conditional of X is

$$g_{X|}(x | u_1, \ldots, u_k) \propto B$$

on support $D = \{x : u_i \le f_i(x) \, \forall i, \, x \in \text{support}(f)\}$. The Gibbs sampling scheme is therefore

$$u_1 \sim U(0, f_1(x))$$
$$u_2 \sim U(0, f_2(x))$$
$$\cdots$$

$$\cdots$$
$$u_k \sim U(0, f_k(x))$$
$$x \sim U(D)$$

As an example, consider a truncated gamma distribution consisting of that part of a full gamma density with values greater than a. This has density

$$f(x) \propto x^{\alpha-1} e^{-x}$$

on support $[a, \infty)$ where $\alpha > 1$. If a is large it is not efficient to sample from the full gamma, rejecting those values that do not exceed a, so slice sampling will be tried. Now $f(x) \propto f_1(x) f_2(x)$ where $f_1(x) = x^{\alpha-1}$ and $f_2(x) = e^{-x}$. Therefore,

$$D = \left\{ x : u_1 \leq x^{\alpha-1}, u_2 \leq e^{-x}, x > a \right\}$$
$$= \left\{ x : x \geq u_1^{1/(\alpha-1)}, x \leq -\ln u_2, x > a \right\}$$
$$= \left\{ x : \max \left(u_1^{1/(\alpha-1)}, a \right) \leq x \leq -\ln u_2 \right\}.$$

The corresponding Gibbs sampling scheme is

$$u_1 \sim U\left(0, x^{\alpha-1}\right)$$
$$u_2 \sim U\left(0, e^{-x}\right)$$
$$x \sim U\left(\max \left(u_1^{1/(\alpha-1)}, a \right), -\ln u_2 \right)$$

In Appendix 8.5 the procedure 'truncatedgamma' implements this for $\alpha = 3$, $a = 9$, with a burn-in of 100 iterations. The efficiency is compared with the naive approach of rejecting gamma variates that are less than 9. Slice sampling becomes worthwhile if a is large. A method that is better than both of these is to use envelope rejection with an exponential envelope on support $[a, \infty)$, as in Dagpunar (1978).

As a further example, the estimation of the 10 pump failure rates in Section 8.4.1 will be revisited. The joint posterior density is

$$\pi\left(\{\lambda_i\}, \beta\right) \propto \frac{\delta^\gamma \beta^{\gamma-1} e^{-\delta\beta}}{\Gamma(\gamma)} \prod_{i=1}^{10} \frac{e^{-\lambda_i t_i} (\lambda_i t_i)^{x_i}}{x_i!} \frac{e^{-\beta\lambda_i} \lambda_i^{\alpha-1} \beta^\alpha}{\Gamma(\alpha)}$$
$$\propto \beta^{\gamma-1} e^{-\delta\beta} \prod_{i=1}^{10} e^{-\lambda_i(\beta+t_i)} \lambda_i^{x_i+\alpha-1} \beta^\alpha.$$

Integrating out the $\{\lambda_i\}$ gives

$$\pi(\beta) \propto \beta^{10\alpha+\gamma-1} e^{-\delta\beta} \prod_{i=1}^{10} (\beta+t_i)^{-(x_i+\alpha)}$$

If we can sample from this density, then there is no need for Gibbs sampling, as it only remains to sample $\lambda_i|\beta \sim \text{gamma}(\alpha+x_i, \beta+t_i)$, a standard distribution. One possibility is to use slice sampling as follows:

$$U_i \sim U\left[0, (\beta+t_i)^{-(x_i+\alpha)}\right], \quad i = 1, \ldots, 10$$
$$U_{11} \sim U\left[0, \beta^{10\alpha+\gamma-1}\right]$$
$$U_{12} \sim U\left[0, e^{-\delta\beta}\right]$$
$$\beta \sim U[D]$$

where

$$D = \left\{ \beta : \left(\frac{1}{\beta + t_i}\right)^{x_i + \alpha} \geq u_i, i = 1, \ldots, 10 : \beta^{10\alpha + \gamma - 1} \geq u_{11}, e^{-\delta\beta} \geq u_{12} \right\}.$$

Sampling a β variate will therefore require 12 uniform random numbers, $R_1, \ldots, R_{12}$. If the last sampled variate is β_0 then $u_i = R_i \left(1/(\beta_0 + t_i)\right)^{x_i + \alpha}$, $i = 1, \ldots, 10$, and the condition $(1/(\beta + t_i))^{x_i + \alpha} \geq u_i$ becomes $(1/(\beta + t_i))^{x_i + \alpha} \geq R_i \left(1/(\beta_0 + t_i)\right)^{x_i + \alpha}$, that is

$$\beta \leq (\beta_0 + t_i) \left(\frac{1}{R_i}\right)^{1/(x_i + \alpha)} - t_i.$$

Similarly, providing $10\alpha + \gamma - 1 > 0$, as in the hyperparameters of (8.22), $\beta^{10\alpha + \gamma - 1} \geq u_{11}$ leads to $\beta^{10\alpha + \gamma - 1} \geq R_{11} \beta_0^{10\alpha + \gamma - 1}$, that is

$$\beta \geq \beta_0 R_{11}^{1/(10\alpha + \gamma - 1)}.$$

Finally, $e^{-\delta\beta} \geq u_{12}$ leads to $e^{-\delta\beta} \geq R_{12} e^{-\delta\beta_0}$, that is

$$\beta \leq \beta_0 - \frac{\ln R_{12}}{\delta}.$$

So the next variate is $\beta \sim U(\beta_{\text{lower}}, \beta_{\text{upper}})$ where

$$\beta_{\text{lower}} = \beta_0 R_{11}^{1/(10\alpha + \gamma - 1)}$$

and

$$\beta_{\text{upper}} = \min \left\{ \min_{i=1,\ldots,10} \left[(\beta_0 + t_i) \left(\frac{1}{R_i}\right)^{1/(x_i + \alpha)} - t_i \right], \beta_0 - \frac{\ln R_{12}}{\delta} \right\}.$$

This is performed in the procedure 'slice' in Appendix 8.6. It is left as an exercise to determine empirically whether or not this is more efficient than the original Gibbs sampling in Appendix 8.3.

8.5.2   Completions

Let the joint density of X and Z be $g(x, z)$, where g is said to be a *completion* of a density f if

$$\int_{-\infty}^{\infty} g(x, z) dz = f(x). \tag{8.33}$$

Using Gibbs sampling of X and Z from g gives a method for sampling (nonindependent) variates from f. Note that the slice sampler is a special case of a completion.

For example, suppose

$$g(x, z) = x^{\alpha - 1} e^{-z}$$

on support $z > x > 0$ where $\alpha > 0$. Then the marginal densities of X and Z are gamma$(\alpha, 1)$ and gamma$(\alpha + 1, 1)$. Therefore g is a completion of both these distributions.

A second example is due to Robert and Casella (2004, p. 487). Suppose $Y_1, \ldots, Y_n$ are i.i.d. Cauchy distributed with density

$$f_{Y_i}(y) \propto \frac{1}{1 + (y - \theta)^2}$$

on support $(-\infty, \infty)$, where the prior density of $\theta \sim N(0, \sigma^2)$. Then the posterior density is

$$\pi(\theta) \propto e^{-\theta^2/(2\sigma^2)} \prod_{i=1}^{n} \frac{1}{1 + (y_i - \theta)^2}.$$

Now the density

$$g(\theta, x_1, \ldots, x_n) \propto e^{-\theta^2/(2\sigma^2)} \prod_{i=1}^{n} e^{-x_i\left[1 + (y_i - \theta)^2\right]}$$

on support $|\theta| < \infty, x_i \geq 0, i = 1, \ldots, n$, is a completion of π. It is left as an exercise to show that the full conditionals are

$$X_i | \{x_j, j \neq i\} \sim \text{Exp}\left(1 + (y_i - \theta)^2\right)$$

for $i = 1, \ldots, n$ and

$$\theta | \{x_j, j = 1, \ldots, n\} \sim N\left(\frac{2\sum x_i y_i}{\sigma^{-2} + 2n\bar{x}}, \frac{1}{\sigma^{-2} + 2n\bar{x}}\right)$$

where $\bar{x}$ is the sample mean of $\{x_i\}$.

8.6 Problems

1. It is required to sample from a folded normal probability density function

$$f(x) = \sqrt{\frac{2}{\pi}} \exp(-0.5x^2) \quad (x > 0)$$

using a Metropolis–Hastings algorithm with a proposal density

$$q_{Y|x}(y) = e^{-y} \quad (y > 0).$$

Use the $U(0, 1)$ random numbers below to sample five candidate variates. Indicate clearly which candidate variates are accepted (state of Markov chain changes) and which are rejected (state unchanged). Start with $x = 0$.

R_1 (for candidate variate)	0.52	0.01	0.68	0.33	0.95
R_2 (for acceptance test)	0.62	0.64	0.03	0.95	0.45

2. Let $P(\theta)$ denote the prior density of a vector parameter $\theta \in S$. Let $P(\theta|D)$ denote the posterior density after observing data D. Consider an MCMC algorithm for sampling from $P(\theta|D)$ with an independence sampler in which, at each iteration, the candidate point is θ with a proposal density $P(\theta)$ for all $\theta \in S$. Given that the current point is θ, show that the acceptance probability for a candidate point ϕ is $\min[1, L(\phi)/L(\theta)]$, where L is the likelihood function.

3. Refer to Appendix 8.2. Modify the procedure 'fail' to obtain an MCMC point estimate of the posterior expected time to failure for a component.

4. Discuss the merits or otherwise of simulating from the posterior density (8.13) using envelope rejection with a proposal density which is the prior shown in (8.12).

5. Robert and Casella (2004, p. 303) consider a Metropolis–Hastings algorithm to sample variates from a density f with support on $[0,1]$ where

$$f(x) \propto \frac{g(x)}{1-\rho(x)},$$

g is a probability density function on $[0, 1]$, and $0 < \rho(x) < 1\ \forall x$. The proposal density $q(y|x)$ is such that if the current point (value) is x and the next candidate point is Y, then with probability $1-\rho(x)$, Y is a variate drawn from g, and with probability $\rho(x)$, $Y = x$.

(a) Define the acceptance probability in the Metropolis–Hastings algorithm to be the proportion of candidate variates (Y variates) that are accepted. Show that in this case the proportion is one.

(b) Write a procedure to simulate variates from f when

$$f(x) \propto \frac{x^{\alpha-1}(1-x)^{\beta-1}}{1-x^5}$$

where $\alpha > 0$, $\beta > 0$, and $0 \le x \le 1$.

6. (From Tierney, 1994.) It is required to generate variates from a probability density function proportional to h using prospective variates from a density g. In the usual envelope rejection method c is chosen such that $h(x)/g(x) \le c\ \forall x \in$ support (g). Now suppose that c is not a uniform bound, that is $h(x)/g(x) > c$ for some x. Let a prospective variate y from g be accepted with probability $\min[1, h(y)/\{cg(y)\}]$.

(a) Show that the probability density function of accepted variates is now r (rather than proportional to h) where

$$r(y) \propto \min[h(y), cg(y)].$$

(b) Consider a Metropolis–Hastings algorithm to sample from a density proportional to h, using a proposal density r. Show that a candidate variate y from r is accepted with probability

$$\alpha(x, y) = \begin{cases} \min\left[1, \dfrac{cg(x)}{h(x)}\right] & \text{if } \dfrac{h(y)}{g(y)} < c, \\[2ex] \min\left[1, \dfrac{h(y)g(x)}{g(y)h(x)}\right] & \text{otherwise.} \end{cases}$$

(c) Discuss the merits of an algorithm based upon (b).

7. Consider the following model.

$$y_i \overset{\text{i.i.d.}}{\sim} N\left(\mu, \tau^{-1}\right) \quad (i = 1, \ldots, n)$$

where

$$\mu \sim N\left(0, \beta^{-1}\right),$$
$$\beta \sim \text{gamma}(2, 1),$$
$$\tau \sim \text{gamma}(2, 1).$$

Devise a Gibbs sampling scheme that can be used for estimating the marginal posterior densities of μ, β, and τ given $y_1, \ldots, y_n$.

8. Construct a Gibbs sampling scheme for sampling from the bivariate density

$$f(x_1, x_2) \propto \exp\left[-\frac{1}{2}x_1\left(1 + x_2^2\right)\right] \quad (x_1 > 0, \infty > x_2 > -\infty).$$

Now find a direct method for sampling from the bivariate density which first samples from the marginal density of X_2, and then the conditional density of X_1 given X_2. Similarly, find a direct method that first samples from the marginal density of X_1, and then the conditional density of X_2 given X_1. Which of the three methods would you choose to implement?

9. Modify the 'capture–recapture' Gibbs sampling scheme of Section 8.4.2 and Appendix 8.4 so that the prior density of p is beta(α, γ). Run the modified procedure to estimate the posterior expectation of N and p when $\alpha = 3$ and $\gamma = 7$ and the other data and parameters are as before.

10. (From Robert and Casella (2004, p. 372)) Write a Maple procedure to perform Gibbs sampling from the trivariate density (Equation (8.15))

$$f(x_1, x_2, x_3) = \exp\left[-(x_1 + x_2 + x_3) - \theta_{12}x_1x_2 - \theta_{23}x_2x_3 - \theta_{31}x_3x_1\right]$$

where $x_i \geq 0$ for all i and where $\{\theta_{ij}\}$ are known positive constants. Suppose $\theta_{12} = 1$, $\theta_{23} = 2$, and $\theta_{31} = 3$. Plot the empirical marginal density, $f_{X_1}(x_1)$, and estimate $\mu_1 = E(X_1)$ and $\sigma_1 = \sigma(X_1)$. Use a burn-in of 100 iterations and a further 1000 iterations for analysis. For μ_1, perform 50 independent replications using estimates in the form of (a) and (b) below. Carry out the corresponding analysis for estimating σ_1. Compare the precision of methods (a) and (b).

(a) $\dfrac{1}{1000} \sum_{i=1}^{1000} x_1^{(i)}$;

(b) $\dfrac{1}{1000} \sum_{i=1}^{1000} \left(1 + \theta_{12} x_2^{(i)} + \theta_{31} x_3^{(i)}\right)^{-1}$.

11. Use slice sampling to derive algorithms for sampling from the following distributions:

(a) $f(x) \propto x^{\alpha-1} (1-x)^{\beta-1}$ for $0 < x < 1, 1 > \alpha > 0, 1 > \beta > 0$ (note carefully the ranges of the parameters);

(b) $f(x, y) = (x + y) \exp\left[-\frac{1}{2}(x^2 + y^2)\right]$ for $(x, y) \in [0, 1]^2$.

12. Let $\alpha > 1$ and $\beta > 0$. Suppose the joint density of X and Z is

$$g(x, z) \propto \begin{cases} (1-x)^{\beta-1} z^{\alpha-2} & (0 < z < x < 1), \\ 0 & \text{(elsewhere)}. \end{cases}$$

(a) Show that the marginal densities of X and Z are beta (α, β) and beta $(\alpha - 1, \beta + 1)$ respectively.

(b) Use the result in (a) to derive a Gibbs algorithm to sample variates from these beta densities.

13. Consider a Brownian motion $\{X(t), t \geq 0\}$ with drift $\mu (\geq 0)$, volatility σ, and $X(0) = 0$. Let a be a known positive real number. It can be shown that the time $T = \inf\{t : X(t) = a\}$ has density (an 'inverse Gaussian')

$$f(t) = \frac{a}{\sigma\sqrt{2\pi t^3}} \exp\left[\frac{-(a - \mu t)^2}{2\sigma^2 t}\right].$$

(a) Show that $f(t) \propto t^{-3/2} \exp\left[-\frac{1}{2}(\chi t^{-1} + \psi t)\right] (t > 0)$, where $\chi = a^2/\sigma^2$ and $\psi = \mu^2/\sigma^2$.

(b) Use slice sampling to construct a Gibbs algorithm for sampling from f.

(c) Show that g is a *completion* of the density f, where

$$g(t, y, z) \propto \begin{cases} t^{-3/2} e^{-z-y} & (y > \chi/(2t), z > \psi t/2, t > 0), \\ 0 & \text{(elsewhere)}. \end{cases}$$

14. A new component is installed in a computer. It fails after a time X has elapsed. It is then repaired and fails again after a further time Y has elapsed. On the second failure the component is scrapped and a new component is installed. The joint density of X_i and Y_i (lifetimes for the ith such component) is f_{X_i,Y_i} where

$$f_{X_i,Y_i}(x, y) \propto \frac{x(\theta x+y) e^{-\lambda(x+y)}}{1+\lambda\theta x} \qquad (x > 0, y > 0)$$

and where $\lambda\,(\geq 0)$ and $\theta\,(0 \leq \theta \leq 1)$ are unknown parameter values. The priors of θ and λ are independently distributed as beta (a, b) and gamma $(\beta+1, \gamma)$ respectively where a, b, β, and γ are known positive values. Let $(x_1, y_1), \ldots, (x_m, y_m)$ denote m independent realizations of (X, Y). Let $s = \sum_{i=1}^{m}(x_i+y_i)$.

(a) Derive an expression for the posterior density of λ and θ given the observed data.

(b) Deduce that the full conditional density of λ given the data is proportional to

$$\frac{e^{-\lambda(s+\gamma)}\lambda^{\beta}}{\prod_{i=1}^{m}(1+\lambda\theta x_i)}.$$

(c) Derive the corresponding full conditional density of θ given λ.

(d) Choosing an envelope that is proportional to $e^{-\lambda(s+\gamma)}\lambda^{\beta}$, derive a rejection procedure for generating a variate λ from the full conditional of λ given θ. You may assume that procedures are available for sampling from standard densities.

(e) Show that for known $t\,(\geq 0)$, x, θ, and λ,

$$P(Y > t \mid X = x) = e^{-\lambda t}\left(1 + \frac{\lambda t}{1+\lambda\theta x}\right).$$

Suppose that N equilibrium realizations $(\theta^{(1)}, \lambda^{(1)}), \ldots, (\theta^{(N)}, \lambda^{(N)})$ are available from Gibbs sampling of the posterior density in (a). Show how the posterior predictive probability of Y exceeding t, given that $X = x$, may be obtained.

15. Construct completion sampling algorithms for:

(a) the von Mises distribution: $f(\theta) \propto e^{k\cos\theta}, 0 < \theta < \pi$;

(b) the truncated gamma: $g(x) \propto x^{\alpha-1}e^{-x}\,(x > t > 0)$.

16. (a) Let $X_1, \ldots, X_n$ denote the lifetimes of n identical pieces of electrical equipment, under specified operating conditions. $X_1, \ldots, X_n$ are identically and independently distributed with the probability density function

$$f_{X_i}(x) = \frac{\lambda^{\beta}x^{\beta-1}e^{-\lambda x}}{\Gamma(\beta)} \qquad (x > 0, i = 1, \ldots, n)$$

where β is a (single) realization of a random variable from the $U(1, 3)$ probability density function, and λ is a (single) realization from the conditional probability density function

$$g(\lambda|\beta) = \frac{5}{\beta}e^{-5\lambda/\beta} \quad (x > 0).$$

If $X_i = x_i$ $(i = 1, \ldots, n)$, show that:

(i) The full conditional posterior density of β is $\pi(\beta \mid \lambda)$ where

$$\pi(\beta|\lambda) \propto \frac{(\lambda^n x_1 \cdots x_n)^{\beta-1}e^{-5\lambda/\beta}}{[\Gamma(\beta)]^n \beta}$$

on support $(1, 3)$. Derive and name the full conditional density of λ given $x_1, \ldots, x_n$ and β.

(ii) Outline three approaches that might be pursued for the design of an algorithm to generate variates from the posterior full conditional density of β. Discuss any difficulties that might be encountered.

(b) Three continuous random variables U_1, U_2, and X have a probability density function that is constant over the region

$$\{(u_1, u_2, x) | 0 < u_1 < x, 0 < u_2 < e^{-x^3}, 0 < x < \infty\}$$

and zero elsewhere.

(i) Show that the marginal density of X is

$$f_X(x) = \frac{xe^{-x^3}}{\int_0^\infty ve^{-v^3}dv} \quad (x > 0).$$

(ii) Use a Gibbs sampling method to design an algorithm that samples $x^{(1)}, x^{(2)}, \ldots$ from f_X.

(iii) Use the $U(0, 1)$ random numbers R_1, R_2, and R_3 in the table below to generate variates $x^{(1)}$, and $x^{(2)}$, given that $x^{(0)} = \left(\frac{1}{3}\right)^{1/3}$.

t	R_1 (for $U_1^{(t)}$)	R_2 (for $U_2^{(t)}$)	R_3 (for $X^{(t)}$)
1	0.64	0.40	0.56
2	0.72	0.01	0.90

17. Suppose $X_1, X_2, \ldots$ are i.i.d. with density

$$f(x) = \frac{\lambda^\alpha x^{\alpha-1}e^{-\lambda x}}{\Gamma(\alpha)}$$

on support $(0, \infty)$, where the prior joint density of $\alpha(> 0)$ and $\lambda(> 0)$ is

$$g(\alpha, \lambda) = g_\alpha(\alpha) g_{\lambda|\alpha}(\lambda|\alpha)$$

$$= g_\alpha(\alpha) \left[\frac{\mu(b-1)}{\alpha} \right]^b \frac{\lambda^{b-1}}{\Gamma(b)} e^{-\mu(b-1)\lambda/\alpha}$$

where $g_\alpha(\alpha)$ is a known density on support $(0, \infty)$ and $\mu(> 0)$ and $b(> 2)$ are known constants.

(a) Derive expressions for the prior expectation and variance of X_i.

(b) Given a realization $x_1, \ldots, x_n$, show that the posterior density is

$$\pi(\alpha, \lambda) \propto g_\alpha(\alpha) \alpha^{-b} \lambda^{b-1} e^{-\mu(b-1)\lambda/\alpha} \frac{(\lambda^n x_1 \cdots x_n)^\alpha \exp(-\lambda \sum_{i=1}^n x_i)}{[\Gamma(\alpha)]^n},$$

that the full conditional of λ is a gamma $(n\alpha + b, \sum_{i=1}^n x_i + \mu(b-1)/\alpha)$ density, and that the full conditional of α is

$$\pi(\alpha|\lambda) \propto \frac{g_\alpha(\alpha)}{[\Gamma(\alpha)]^n} \alpha^{-b} e^{-\mu(b-1)\lambda/\alpha} (\lambda^n x_1 \cdots x_n)^\alpha.$$

Find a condition for the latter to be log-concave.

18. A gamma distributed time to failure with a shape parameter greater than one is useful when it is known that the failure rate is increasing with age to a *finite* limit (in contrast to an increasing failure rate for a Weibull density, which always increases without bound). Accordingly, suppose the times to failure $X_1, \ldots, X_n$ of n similar components are i.i.d. with density

$$f(x) = \frac{\lambda^\alpha x^{\alpha-1} e^{-\lambda x}}{\Gamma(\alpha)}$$

where $\alpha > 1$, $\lambda > 0$, that is $X_i \sim \text{gamma}(\alpha, \lambda)$. Percy (2002) has remarked on the use of a *conjugate prior* density of the form

$$g(\alpha, \lambda) \propto \frac{(a\lambda)^{b\alpha} e^{-c\lambda}}{[\Gamma(\alpha)]^d}$$

where the support here has been modified to $\alpha > 1$, $\lambda > 0$ for specified $a, b, c, d > 0$. An advantage of any conjugate prior is that the posterior density comes from the same family of densities.

(a) Show that the joint posterior density is

$$\pi(\alpha, \lambda) \propto \frac{(a^b \lambda^{n+b} x_1 \cdots x_n)^\alpha \exp[-\lambda(c + \sum_{i=1}^n x_i)]}{[\Gamma(\alpha)]^{d+n}}.$$

(b) Show that the posterior marginal density of α is

$$\pi\left(\alpha\right) \propto \frac{\left[a^b x_1 \cdots x_n / \left(c + \sum_{i=1}^n x_i\right)^{n+b}\right]^\alpha \Gamma\left[\alpha\left(n+b\right)+1\right]}{\left[\Gamma\left(\alpha\right)\right]^{d+n}}.$$

(c) Show that the full conditionals are

$$\pi\left(\alpha \mid \lambda\right) \propto \frac{\left(a^b \lambda^{n+b} x_1 \cdots x_n\right)^\alpha}{\left[\Gamma\left(\alpha\right)\right]^{d+n}}$$

and

$$\pi\left(\lambda \mid \alpha\right) \propto \lambda^{\alpha(n+b)} \exp\left[-\lambda\left(c + \sum_{i=1}^n x_i\right)\right].$$

How amenable is $\pi\left(\alpha, \lambda\right)$ to Gibbs sampling?

(d) Show that there is no need to use simulation to derive the posterior expectation of X, since this is simply $(c + \sum_{i=1}^n x_i)/(n+b)$.

(e) Given Gibbs sampling yields equilibrium values of $\left\{\left(\alpha^{(i)}, \lambda^{(i)}\right), i = 1, \ldots, m\right\}$. How would you estimate the Bayes estimates of λ, α? How would you estimate the posterior predictive probability that the component survives to at least age x?

9

Solutions

9.1 Solutions 1

Solutions to the problems in Chapter 1 appear in Appendix 1.

9.2 Solutions 2

1. (a) $a = 5, m = 16, c = 3$

    $X_1 = (5 \times 5 + 3) \bmod 16 = 28 \bmod 16 = 12$
    $X_2 = (12 \times 5 + 3) \bmod 16 = 15$
    $\vdots$
    $X_{16} = 5 = X_0$
    Period is 16. Since $m = 16$, the generator is full period. This agrees with theory since $c = 3$ and $m = 16$ are relatively prime, $a - 1 = 4$ is a multiple of 2, which is the only prime factor of $m = 16$, and $a - 1$ is a multiple of 4.

    (b) $\lambda = 16$, period is unaffected by choice of seed.

    (c) $\lambda = 4$, not full period, since $a - 1 = 7 - 1 = 6$, which is not a multiple of 4.

    (d) $\lambda = 2$, not full period since $c = 4$ and $m = 16$ are not relatively prime.

    (e) $\lambda = 16 = m/4$, since m is a power of 2 and $a = 5 \bmod 8$ and X_0 is odd.

    (f) $\lambda = 4 < m/4$ since X_0 is not odd.

2. The code below finds the smallest n for which $X_n = X_0$. Note the use of the colon rather than the semicolon to suppress output of every X_n. To obtain a random number in $[0,1)$ insert $R := x/61$. Answers: (a) $\lambda = 60$, (b) $\lambda = 30$.

```
> rn := proc()globalseed;
  seed := (seed*49) mod (61);
  end proc;

> seed := 1;
  x0 := seed;
  x := rn();
  n := 1:
  while x <> x0 do:
  x := rn():
  n := n+1:
  end do:
  n;
```

The period of 60 in (a) suggests that 7 is a primitive root of 61 as will now be shown. The prime factors of $m-1=60$ are 5, 3, and 2. Therefore, $7^{60/5}-1, 7^{60/3}-1, 7^{60/2}-1$ should not be divisible by 61. Now $(7^{12}-1) \bmod 61 = (7^6-1)(7^6+1) \bmod 61$. However, $7^6 \bmod 61 = 7^3 7^3 \bmod 61 = (7^3 \bmod 61)^2 = 38^2 \bmod 61 = 41$. Thus $(7^{12}-1) \bmod 61 = (40 \times 42) \bmod 61 = 33$. Similarly, $(7^{20}-1) \bmod 61 = 46$ and $(7^{30}-1) \bmod 61 = 59$, showing that none of these is divisible by 61, so the generator is a maximum period prime modulus one with $\lambda = m-1 = 60$.

(b) 49 is not a primitive root of 61 so the period is less than 60.

4. $a = 1,000,101, X_0 = 53,547,507,752$. Hand calculation therefore gives

$$aX_0 = 53,552,916,050,282,952$$

$$= 53552 \times 10^{12} + 916,050,282,952.$$

Therefore, $aX_0 \bmod (10^{12} - 11) = 11 \times 53552 + 916,050,282,952 = 916,050,872,024$.

5. (a) $X_{i+1} = aX_i \bmod m$ gives an almost full period generator of period $m-1$, since a is a primitive root of m. Therefore, the sequence is $\{X_0, \ldots, X_{m-2}, X_0, \ldots\}$ where $X_0, \ldots, X_{m-2}$ is a permutation of $1, 2, \ldots, m-1$. Over the entire cycle $E(X_i) = [1/(m-1)]\sum_{i=0}^{m-2} X_i = [1/(m-1)](1+2+\cdots+m-1) = m/2$. Also, $E(X_i^2) = m(2m-1)/6$ and hence gives the result for variance. Since $R_i = X_i/m$, $E(R_i) = (1/m)E(X_i) = \frac{1}{2}$ and $Var(R_i) = (1/m^2)Var(X_i) = (m-2)/(12m)$, which is almost $\frac{1}{12}$ when m is large.

6. First show 2 is a primitive root of 13. The remaining primitive roots are $\{2^j : j < 12, \text{where } j \text{ and } 12 \text{ are relatively prime}\} = \{2^5, 2^7, 2^{11}\} \bmod 13 = \{6, 11, 7\}$. Therefore, when $m = 13$, the only multiplicative generators having period 12 are those where $a = 2, 6, 7, 11$.

7. The multiplicative generator $X_{i+1} = 13X_i \bmod 2^5$ with $X_0 = 3$ gives $\{X_i\} = \{3, 7, 27, 31, 19, 23, 11, 15\}$. Put $R_i = X_i/m$; then $\{R_i\} = \{3/32, 7/32, \ldots, 15/32\}$. The mixed generator $X_{i+1}^* = [13X_i^* + (3 \bmod 4)(13-1)/4] \bmod 2^3$ with $X_i^* = 0$ gives $\{X_i^*\} = \{0, 1, 6, 7, 4, 5, 2, 3\}$. Put $R_i^* = X_i^*/8$; then $R_i^* = \{0, \frac{1}{8}, \ldots, \frac{3}{8}\}$. Now observe that $R_i - R_i^* = 3/32$ as theory predicts.

8. (a) The period is λ where λ is the smallest positive integer for which $(X_0, X_1) = (X_\lambda, X_{\lambda+1})$. The number of pairs (i, j) where $i, j \in [0, m-1]$ is m^2. Clearly, $X_0 = X_1 = 0$ is not feasible, so the number of pairs to a repeat cannot exceed $m^2 - 1$.

   (b) $X_0 = 0$ and $X_1 = 1$ gives $\{0, 1, 1, 2, 3, 0, 3, 3, 1, 4, 0, 4, 4, 3, 2, 0, 2, 2, 4, 1, 0, 1, \dots\}$ and the other one is $\{1, 3, 4, 2, 1, 3, \dots\}$, having periods of 20 and 4 respectively, compared with the upper bound of $5^2 - 1 = 24$.

9. (a) Periods are 12 (X) and 16 (Y) respectively. Plots (see figures 9.1 and 9.2) of overlapping pairs suggests that the X generator has a somewhat more uniform distribution of points. Observe that in the basic lattice cells, the ratio of side lengths $\left(r_2 = \sqrt{34/5}\right)$ is smaller in the X generator compared to the Y one $(r_2 = 8)$.

   (b) A combined generator is $R_n = (X_n/13 + Y_n/16) \bmod 1$.

10. (b) The periods of $\{X_n\}$ and $\{Y_n\}$ are 8 and 6 respectively. Therefore, $\{R_n\}$ will repeat after a period equal to the lowest common multiple of 8 and 6, which is 24.

11. Using the algorithm in Section 2.3, initially $T(0) = 0.69$, $T(1) = 0.79$, $T(2) = 0.10$, $T(3) = 0.02$, and $T(4) = 0.43$. Now $N = \lfloor 4T(4) \rfloor = 1$. Therefore, $T(4) := T(1) = 0.79$ and $T(1) := 0.61$, which is the next number to enter the buffer. Continuing in this manner the shuffled sequence becomes 0.43, 0.79, 0.02, 0.69, 0.10, 0.66, ...

13. $X^2 = 57.31$ compared with a 1 % critical value of $\chi^2_{9,0.01} = 21.666$. Assuming these data are independent, we can confidently reject the null hypothesis that these are $U(0, 1)$.

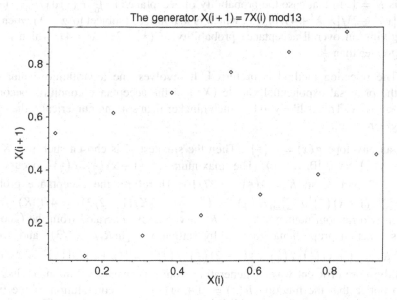

Figure 9.1 The lattice plot for the $\{X\}$ generator, problem 2.9

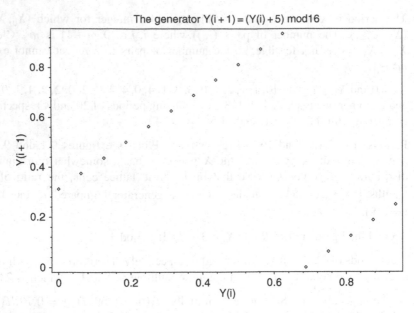

Figure 9.2 The lattice plot for the {Y} generator, problem 2.9

9.3 Solutions 3

6. (a) Suppose the envelope is $Ke^{-\mu x}$ for some positive μ. Clearly, if $\mu > \lambda$ then it is impossible to envelope $h(x)$. If $\mu \le \lambda$ the smallest K allowing enveloping of h is $K = 1$. In that case the probability of acceptance is $\int_0^\infty h(x)\,dx / \int_0^\infty g(x)\,dx = \{(\lambda^2 + 2)/[\lambda(\lambda^2 + 4)]\}/\mu^{-1}$, which is maximized (subject to $\mu \le \lambda$) when $\mu = \lambda$, giving an overall acceptance probability of $(\lambda^2 + 2)/(\lambda^2 + 4)$, which is always greater than $\frac{1}{2}$.

   (c) The rejection method is preferred. It involves one logarithmic evaluation for the proposal exponential variate (X) and the acceptance condition becomes $R_2 < \cos^2 X$. This is likely to be much quicker than solving numerically the equation given in (c).

7. Use an envelope $g(x) = K \left(\frac{2}{3}\right)^x$. Then the smallest K is chosen such that $K \left(\frac{2}{3}\right)^x \ge (1+x)\left(\frac{1}{2}\right)^x \forall x \in [0, 1, \dots)$. The maximum of $(1+x)\left(\frac{1}{2}\right)^x / \left(\frac{2}{3}\right)^x$ occurs jointly at $x = 2$ and 3, so $K = 3\left(\frac{3}{4}\right)^2 = 27/16$. Therefore, the acceptance probability is $\sum_{x=0}^\infty (1+x)\left(\frac{1}{2}\right)^x / \sum_{x=0}^\infty K \left(\frac{2}{3}\right)^x = \left(1 - \frac{1}{2}\right)^{-2} / [K/(1 - 2/3)] = 4/(3K) = 64/81$. Given two random numbers R_1 and R_2, a variate is generated from the (geometric) mass function proportional to $\left(\frac{2}{3}\right)^x$ by setting $x = \lfloor \ln R_1 / \ln(2/3) \rfloor$ and accepting it if $R_2 < (1+x)\left(\frac{1}{2}\right)^x / \left(\frac{2}{3}\right)^x / \left[(1+2)\left(\frac{1}{2}\right)^2 / \left(\frac{2}{3}\right)^2 \right] = [(1+x)/3]\left(\frac{3}{4}\right)^{x-2}$. Actually, a slightly more efficient way of generating from this (negative binomial) distribution is to notice that the function $h(x) = (1+x)\left(\frac{1}{2}\right)^x$ is a convolution of the function $\left(\frac{1}{2}\right)^x$ with itself. Thus, $h(x) = \sum_{v=0}^x \left(\frac{1}{2}\right)^{x-v}\left(\frac{1}{2}\right)^v$, and therefore the required variate is

the sum of two independent geometric variates. In that case set $x = \lfloor \ln R_1 / \ln(1/2) \rfloor + \lfloor \ln R_2 / \ln(1/2) \rfloor$.

10. (a) Applying the ratio method to C_θ let $Y = S/R$ where (R, S) is uniformly distributed over C_θ. Then $Y = y \leftrightarrow s/r = y \leftrightarrow (v - u \tan \theta)/u = y$ where

$$\begin{pmatrix} u \\ v \end{pmatrix} \text{ maps to } \begin{pmatrix} r \\ s \end{pmatrix}.$$

This is true $\leftrightarrow x - \tan \theta = y$ where x is a random variate from a density proportional to $h(x)$. Thus $Y = X - \tan \theta$ and the density of Y is $h(y + \tan \theta)$.

(b) Note that

$$A = \begin{pmatrix} \cos \theta & -\sin \theta \\ 0 & \frac{1}{\cos \theta} \end{pmatrix} \begin{pmatrix} \cos \theta & \sin \theta \\ -\sin \theta & \cos \theta \end{pmatrix}$$

which represents the rotation followed by a deformation.

(c) By definition, for any $(u, v) \in C$,

$$u \le \sqrt{h\left(\frac{v}{u}\right)} \le \max \sqrt{h(x)} = u^+.$$

Also by definition, $h(x)$ is maximized at $x = m$ so $u^+ = \sqrt{h(m)}$. Therefore, the boundary of C passes through $(u^+, u^+ m)$ and the line $u = u^+$ touches the boundary at this point.

(d) If the random variable X has a small coefficient of variation, C will be a thin region about the line $v/u = m$, and will occupy a small proportion of the minimal rectangle enclosing C. Using the ratio method on a random variable $X - \tan \theta$, where θ is chosen suitably, will allow the region C_θ to be distributed roughly evenly on either side of the line $s = 0$. The region C_θ will occupy a larger proportion of the minimal enclosing rectangle for C_θ, resulting in a larger probability of acceptance.

(e) (i) $\theta = 0$, acceptance probability $= 0.45486$; (ii) $\tan \theta = 2/9$, acceptance probability $= 0.71943$; (iii) $\tan \theta = 1826/9000$, acceptance probability $= 0.72297$.

11. In the usual notation it is shown that $u_+ = \left(\frac{1}{2}\right)^{\alpha-1}$, $v_+ = -v_- = (1 - \alpha^{-1})^{(\alpha-1)/2} / (2^\alpha \sqrt{\alpha})$. The acceptance probability is $\frac{1}{2} \int_{-1/2}^{1/2} h(y) \, dy / (2u_+ v_+)$. Using Maple $\int_{-1/2}^{1/2} h(y) \, dy = 2^{1-2\alpha} \sqrt{\pi} \Gamma(\alpha) / \Gamma(\alpha + 1/2)$ and the result follows. Using Maple, the limit of this as $\alpha \to \infty$ is as given in the question.

9.4 Solutions 4

1. We have $x = \sqrt{-2 \ln(j/m)} \sin(2\pi [(aj) \bmod m]/m)$. The log term ensures that the largest values of x are for small j. For such j, $x = \sqrt{-2 \ln(j/m)} \sin(2\pi a j/m)$,

and the first maximum occurs at the smallest j such that $\tan(2\pi aj/m) = -2(2\pi aj/m)\ln(j/m)$. This is approximately $2\pi aj/m = \pi/2$, giving $x \approx \sqrt{2\ln(4a)} = \sqrt{2\ln 524} = 3.54$. Similarly, the largest negative value is at approximately $2\pi aj/m = 3\pi/2$, giving $x \approx -\sqrt{2\ln(4a/3)} = -3.21$. Using Maple the expected frequencies in 2^{30} variates are (i) 214817 and (ii) 712615 respectively.

2. The joint density of U and V is $f_{U,V}(u,v) = 1/\pi$ over support $u^2 + v^2 \le 1$. The Jacobian of the transformation is

$$J = \begin{vmatrix} \dfrac{\partial y}{\partial u} & \dfrac{\partial y}{\partial v} \\[2mm] \dfrac{\partial \theta}{\partial u} & \dfrac{\partial \theta}{\partial v} \end{vmatrix} = \begin{vmatrix} 2u & 2v \\[2mm] \dfrac{-v/u^2}{1+v^2/u^2} & \dfrac{1/u}{1+v^2/u^2} \end{vmatrix} = 2.$$

Therefore, the joint density of Y and Θ is $f_{Y,\Theta}(y,\theta) = (1/J)f_{U,V}(u[y,\theta],v[y,\theta]) = (1/\pi)/2 = 1/(2\pi)$ over support $\theta \in [0, 2\pi)$ and $Y \in [0, 1)$. Since the joint density is evidently the product of the two marginal densities it is concluded that Y and Θ are independent.

4. (b) This should be close to the theoretical value, which can be shown to be 11.2 pence.

5. $h(x) = e^{-x}x^{\alpha-1}$ on support $[0, \infty)$. Choose an envelope $g(x) = Ke^{-\lambda x}$ where λ is to be selected. The probability of acceptance is $\int_0^\infty h(x)\,dx / \int_0^\infty g(x)\,dx = \Gamma(\alpha)/(K/\lambda)$. Therefore, for given λ we select the smallest K s.t. $Ke^{-\lambda x} \ge e^{-x}x^{\alpha-1} \forall x \in [0, \infty)$. Accordingly, we set $K = \max_{x>0}(e^{-x}x^{\alpha-1}/e^{-\lambda x})$. If $\lambda \ge 1$ then K is infinite, leading to zero efficiency. If $\lambda < 1$ then $K = e^{-(\alpha-1)}\left[\frac{\alpha-1}{1-\lambda}\right]^{\alpha-1}$. The resulting probability of acceptance is $\lambda\Gamma(\alpha)e^{\alpha-1}\left[\frac{\alpha-1}{1-\lambda}\right]^{-\alpha+1}$. This is maximized when $\lambda = \alpha^{-1}$, giving an acceptance probability of $\Gamma(\alpha)e^{\alpha-1}/\alpha^\alpha$ and an optimal exponential envelope that is proportional to $\exp(-x/\alpha)$. Note that for large α, using Stirling's approximation, this acceptance probability behaves as $\sqrt{2\pi}(\alpha-1)^{\alpha-1/2}/\alpha^\alpha \sim \sqrt{2\pi}e^{-1}/\sqrt{\alpha-1} \sim e^{-1}\sqrt{2\pi/\alpha}$. This is poor (e.g. 0.206 when $\alpha = 20$) and tends to zero as $\alpha \to \infty$. The method cannot be used when $\alpha < 1$ as the gamma density is unbounded as $x \to 0$. It is therefore impossible to envelope it with any negative exponential function.

6. The first part is easy. For the second part, $h(w) = \exp(-w^{1/\alpha})$, so $u^+ = \max_{w>0}\left[\exp(-\frac{1}{2}w^{1/\alpha})\right] = 1$ and $v^+ = \max_{w>0}\left[w\exp(-\frac{1}{2}w^{1/\alpha})\right] = (2\alpha/e)^\alpha$. The acceptance probability is $\frac{1}{2}\int_0^\infty h(w)\,dw/(u^+v^+) = \frac{1}{2}\int_0^\infty e^{-x}\alpha x^{\alpha-1}dx/(2\alpha/e)^\alpha = (e/2\alpha)^\alpha[\Gamma(\alpha+1)]/2$. This is bounded (e.g $\frac{1}{2}$ as $\alpha \to 0$, and $e/4$ at $\alpha = 1$) for all $\alpha \in (0, 1)$ (see figure 9.3 and Maple procedure below). Each prospective variate requires one exponentiation and one logarithmic evaluation, which is not unusually costly (see below). It is therefore a good method. The method is still valid when $\alpha \ge 1$, but leads to steadily decreasing acceptance probabilities in this range.

```
> gammaalphaless1:=proc(alpha) local r1,r2,x;
  do;
  if alpha>=1 then ERROR("alpha should be less than 1") end if;
  r1:=rand()/10^12;
  r2:=rand()/10^12;
```

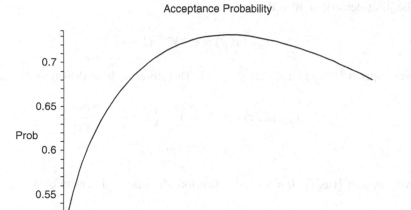

Figure 9.3   Acceptance probability, problem 4.6

```
x:=evalf((r2/r1)^(1/alpha)*(2*alpha/2.718281828));
if evalf(−ln(r1)−x/2) > 0 then break end if;
end do;
end proc;
```

7. The joint density of R_1 and R_2 is uniform over

$$\left\{ (r_1, r_2) : 0 < r_1, 0 < r_2, r_1^{1/\alpha} + r_2^{1/\beta} \leq 1 \right\}.$$

Let $X = R_1^{1/\alpha} / \left(R_1^{1/\alpha} + R_2^{1/\beta} \right)$. Then the joint density of R_1 and X is

$$f_{R_1, X}(r_1, X) \propto \left| \frac{\partial r_2}{\partial x} \right| = \beta x^{-\beta-1} (1-x)^{\beta-1} r_1^{\beta/\alpha}$$

over support $\{ (r_1, x) : 0 < r_1 < x^\alpha, 0 < x < 1 \}$. Integrating over r_1 the marginal density of X is

$$f_X(x) \propto x^{-\beta-1} (1-x)^{\beta-1} \int_0^{x^\alpha} r_1^{\beta/\alpha} \, dr_1$$

$$\propto x^{-\beta-1} (1-x)^{\beta-1} x^{\alpha+\beta}$$

$$= x^{\alpha-1} (1-x)^{\beta-1},$$

as required.

8. The joint density of W and Y is

$$f_{W,Y}(w, y) \propto e^{-w} y^{\alpha-1} (1-y)^{-\alpha}$$

over support $\{(w, y) : 0 \leq w, 0 < y < 1\}$. Therefore the joint density of W and X is

$$f_{W,X}(w, x) \propto e^{-w} \left(\frac{x}{w}\right)^{\alpha-1} \left(1 - \frac{x}{w}\right)^{-\alpha} \left|\frac{\partial y}{\partial x}\right|$$

$$= e^{-w} x^{\alpha-1} (w - x)^{-\alpha}$$

over support $\{(w, x) : 0 < x < w\}$. Therefore the marginal density of X is

$$f_X(x) \propto x^{\alpha-1} \int_x^{\infty} e^{-w} (w - x)^{-\alpha} dw$$

$$\propto x^{\alpha-1} e^{-x},$$

the required gamma density.

9. The density of X is

$$f_X(x) \propto \left(\frac{x}{\rho(1-x)}\right)^{\alpha-1} \left(\frac{1}{1-x}\right)^{-\alpha-\beta} \left|\frac{dy}{dx}\right|$$

$$\propto \left(\frac{x}{1-x}\right)^{\alpha-1} \left(\frac{1}{1-x}\right)^{-\alpha-\beta+2}$$

$$= x^{\alpha-1} (1-x)^{\beta-1}$$

as required.

10. The joint density is $f_{X,Y}(x, y) = f\left(\sqrt{x^2+y^2}\right) / (2\pi\sqrt{x^2+y^2})$ on domain $\mathbb{R}^2$. The marginal density of X is $\int_{-\infty}^{\infty} f\left(\sqrt{x^2+y^2}\right) / (2\pi\sqrt{x^2+y^2}) dy = \int_0^{\infty} f\left(\sqrt{x^2+y^2}\right) / (\pi\sqrt{x^2+y^2}) dy$ and similarly for Y.

(a) Here, $f_{X,Y}(x, y) = [1/(2\pi)] \exp(-x^2/2 - y^2/2)$, which is separable in x and y, showing that X and Y are independently distributed as $N(0, 1)$.

(b) Here, $f_X(x) \propto \int_0^{\sqrt{1-x^2}} (1 - x^2 - y^2)^{c-1} dy \propto (1 - x^2)^{c-1/2}$. Put $T = \sqrt{n}X/\sqrt{1-X^2}$ and $S = \sqrt{n}Y/\sqrt{1-Y^2}$. Then T is monotonic increasing in X in $(-1, 1)$. Therefore, $f_T(t) \propto [n/(n+t^2)]^{n/2-1} |dx/dt| \propto [1/(n+t^2)]^{n/2-1} [1/(n+t^2)]^{3/2} \propto (1+t^2/n)^{-(n+1)/2}$. The joint density of S and T is not separable in s and t and therefore these two random variables are not independent. Accordingly, a simulation should use either the cosine form or the sine form but not both. (Recall that in the standard Box–Müller method for normal variates, the sine and cosine forms are independent.) The result (4.15) is obtained by using the inversion method to sample from the density of R.

9.5   Solutions 5

1. (a) Using 5000 replications, each replication generating a primary and antithetic waiting time for five customers, the mean such time was 1.287 and the standard error was 0.00777. The estimated variance reduction ratio is approximately 2.5.

   (b) Using the same seed, the corresponding results were 5.4751, 0.00820, and 14. The v.r.r. is now better as W_i is now more linear in A_i and S_{i-1} than in (a). Why?

4. (b) (i) $t = 2.8386$.

5. The conditional probability density function is

$$f_{X|X \in (a_{i-1}, a_i)}(x) = -N \ln(x)$$

on support (a_{i-1}, a_i), since for equiprobable intervals, $P(X \in (a_{i-1}, a_i)) = 1/N$. For large N, there is little variation in f over (a_{i-1}, a_i), except for $i = 1$, so a uniform envelope is appropriate. Therefore, for $i \geq 2$,

> 1. generate $X \sim U(a_{i-1}, a_i)$
>
> generate $R \sim U(0, 1)$
>
> if $R < \dfrac{-\ln X}{-\ln(a_{i-1})}$ deliver X else goto 1

For $i = 1$, we could continue to use inversion. For large n this will not degrade the performance much.

10. Let f denote the p.d.f. of points uniformly distributed over D. Since

$$\int \cdots \int \prod_{j=1}^{m} dx_j = \frac{1}{m!}$$

it follows that

$$f(x_{(1)}, \ldots, x_{(m)}) = m! \quad (x_{(1)}, \ldots, x_{(m)}) \in D$$

Therefore,

$$\int \cdots \int_D \left[\sum_{j=1}^{m} (m-j+1)^2 x_j \right] dx = \frac{1}{m!} E_f \left[\sum_{j=1}^{m} (m-j+1)^2 x_j \right].$$

Points are sampled uniformly over D by sampling uniformly over $(0, 1)^m$ and then sorting the coordinates so that $0 < x_{(1)} < \cdots < x_{(m)} < 1$. The Maple code below shows the numerical derivation of a 95% confidence interval for the integral. Note that the naive approach of sampling points uniformly over $(0, 1)^m$ and accepting only those that lie in D would be hopelessly inefficient.

```
> restart;
> with(stats);
[anova, describe, fit, importdata, random, statevalf, statplots, transform]


> randomize(96341641);
> n := 10000; m := 10;


n := 10000
m := 10


> for i from 1 to n do;
   for j from 1 to m do;
   a[j]:=evalf(rand()/10^12);
   end do;
   b:=[seq(a[j],j=1..m)];
   c:=transform[statsort](b);
   d:=0;
   for j from 1 to m do;
   d := d+ (m −j+ 1)^2*c[j];
   end do;
   end do:
   f:=[seq(e[i],i=1..n)]:
   g1:=describe[mean](f)/m!;
   g2:=describe[standarddeviation[1]](f)/m!;
   interval:=evalf([g1-1.96*g2/sqrt(n),g1+1.96*g2/sqrt(n)]);


g1 := .00003042677684
g2 := .9879697859e-5
interval := [.00003023313476, .00003062041892]
```

9.6   Solutions 6

1. Put call parity gives $c(t) + Ke^{-r(T-t)} = p(t) + x(t) e^{-r_f(T-t)}$. Using the result derived for the delta of the corresponding call option,

$$\frac{\partial p(t)}{\partial x(t)} = \frac{\partial c(t)}{\partial x(t)} - e^{-r_f(T-t)}$$

$$= e^{-r_f(T-t)} \Phi\left(d_{r_f}\right) - e^{-r_f(T-t)}$$

$$= -e^{-r_f(T-t)} \left[1 - \Phi\left(d_{r_f}\right)\right]$$

$$= -e^{-r_f(T-t)} \Phi\left(-d_{r_f}\right)$$

2. Use put call parity with $r_f = 0$. This gives (a) $p = £120.86$, (b) $p = £186.38$, and (c) $p = £268.94$. The seller of a put will initially hedge his or her position by having

a portfolio consisting of -1 put and Δ blocks of shares, where $\Delta = -\Phi(-d)$ (see Problem 1) and $d = \{(r + \sigma^2/2)(T - t) + \ln[x(t)/K]\}/(\sigma\sqrt{T - t})$. The number of shares shorted initially is (a) 328, (b) 440, and (c) 553.

3. (a) Let $P(X(T)|t, x(t))$ denote the payoff for the bond holder at time T, given that the current FTSE is at $x(t)$. Then

$$P(X(T)|t, x(t)) = \begin{cases} 1, & \frac{X(T)}{X(0)} < 1, \\ 1 + \frac{1}{2}\left[\frac{X(T)}{X(0)} - 1\right], & 1 \le \frac{X(T)}{X(0)} < 1.6, \\ 1 + 0.3, & 1.6 \le \frac{X(T)}{X(0)}. \end{cases}$$

In a risk-neutral world,

$$X(T) = x(t)\exp\left[(r - 0.5\sigma^2)(T - t) + \sigma\sqrt{T - t}Z\right]$$

where $Z \sim N(0, 1)$. Let $Q(Z|t, x(t))$ denote the payoff as a function of Z rather than of $X(T)$ and let

$$d_1 = \frac{\ln[x(0)/x(t)] - (r - 0.5\sigma^2)(T - t)}{\sigma\sqrt{T - t}},$$

$$d_2 = \frac{\ln[1.6x(0)/x(t)] - (r - 0.5\sigma^2)(T - t)}{\sigma\sqrt{T - t}}.$$

Then

$$Q(Z|t, x(t)) = \begin{cases} 1, & Z < d_1, \\ 1 + \frac{1}{2}\left[\frac{X(T) - X(0)}{X(0)}\right], & d_1 \le Z < d_2, \\ 1 + 0.3, & d_2 \le Z, \end{cases}$$

and so

$$V(x(t), t) = e^{-r(T - t)}$$
$$\times \left\{1 + \frac{1}{2}\int_{d_1}^{d_2}\left[\frac{x(t)}{x(0)}e^{(r - 0.5\sigma^2)(T - t) + \sigma\sqrt{T - t}z} - 1\right]\phi(z)\,dz \right.$$
$$\left. + 0.3\Phi(-d_2)\right\}$$

(b) The following Maple procedure computes this for given t and xt, where $xt = x(t)$ is expressed in terms of $x0$.

```
> capped_bond:=proc(xt,t) local r,sigma,T,b,d1,d2,P,price;
  r := 0.04;
  sigma := 0.2;
  T := 4;
  b := xt/x0;
  d1 := (ln(1/b)-(r-0.5*sigma^2)*(T-t))/sqrt(T-t)/sigma;
```

```
d2 := (ln(1.6/b)-(r-0.5*sigma^2)*(T-t))/sqrt(T-t)/sigma;
P := 0.5*(b* exp((r-0.5*sigma^2)*(T-t))
  +sigma*sqrt(T-t)*z)-1)* exp(-0.5*z^2)/sqrt(2*Pi);
price := exp(-r*(T-t))*((1+int(P, z = d1..d2))
  +0.3*statevalf[cdf,normald](-d2));
print("price at time",t,"is £",price);
end proc;
```

It is found that $V(x(0), 0) = £0.9392$ and

(c) that $V(1.8x(0), 2) = £1.1678$.

4. (a) At time T portfolio A has the value $K + V(x(T), T) = K + x(T) - K = x(T)$, which is the same as the value of portfolio B. Therefore, the two portfolios must have identical values at any time $t \in [0, T]$ otherwise a riskless profit could be made by investing in the cheaper portfolio. At time t, the cash in A has grown to $Ke^{-rT}e^{rt}$ so the value of portfolio A is $Ke^{-r(T-t)} + V(x(t), t)$ while that of B is $x(t)$. Equating these two gives the desired result.

(b) We have $\partial V/\partial x(t) = 1$, $\partial V/\partial t = r[V(x(t), t) - x(t)]$, and $\partial^2 V/\partial x(t)^2 = 0$; the result follows by substitution.

(c) A portfolio that is a long one forward contract and a short one share will have the value at time t given by $V(x(t), t) - x(t) = -Ke^{-r(T-t)}$. Since there is no uncertainty in $-Ke^{-r(T-t)}$ for all $t \in [0, T]$, the hedge is perfect. In practice the hedge involves, at time zero, selling short one share for $x(0)$. A forward contract is purchased for $V(x(0), 0) = x(0) - Ke^{-rT}$, leaving an amount of cash Ke^{-rT}. This grows to K at time T, which is used at that time to meet the obligations on the forward contract and to the initial lender of the share.

(d) The delivery price K is now such that $V(x(0), 0) = 0$. The contract at time zero has zero value. No money passes between A and B at that time. This is a standard forward contract.

5. At time zero, the writer of the option sets up a portfolio consisting of a -1 call option and $\Delta(0)$ shares. The share purchase is financed by borrowing $\Delta(0)X(0)$. At time T this has grown to $\Delta(0)X(0)e^{rT}$. At time T, if $X(T) > K$, then a further $1 - \Delta(0)$ shares are purchased at a cost of $[1 - \Delta(0)]X(T)$. Since the customer will exercise the option, the writer will sell the one share for K. The total cost at time T of writing and hedging the option in this case is

$$\Delta(0)X(0)e^{rT} + [1 - \Delta(0)]X(T) - K.$$

Otherwise, if $X(T) \leq K$ at time T, the writer will sell the existing $\Delta(0)$ shares, obtaining $\Delta(0)X(T)$. The total cost of writing and hedging the option in this case is

$$\Delta(0)X(0)e^{rT} - \Delta(0)X(T).$$

Bringing these two results together, the total cost is

$$\Delta(0)X(0)e^{rT} - \Delta(0)X(T) + [X(T) - K]^+.$$

The present value of this is

$$C = \Delta(0) X(0) - \Delta(0) X(T) e^{-rT} + e^{-rT} [X(T) - K]^{+}.$$

Therefore,

$$
\begin{aligned}
E(C) - c &= E\left[\Delta(0) X(0) - \Delta(0) X(T) e^{-rT} + e^{-rT} [X(T) - K]^{+}\right] - c \\
&= \Delta(0) X(0) - E\left[\Delta(0) X(T) e^{-rT}\right] \\
&\quad + e^{-rT} \left[X(0) \Phi(d_\mu) e^{\mu T} - K\Phi\left(d_\mu - \sigma\sqrt{T}\right)\right] \\
&\quad - e^{-rT} \left[X(0) \Phi(d) e^{rT} - K\Phi\left(d - \sigma\sqrt{T}\right)\right] \\
&= \Delta(0) X(0) - \Delta(0) X(0) e^{(\mu-r)T} \\
&\quad + e^{-rT} X(0) \Phi(d_\mu) e^{\mu T} - X(0) \Phi(d) \\
&\quad + K e^{-rT} \left[\Phi\left(d - \sigma\sqrt{T}\right) - \Phi\left(d_\mu - \sigma\sqrt{T}\right)\right] \\
&= X(0) e^{(\mu-r)T} \left[\Phi(d_\mu) - \Phi(d)\right] \\
&\quad + K e^{-rT} \left[\Phi\left(d - \sigma\sqrt{T}\right) - \Phi\left(d_\mu - \sigma\sqrt{T}\right)\right]
\end{aligned}
$$

and this is plotted in Figure 9.4.

Notice that when $\mu = r$, there is no difference between the *expected* cost of writing and hedging the option in this way and the Black–Scholes price.

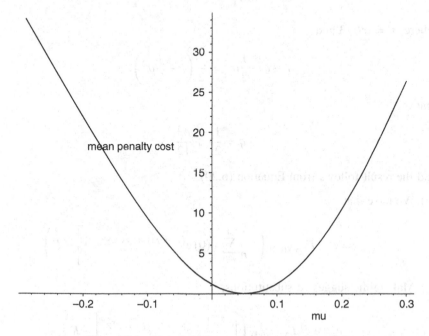

Figure 9.4 Plot of $E(C) - c$ against μ, problem 6.5

6. From Equation (6.30),

$$c_g = e^{-rT} E_{N(0,1)} \left[X_0 e^{(r-\sigma^2/2)[(n+1)/2]h+(\sigma\sqrt{h}/n)\sum_{i=1}^{n}(n-i+1)Z_j} - K \right]^+$$

Put

$$Z = \frac{\sum_{i=1}^{n}(n-i+1)Z_j}{\sqrt{\sum_{i=1}^{n}(n-i+1)^2}}$$

$$= \frac{\sum_{i=1}^{n}(n-i+1)Z_j}{\sqrt{n(n+1)(2n+1)/6}}$$

$$= \frac{\sum_{i=1}^{n}(n-i+1)Z_j}{\sqrt{a}}$$

$$\sim N(0,1).$$

Then

$$c_g = e^{-rT} E_{N(0,1)} \left[X_0 e^{(r-\sigma^2/2)[(n+1)/2]h+\sigma\sqrt{ah}Z/n} - K \right]^+$$

and

$$\lim_{n\to\infty} c_g = e^{-rT} E_{N(0,1)} \left[X_0 e^{(r-\sigma^2/2)(T/2)+\sigma\sqrt{T}Z/\sqrt{3}} - K \right]^+$$

$$= e^{-rT} E_{N(0,1)} \left[X_0 e^{(r-q-\sigma'^2/2)T+\sigma'\sqrt{T}Z} - K \right]^+$$

where $\sigma' = \sigma/\sqrt{3}$ and

$$r - q - \frac{1}{2}\sigma'^2 = \frac{1}{2}\left(r - \frac{1}{2}\sigma^2 \right),$$

that is

$$q = \frac{1}{2}r + \frac{\sigma^2}{12},$$

and the result follows from Equation (6.20).

9. (a) We have that

$$c = e^{-rT} E_{Y\sim N(0,I)} \left(\left[\frac{1}{n}\sum_{j=1}^{n} x(0) e^{(r-\sigma^2/2)jh+\sigma\sqrt{h}\sum_{i=1}^{j} Y_i} \right] - K \right)^+$$

Making the suggested substitutions,

$$c = e^{-rT} E_{Y\sim N(0,I)} \left(\left[\sum_{j=1}^{n} \frac{1}{n} x_j e^{(r-\sigma_j^2/2)T+\sigma_j\sqrt{T}Z_j} \right] - K \right)^+$$

where $Z_j = \sum_{i=1}^{j} Y_i / \sqrt{j} \sim N(0,1)$ for $j = 1, \ldots, n$. This is clearly the price of a basket option with quantities $1/n$ of each asset, where the correlation on the returns between assets j and $m (\geq j)$ is

$$\mathrm{Cov}(Z_j, Z_m) = \mathrm{Cov}\left(\frac{\sum_{i=1}^{j} Y_i}{\sqrt{j}}, \frac{\sum_{i=1}^{m} Y_i}{\sqrt{m}}\right)$$

$$= \frac{1}{\sqrt{jm}} \sum_{i=1}^{j} \mathrm{Var}(Y_i)$$

$$= \sqrt{\frac{j}{m}}$$

(b) The following Maple execution group makes the necessary substitutions and creates the correlation matrix of returns between the 16 equivalent assets.

```
> n: = 16; r: = 0.05; x: = Vector(n); sigma: = Vector(n); q: = Vector(n);
  rho: = Matrix(n); T: = 1; t: = 0; m: = 20;
  npath: = 400; p: = 100; K: = 50; upper: = 200;
  T: = 1; sig: = 0.3; x0: = 50;
  h: = evalf(T/n);
  f: = (i, j)- > if i < j then evalf(sqrt(i/j)) else evalf(sqrt(j/i)) end if:
  rho: = Matrix(n, f);
  for j from 1 to n do;
  x[j]: = x0* exp(r*(j*h − T));
  q[j]: = evalf(1/n);
  sigma[j]: = sig*sqrt(j*h/T);
  end do;
  spot:=Transpose(x).q;
```

From the resulting 100 replications, using basketimppoststrat (seed = 9624651), each consisting of 400 payoffs over 20 strata, a price of 4.1722 with a standard error of 0.0014 was obtained. This compares with a value of 4.1708 and standard error of 0.00037 using 100 replications, each consisting of 2500 replications over 100 strata, as given in Table 6.2. Bearing in mind the different sample sizes and the sampling error, there is little to choose between the two approaches.

10. (a) Suppose $\lambda_Y = 0$. In the usual notation, an Euler approximation scheme is (in a risk-neutral world)

$$X(T) \approx x_0 e^{\sum_{j=1}^{n} (r - \sigma_j^2/2) h + \sigma_j \sqrt{h}\left(\sqrt{1-\rho^2} Z_j + \rho W_j\right)}$$

where

$$\sigma_j = e^{Y_j},$$

$$Y_j = Y_{j-1} e^{-\alpha h} + m\left(1 - e^{-\alpha h}\right) + \nu \sqrt{1 - e^{-2\alpha h}} W_j$$

for $j = 1, \ldots, n$, with $Y_0 = \ln[\sigma(0)]$. $\{W_j\}$ and $\{Z_j\}$ are independent sequences of $N(0, 1)$ random variables.

(b) (i) Antithetics can be implemented with ease at little extra cost, but the improvement may be unimpressive in many cases.

(ii) Stratification may be promising for certain parameter values. The distribution of Y_n given Y_0 is normal, so post-stratified sampling on Y_n may be of some value for slow mean reverting processes.

(iii) However, the method likely to be most successful is to note that conditional on $W_1, \ldots, W_n, \sigma_1, \ldots, \sigma_n$,

$$X_n = X_0 e^{\sum_{j=1}^n \left[(r - \sigma_j^2/2)h + \sigma_j \sqrt{h} \left(\sqrt{1-\rho^2} Z_j + \rho W_j \right) \right]}.$$

Put $Z = \sum_{j=1}^n \sigma_j \sqrt{h} Z_j / (\sum_{j=1}^n \sigma_j^2 h)^{1/2} \sim N(0,1)$. Then

$$X_n = X_0 e^{\rho \sqrt{h} \sum_{j=1}^n \sigma_j W_j} e^{(r - \Delta^2/2)T + \Delta\sqrt{T(1-\rho^2)} Z}$$

where $\Delta^2 = (1/n)\sum_{j=1}^n \sigma_j^2$. Put $\Omega^2 = \Delta^2(1-\rho^2)$ and $S_0 = X_0 \exp\left(\rho\sqrt{h} \sum_{j=1}^n \sigma_j W_j\right)$. Then

$$X_n = S_0 e^{[r - (\rho^2/2n)\sum_{j=1}^n \sigma_j^2 - \Omega^2/2]T + \Omega\sqrt{T} Z}$$

Let $BS(S_0, s, T, K, r, q)$ denote the price at time zero of a European call with strike price K and exercise time T, on an underlying asset having an initial price S_0 which earns interest continuously at rate q with *constant* volatility s, and where the risk-free interest rate is r. Then

$$\widehat{c} = BS\left(X_0 \exp\left(\rho\sqrt{h} \sum_{j=1}^n \sigma_j W_j \right), \Omega, T, K, r, \frac{1}{2}\Delta^2\rho^2 \right)$$

This is an example of conditional Monte Carlo.

9.7 Solutions 7

1. (c) Each state variable, $X(t), t \geq 0$, is a mixture of discrete and continuous, since there is a positive probability that none has been consumed in $[0, t]$. The time variable is continuous. The system is a continuous event system during periods of consumption and is trivially a discrete state continuous time system during periods of abstinence.

2. (b) The ith event is at $T_{(i)} = (1/\beta)\ln(1 + \beta\tau_{(i)} e^{-\alpha})$ where $\{\tau_{(i)}\}$ follow a simple Poisson process of rate one. Stop at event number $\max\{i : \tau_{(i)} < (1/\beta)e^\alpha (e^{\beta t_0} - 1)\}$.

3. It is probably easiest and quite efficient to use rejection of points falling in the rectangle $\{(x,y) : x \in [-a,a], y \in [-b,b]\}$, but not within the ellipse.

7. To obtain the theoretical result, let m_i denote the mean time to absorption given the current state is i. Then $m_i = 1 + \sum_{j=i}^4 p_{ij} m_j$ for $i = 1, \ldots, 4$. The expected life of the equipment is m_1.

8. Since $N(t) + K(t) + D(t)$ is a constant for all $t \geq 0$ the state is uniquely represented by any two state variables, say $(N(t), K(t))$. The 'leaving rate' for state (n, k) is $\lambda = nk\beta p + n\mu_2 + k\mu_1$ and so, given that $(N(t), K(t)) = (n, k)$, the next event is at time $t - \lambda^{-1} \ln R$, where $R \sim U(0, 1)$, and the state immediately after that event will be either $(n-1, k+1)$ or $(n-1, k)$ or $(n, k-1)$ with probabilities $nk\beta p/\lambda, n\mu_2/\lambda, k\mu_1/\lambda$ respectively. Simulate realizations of the epidemic by setting $(N(0), K(0)) = (N-1, 1)$ say.

12. (a) Given that there is an accident in $(t, t+\delta t)$, the conditional distribution of R, the distance of occurrence from the hospital, is given by

$$P(R \leq r \mid R \leq 5) = \left(\frac{r}{5}\right)^2$$

Using inversion of this cumulative distribution function gives $(r/5)^2 = U$ or $r = 5\sqrt{U}$, so $x = 2\left(5\sqrt{U}/v\right)$.

(b) This is a five server (ambulance) queueing system with waiting customers (patients) forming one queue. Bound state changes are customer arrivals (emergencies) and customer departures (patients deposited at the hospital). Conditional state changes are starts of service (ambulances despatched to patients). For each of the five ambulances it is necessary to store (i) the time at which it next deposits a patient at the hospital and (ii) the time at which the patient incurred the emergency. Suppose these are called $TD[j]$ and $TA[j]$ respectively for $j = 1, \ldots, 5$. Let b = number of busy servers, q = number of patients waiting for an ambulance, A = time of next emergency, $clock$ = present time, and $simtim$ = duration of simulation. Let $A[j]$ = time of arrival of the patient who is currently jth in the queue for an ambulance. Then, assuming ambulances are despatched to emergencies on a 'first come, first served' protocol, the core part of a simulation might be based around the following algorithm:

```
While clock < simtim do
clock := min (A, TD[1], . . . , TD[5])
If event is arrival (clock = A) then
q := q + 1
A := clock + interarrivaltime
A[q] := clock
end if
If event is departure then
identify j the server involved
print clock − TA[j]
TA[j] := 0 (arbitrary value)
```

$TD[j] := \infty$
$b := b - 1$
end do
If $q > 0$ and $b < 5$ then
Identify j, a free server
$TD[j] := clock + service\ duration$
$TA[j] := A[1]$
$A[1] := \infty$
sort $\{A[k], k = 1, \ldots, q\}$
$q := q - 1$
$b := b + 1$
end if
End do

13. The bound state changes are machine breakdowns, completion of machine repairs, and completion of machine tests. Conditional state changes are repair starts and testing starts. Let the time of the next state change for machine j be $T[j]$. Let the state of machine j be $S[j] = 1, 2, 3, 4, 5$ according to whether it is working, in the repair queue, being repaired, in the testing queue, or being tested respectively. Let nw, nr, and nt denote the number of machines working, the number of repairmen who are free, and the number of testers who are free respectively. Let qr and qt denote the number of machines in the repair queue and testing queue respectively. Let *clock* and *clockprev* denote the current simulation time and time at the previous event respectively.

    If working periods happen to be exponentially distributed, then regeneration points would exist at those events where the number of working machines changes from $m - 1$ to m. Otherwise, regenerative analysis cannot be used, since the only regenerative points (assuming steady state behaviour) are those instants at which all machines are returned simultaneously to the working state – an impossibility with continuous random variables. Therefore, a proper analysis should plot nw against *clock* and identify a burn-in period, tb. Then a point estimate of the long-run average utilization will be $\{1/[m(simtim - tb)]\} \int_{tb}^{simtim} nw(t)\,dt$. A confidence interval may be obtained by replicating a large number of realizations over identical values of *simtim* and tb, preferably starting in different states.

14. (a) Bound events are customer arrivals, customer departures from window A, and customer departures from window B. Conditional events are the start of service at window A, the start of service at window B, a queue-hop from A to B, and a queue-hop from B to A.

    (c) Regeneration points are when one server is idle and the other one becomes idle. If the traffic intensity is not too close to 1, these will occur frequently enough to allow a regenerative analysis.

15. This is similar to Problem 13, but here the individual waiting times need to be recorded.

9.8   Solutions 8

1. If x is the current point and y is a prospective variate then the acceptance probability is

$$\alpha(x, y) = \min\left(1, \frac{e^{-y^2/2}\,e^{-x}}{e^{-x^2/2}e^{-y}}\right)$$

and it is found that $\{x_i\}$ becomes $\{0, 0.6539, 0.6539, 0.3587, 1.1087, 0.0513, \dots\}$.

4. The overall acceptance rate is very low, being of the order of 2.5%. This is because the posterior is so different from the prior which provides a very poor envelope. On the other hand, inspection of the plots for the sequence $\{\beta^{(i)}\}$ say, (Appendix 8.2), for the MCMC independence sampler shows that the acceptance rate, although poor, is not quite so low (approximately 7.5 %). In the independence sampler the probability of acceptance of a candidate point (α_c, β_c) is $\min(1, L(\alpha_c, \beta_c)/L(\alpha, \beta))$ where (α, β) is the current point. This is always greater than $L(\alpha_c, \beta_c)/L_{\max}$, the acceptance probability using envelope rejection. Offset against this is the fact that envelope rejection gives independent variates, whereas MCMC does not.

6. (c) Sometimes it is difficult to find a uniform bound c for $h(x)/g(x)$ in standard envelope rejection. Part (b) suggests that variates from a density proportional to h can be sampled using Metropolis–Hastings with the usual disadvantage that variates will not be independent. However, each time $h(y)/g(y) > c$, we can update c to $h(y)/g(y)$. In the limit (after many proposal variates), c becomes a uniform bound and the variates produced by Metropolis–Hastings become independent.

7. Let $\bar{y}$ and s_y^2 denote the sample mean and variance of $\{y_1, \dots, y_n\}$. Gibbs sampling is based upon the following full conditionals:

$$\beta \,|\, \mu, \tau \sim \text{gamma}\left(\frac{5}{2}, 1 + \frac{1}{2}\mu^2\right)$$

$$\tau \,|\, \mu, \beta \sim \text{gamma}\left(\frac{n}{2} + 2, 1 + \frac{1}{2}(n-1)s_y^2 + \frac{1}{2}n(\bar{y} - \mu)^2\right)$$

$$\mu \,|\, \tau, \beta \sim N\left(\frac{n\bar{y}\tau}{n\tau + \beta}, \frac{1}{n\tau + \beta}\right)$$

8. The full conditionals are $x_1 \,|\, x_2 \sim \text{Exp}\left[\frac{1}{2}(1 + x_2^2)\right]$ and $x_2 \,|\, x_1 \sim N(0, 1/x_1)$. One direct method is to sample from the marginal of X_1 which has a gamma $\left(\frac{1}{2}, \frac{1}{2}\right)$ density, that is a chi-squared variate with one degree of freedom. Then put $x_1 = z_1^2$ where $z_1 \sim N(0, 1)$ and $x_2 \sim N(0, 1/x_1)$. The other is to sample from the marginal of X_2 which has a Cauchy density. Therefore $x_2 = \tan\left[(\pi/2)(2R_2 - 1)\right]$ where $R_2 \sim U(0, 1)$, and then put $x_1 = -2\ln R_1/(1 + x_2^2)$ where $R_1 \sim U(0, 1)$. Either of the direct methods is better than Gibbs sampling, for the usual reason that the former produces sequences of independent variates. The first direct

method is probably preferred, due to the prevalence of efficient standard normal generators.

10. (a) $\widehat{\mu}_1 = 0.4597$, e.s.e. $(\widehat{\mu}_1) = 0.0219$.

(b) $\widehat{\mu}_1 = 0.4587$, e.s.e $(\widehat{\mu}_1) = 0.0073$.

11. (a) If $x^{(i)}$ is the current point then

$$x^{(i+1)} \sim U\left(\max\left[1 - (1-x^{(i)})\,R_2^{1/(\beta-1)}, 0\right], \min\left[1, x^{(i)}\,R_1^{1/(\alpha-1)}\right]\right)$$

where $R_1, R_2 \sim U(0, 1)$.

(b) $x^{(i+1)} \sim U\left(\max\left[R_3\left(x^{(i)}+y^{(i)}\right) - y^{(i)}, 0\right], \min\left[1, \sqrt{x^{(i)^2} - 2\ln R_1}\right]\right)$,

$y^{(i+1)} \sim U\left(\max\left[R_3\left(x^{(i+1)}+y^{(i)}\right) - x^{(i+1)}, 0\right], \min\left[1, \sqrt{y^{(i)^2} - 2\ln R_2}\right]\right)$,

where $R_1, R_2, R_3 \sim U(0, 1)$.

12. (b) $z^{(i+1)} = x^{(i)}\,R_2^{1/(\alpha-1)}$ and $x^{(i+1)} = 1 - \left(1 - z^{(i+1)}\right)R_1^{1/\beta}$ where $R_1, R_2 \sim U(0, 1)$.

15. (a) Let $g(\theta, y)$ denote the joint density of θ and y where

$$g(\theta, y) \propto e^k\,e^{-ky}$$

on support $\infty > y > 1 - \cos\theta, 0 < \theta < \pi$. Then g is a completion of f. One full conditional is

$$g(y|\theta) = k\,e^{-k(y-1+\cos\theta)}$$

on support $y > 1 - \cos\theta$. Therefore $y|\theta = 1 - \cos\theta - \ln(R_1)/k$ where $R_1 \sim U(0, 1)$. The other is

$$g(\theta|y) = \text{constant}$$

on support $\cos\theta > 1 - y$. Therefore,

$$\theta|y \sim \begin{cases} U(0, \pi), & y \geq 2, \\ U\left(0, \cos^{-1}(1-y)\right), & 0 < y < 2, \end{cases}$$

or

$$\theta|y \sim \begin{cases} R_2\pi, & y \geq 2, \\ R_2\cos^{-1}(1-y), & 0 < y < 2, \end{cases}$$

where $R_2 \sim U(0, 1)$.

17. (a) μ and $\left[\mu^2/(b-2)\right]\left(1+[b-1]E_{g_\alpha}(1/\alpha)\right)$.

(b) $g_\alpha(\alpha)\,\alpha^{-b}\,(x_1\cdots x_n)^\alpha\,/\,[\Gamma(\alpha)]^n$ should be log-concave.

18. (e) The posterior predictive survival probability is

$$P_{\text{post}}(X > x) = E_{\pi(\alpha,\lambda)}\left[\int_x^\infty \frac{\lambda^\alpha y^{\alpha-1} e^{-\lambda y}}{\Gamma(\alpha)} \, dy\right]$$

$$= E_{\pi(\alpha,\lambda)}\left[\int_{\lambda x}^\infty \frac{t^{\alpha-1} e^{-t}}{\Gamma(\alpha)} \, dt\right]$$

and an estimate of this is

$$\widehat{P}_{\text{post}}(X > x) = \frac{1}{m}\sum_{i=1}^m \text{GAMMA}\left(\alpha^{(i)}, \lambda^{(i)} x\right)$$

where $\text{GAMMA}(\alpha, s)$ is the Maple incomplete gamma function $\int_s^\infty \frac{t^{\alpha-1} e^{-t}}{\Gamma(\alpha)} \, dt$.

Appendix 1: Solutions to problems in Chapter 1

Problem 1.1

Use a Monte Carlo method, based upon 1000 random standard normal deviates, to find a 95 % confidence interval for $\int_{-\infty}^{\infty} \exp(-x^2)|\cos x|\mathrm{d}x$. Use the Maple 'with(stats)' command to load the stats package. The function `stats[random,normald](1)` will generate a random standard normal deviate.

Solution

Let $x = y/\sqrt{2}$. Then the integral becomes

$$I = \frac{1}{\sqrt{2}} \int_{-\infty}^{\infty} e^{(-y^2/2)} \left|\cos\left(\frac{y}{\sqrt{2}}\right)\right| \mathrm{d}y = \sqrt{\pi} E_f\left(\left|\cos\left(\frac{Y}{\sqrt{2}}\right)\right|\right)$$

where f is $N(0,1)$.

```
> restart;with(stats);
```

$[anova, describe, fit, importdata, random, statevalf, statplots, transform]$

```
> randomize(135);n:=1000;
```
$$135$$
$$n := 1000$$

```
> u:=evalf(sqrt(2)):v:=evalf(sqrt(Pi)):
  for i from 1 to n do:
     y:=stats [random,normald](1);
     a[i]:=evalf(abs(cos(y/u)));
  end do:
```

```
> c:=seq(a[i],i=1..n):
```

```
> mu:=v* describe[mean]([c]);
```

$$\mu := 1.389072868$$

```
> s:=v* describe[standarddeviation[1]]([c]);
```

$$s := 0.4421267288$$

```
> interval:= evalf([mu-1.96*s/sqrt(n), mu+1.96*s/sqrt(n)]);
```

$$interval := [1.361669569, \ 1.416476167]$$

Problem   1.2

Use a Monte Carlo method to find a 95% confidence interval for

$$\int_{-\infty}^{\infty} \int_{-\infty}^{\infty} \exp\left\{-0.5\left[x^2 + (y-1)^2 - \frac{x(y-1)}{10}\right]\right\} dx \ dy.$$

Solution

Put $z = y - 1$. Then the integral is $2\pi E_f[\exp(-XZ/20)]$ where X and Z are i.i.d. $N(0,1)$. We will sample 500 values of (X, Z).

```
> restart;
```

```
> with(stats);
```

$[anova, describe, fit, importdata, random, statvalf, statplots, transform]$

```
> seed:=randomize(567);
  u:=evalf(sqrt(2)):n:=500;
  for i from 1 to n do:
    x:=stats[random,normald](1);z:=stats[random,
    normald](1):
    a[i]:=evalf(exp(0.05*x*z));
  end do:
```

$$seed := 567$$
$$n := 500$$

```
> c:=seq(a[i],i=1..n):
```

```
> mu:=describe[mean]([c]);
```

$$\mu := 1.003656281$$

```
> s:=describe[standarddeviation[1]]([c]);
```

$$s := 0.05394465324$$

```
> interval :=evalf([2*Pi*(mu-1.96*s/sqrt(n)),
  2*Pi*(mu+1.96*s/sqrt(n))]);
```

$$interval := [6.276448626,\ 6.335868170]$$

Problem   1.3

A machine tool is to be scrapped 4 years from now. The machine contains a part that has just been replaced. It has a life distribution with a time-to-failure density $f(x) = x\,e^{-x}$ on support $(0, \infty)$. Management must decide upon one of two maintenance strategies. The first is to replace the part whenever it fails until the scrapping time. The second is to replace failures during the first two years and then to make a preventive replacement two years from now. Following this preventive replacement the part is replaced on failures occurring during the second half of the 4 year span. Assume that replacements are instantaneous and cost c_f on failure and c_p on a preventive basis. Simulate 5000 realizations of 4 years for each policy and find a condition on c_p/c_f for preventitive replacement to be the preferred option.

Solution

Firstly, we simulate the number of failures during 5000 of these 4 year periods.

```
> restart;
```

```
> n:=5000;
```
$$n := 5000$$

```
> nf:=0:
  randomize(134):
  for j from 1 to n do;
     t:=0;
     do;
          r1:=evalf(rand()/10^12);# rand() samples integers
     ~U[1,10^12-12]
          r2:=evalf(rand()/10^12);
          x:=-ln(r1*r2);
          t:=t+x;
          if t>4 then break end if;
          nf:=nf+1;
        end do:
     end do:
     printf("nf=%d",nf);
  nf=8851
```

For the second strategy we obtain the number of failures during 10 000 periods, each of 2 years duration.

```
> n:=10000;
```
$$n := 10000$$

```
> nf:=0:
  randomize(134):
  for j from 1 to n do;
     t:=0;
     do;
          r1:=evalf(rand()/10^12);
          r2:=evalf(rand()/10^12);
          x:=-ln(r1*r2);
          t:=t+x;
          if t>2 then break end if;
          nf:=nf+1;
        end do:
     end do:
  printf("nf=%d",nf);
  nf=7577
```

For the first strategy the expected cost over 4 years is $c_f 8851/5000$ and for the second it is $c_p + c_f 7577/5000$. Therefore, it is estimated that preventive replacement is better when $c_p/c_f < 637/2500$.

Problem 1.4

Two points A and B are selected randomly in the unit square $[0, 1]^2$. Let D denote the distance between them. Using Monte Carlo:

(a) Estimate $E(D)$ and $Var(D)$.

(b) Plot an empirical distribution function for D.

(c) Suggest a more efficient method for estimating $P(D > 1.4)$, bearing in mind that this probability is very small.

Solution

(a)

```
> restart;
```

```
> randomize(462695):#specify a seed
```

```
> distance:=proc(n) local j,x1,x2,y1,y2,d;
    for j from 1 to n do;
        x1:=rand()/10^12;
        y1:=rand()/10^12;
        x2:=rand()/10^12;
        y2:=rand()/10^12;
        d[j]:=sqrt((x1-x2)^2+(y1-y2)^2);
    end do;
    seq(d[j],j=1..n);
    end proc:
```

```
> n:=1000;
```
$$n := 1000$$

```
> f:=evalf(distance(n)):
```

```
> with(stats);
```

[*anova, describe, fit, importdata, random, statevalf, statplots, transform*]

```
> mean:=evalf(describe[mean]([f]));
```

$$mean := 0.5300028634$$

```
> stddev:=evalf(describe[standarddeviation[1]]([f]));
```

$$stddev := 0.2439149391$$

```
> std_error_of_mean:=evalf(stddev/sqrt(n));
```

$$std\_error\_of\_mean := 0.007713267629$$

(b)

```
> d:=transform[statsort]([f]):#sorts the data. d[j] is
  now the j th. order statistic for distance
```

```
> for j from 1 to n do:#constructs the empirical
  distribution function
      h[j]:=[d[j],evalf(j/(n+1))]:
  end do:
  e:=seq(h[j],j=1..n):
```

```
> PLOT(CURVES([e]),TITLE("Empirical
  c.d.f."),AXESLABELS("distance","prob."),
  AXESSTYLE(BOX));
```

Empirical c.d.f.

(c)

Since the maximum possible value of D is $\sqrt{2} = 1.41421$, the probability, p say, that a distance exceeds 1.4 will be extremely small. Therefore the standard error of the estimate for p using naive Monte Carlo will be approximately $\sqrt{p/n}$ and the coefficient of variation $\sqrt{1/(np)}$, giving very low precision.

To motivate a more efficient scheme set up Cartesian coordinates at the centre of the square with the square vertices at, $(-1/2, 1/2), (1/2, 1/2), (1/2, -1/2), (-1/2, -1/2)$. Construct a circle of radius 1.4, centre $(-1/2, -1/2)$. It cuts that portion of the square in the first quadrant at $(\sqrt{1.4^2 - 1} - 1/2, 1/2)$ and $(1/2, \sqrt{1.4^2 - 1} - 1/2)$. Repeat for the remaining three vertices of the square. This construction gives four truncated corner sections formed from the four arcs and the square. Observe that a necessary condition for $D > 1.4$ is that A and B lie in opposite truncated corner sections. Now consider an isosceles triangle containing the top right truncated sector and having two sides identical to it. The equation of the third side is $x + y = \sqrt{1.4^2 - 1}$. Similarly, equations for the third side of the three remaining isosceles triangles, constructed in a similar manner, are $x - y = \sqrt{1.4^2 - 1}$, $x + y = -\sqrt{1.4^2 - 1}$, and $x - y = -\sqrt{1.4^2 - 1}$. It follows that $D < 1.4$ whenever either A or B lies in the region $Q = \{(x, y) : |x + y| < \sqrt{1.4^2 - 1}, |x - y| < \sqrt{1.4^2 - 1}\}$. Therefore, we may *condition* the Monte Carlo on that part of the square that does not include Q. This gives a phenomenal reduction in variance.

Problem 1.5

An intoxicated beetle moves over a cardboard unit circle $x^2 + y^2 < 1$. The (x, y) plane is horizontal and the cardboard is suspended above a wide open jar of treacle. In the time interval $[t, t + \delta t)$ it moves by amounts $\delta x = Z_1 \sigma_1 \sqrt{\delta t}$ and $\delta y = Z_2 \sigma_2 \sqrt{\delta t}$ along the x and y axes where Z_1 and Z_2 are independent standard normal random variables and σ_1 and σ_2 are specified positive constants. The aim is to investigate the distribution of time until the beetle arrives in the treacle pot starting from the point (x_0, y_0) on the cardboard.

(a) Write a procedure that simulates n independent times between starting at the point (x_0, y_0) and landing in the treacle. The function `stats [random, normald] (1)` creates a random standard normal deviate.

(b) Plot a histogram showing the distribution of 200 such times when $\sigma_1 = \sigma_2 = 1$, $\delta t = 0.01$, and $x_0 = y_0 = 0$. To create a histogram, load the subpackage 'statplots' using `with(statplots)` and use the function `histogram(a)` where a is a *list* of the 200 times.

Solution

(a)

```
> restart;
```

```
> with(stats);
```

[*anova, describe, fit, importdata, random, statevalf, statplots, transform*]

```
> beetle:=proc (x0,y0,sigma1,sigma2,h,n,seed) local
  x,y,g,i,t,time;
  # 'h' is the time increment (deltat), 'n' is the number
  of realisations (falls), 't' is current time,'time[i]'
  is the time to the ith. fall.
  g:=sqrt(h):randomize(seed):
  for i from 1 to n do:
    t:=0:x:=x0:y:=y0:
    while x^2+y^2<1 do:
      x:=x+sigma1*g*stats[random,normald](1):
      y:=y+sigma2*g*stats[random,normald](1):
      t:=t+h:
    end do:
    time[i]:=t:
  end do:
  seq(time[i],i=1..n):
  end proc:
```

```
> x0:=0:y0:=0:sigma1:=1:sigma2:=1:h:=0.01:n:=200:
  seed:=6345:
```

```
> a:=beetle(x0,y0,sigma1,sigma2,h,n,seed);
a :=0.71, 0.18, 0.19, 0.59, 0.34, 0.97, 1.24, 0.52,
0.31, 0.59, 0.17, 0.45, 0.55, 0.38, 0.14, 0.18, 0.26,
0.16, 0.43, 0.23, 0.38, 0.69, 0.56, 0.68, 0.40, 0.52,
0.35, 0.40, 0.95,
0.23, 0.64, 0.89, 0.48, 0.27, 0.35, 0.43, 0.27, 0.15,
0.28, 0.23, 1.27, 0.13, 0.35, 0.23,
1.79, 0.97, 0.82, 0.34, 0.42, 0.19, 0.31, 0.92, 1.17,
0.52, 0.20, 0.82, 0.43, 0.88, 0.25,
```

```
0.33, 0.44, 0.38, 0.34, 0.62, 0.41, 0.27, 0.37, 0.72,
0.45, 0.59, 0.66, 0.32, 0.28, 1.25,
0.39, 0.36, 0.28, 0.36, 0.19, 0.47, 0.53, 0.59, 0.33,
0.36, 0.10, 0.24, 0.11, 0.20, 0.20,
0.54, 0.72, 0.60, 1.39, 0.32, 0.19, 0.35, 0.39, 0.28,
0.33, 0.30, 1.01, 0.63, 1.47, 0.35,
0.31, 0.34, 0.64, 0.26, 1.51, 0.16, 1.98, 0.44, 0.38,
0.29, 0.84, 0.34, 0.64, 0.36, 0.63,
0.92, 0.28, 0.65, 0.21, 0.18, 0.14, 0.67, 0.25, 0.16,
0.13, 0.77, 0.46, 0.67, 0.47, 0.22,
0.49, 0.16, 0.08, 0.85, 0.38, 0.23, 0.94, 0.83, 0.45,
0.87, 0.22, 0.99, 0.20, 0.64, 1.39,
0.62, 0.22, 0.15, 2.21, 0.76, 0.80, 0.97, 1.06, 0.26,
0.47, 0.13, 0.28, 0.18, 0.57, 1.35,
0.61, 0.48, 0.14, 0.16, 0.32, 0.29, 0.32, 0.48, 1.14,
0.31, 0.40, 0.66, 0.21, 0.91, 0.18,
0.97, 0.48, 0.34, 0.20, 1.60, 1.82, 0.54, 0.65, 0.41,
0.74, 0.26, 0.76, 0.24, 0.61, 0.58,
0.50, 0.33, 0.41, 0.68, 1.26, 0.82
```

(b)

```
> with(statplots):

Warning, these names have been redefined: boxplot,
histogram, scatterplot, xscale, xshift, xyexchange,
xzexchange, yscale, yshift, yzexchange, zscale, zshift
```

```
> histogram([a],title="Histogram of time to
  fall",labels=[time,density],axes=BOXED);#produces
  rectangles of equal area
```

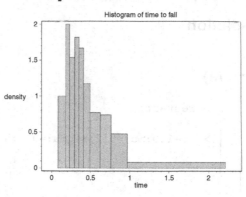

Problem 1.6

The following binomial model is frequently used to mimic share price movements. Let S_i denote the price at time ih where $i = 0, 1, 2, \ldots$ and h is a positive time increment. Let μ and σ denote the growth rate and volatility respectively. Let

$$u = \frac{1}{2}(e^{-\mu h} + e^{(\mu+\sigma^2)h}) + \frac{1}{2}\sqrt{(e^{-\mu h} + e^{(\mu+\sigma^2)h})^2 - 4}, \quad v = u^{-1}, \quad p = \frac{e^{\mu h} - v}{u - v}.$$

Then

$$S_i = X_i S_{i-1}$$

where $X_i, i = 0, 1, \ldots$, are independent Bernoulli random variables with distribution $P(X_i = u) = p$, $P(X_i = v) = 1 - p$ for all i.

(a) Simulate the price at the end of each week during the next year when $S_0 = 100$ pence, $\mu = 0.2$ per annum, $\sigma = 0.3$ per annum, and $h = 1/52$ years.

(b) Now suppose there are 252 trading days in a year. Put $h = 1/252$. For any realization let $S_{\max} = \max(S_j : j = 0, \ldots, 756)$. Let $loss = S_{\max} - S_{756}$. $loss$ denotes the difference between selling the share at the peak value during the next 3 years and selling it after 3 years. Simulate 200 realizations of $loss$ and construct an empirical distribution function for it. You will need to sort the 200 values. Do this by loading the 'stats' package and using the function **transform [statsort] (x)** where **x** is a list of the data to be sorted. Note that if the order statistics to $loss$ are $x_{(1)}, \ldots, x_{(n)}$ then an unbiased estimate of $P(X \le x_{(i)})$ is $i/(n+1)$.

Solution

(a)

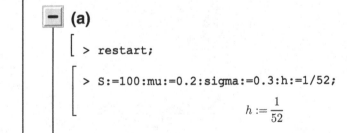

```
> restart;

> S:=100:mu:=0.2:sigma:=0.3:h:=1/52;
```

$$h := \frac{1}{52}$$

[Compute u, ν, p.

```
> u:=0.5*(exp(-mu*h)+exp((mu+sigma^2)*h))
  +0.5*sqrt((exp(-mu*h)+exp((mu+sigma^2)*h))^2-4);
```

$$u := 1.042763680$$

```
> v:=1/u;
```

$$v := 0.9589900561$$

```
> p:=(exp(mu*h)-v)/(u-v);
```

$$p := 0.5355325675$$

```
>
```

[Do loop prints out week, random number, price.

```
> randomize(15725):# set seed
  for j from 1 to 3 do:
    R:=evalf(rand()/10^12):
    if R<p then X:=u else X:=v end if:
    S:=X*S:printf("j=%d R=%f S[j]=%f\n",j,R,S);
    #d prints as integer,f prints in floating point,
    \n gives new line
  end do:

j=1 R=0.174296 S[j]=104.276368
j=2 R=0.465652 S[j]=108.735609
j=3 R=0.630084 S[j]=104.276368
```

[and similarly for the full year.

(b)

```
> n:=200;randomize(5640):
```

$$n := 200$$

```
> h:=1/252;u:=0.5*(exp(-mu*h)+exp((mu+sigma^2)*h))
  +0.5*sqrt((exp(-mu*h)+exp((mu+sigma^2)*h))^2-4);
```

$$h := \frac{1}{252}$$

$$u := 1.019103964$$

```
> v:=1/u;
```

$$v := 0.9812541559$$

```
> p:= (exp(mu*h)-v)/(u-v);
```

$$p := 0.5162459489$$

The following execution group computes loss[i], which is the maximum share price during the three years minus the price on the last day of the three years, on the ith realization, where there are n realizations in all. This represents the loss in selling the share on the last day as opposed to selling it at its maximum price.

```
> for i from 1 to n do:
  S:=100:smax:=S:
  for j from 1 to 756 do:
      R:=evalf(rand()/10^12):
      if R<p then X:=u else X:=v end if:
      S:=X*S:
      if S>smax then smax:=S end if:
  end do:
  loss[i]:=smax-S:
  end do:
```

[Form a sequence a, comprising the n losses.

```
> a:=seq(loss[i],i=1..n):
```

[Sort the losses.

```
> with(stats);d:=transform[statsort]([a]):#sorts the
  data. d[i] is now the i th. order statistic
```

[anova, describe, fit, importdata, random, statevalf, statplots, transform]

Construct a sequence e, each element of the sequence giving the rank order and the cumulative probability.

```
> for i from 1 to n do:#constructs the empirical
  distribution function
      h[i]:=[d[i],evalf(i/(n+1))]:
  end do:
  e:=seq(h[i],i=1..n):
```

```
> PLOT(CURVES([e]),TITLE("Empirical
    c.d.f."),AXESLABELS("loss","prob."),AXESSTYLE(BOX));
```

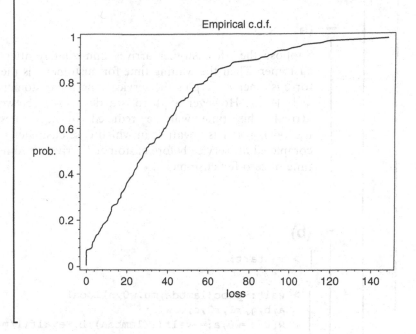

Problem 1.7

Consider a single server queue. Let a_i denote the interarrival time between customer $i - 1$ and customer i, s_i the service time of customer i, and w_i the waiting time in the queue (i.e. the time between arrival and start of service) for customer i.

(a) Show that $w_i = \max(0, \ w_{i-1} - a_i + s_{i-1})$.

(b) Now consider an M/M/1 queue in which the arrival rate is λ and the service rate is μ. Write a procedure that simulates $w_1, \ldots, w_n$ given w_0, λ, and μ.

(c) Experiment with different values for the traffic intensity λ/μ, plotting w_i against i to demonstrate queues that achieve stationary behaviour (i) quickly, (ii) slowly, and (iii) never. In cases (i) and (ii) provide point estimates of the expectation of w_i in the steady state.

Solution

 (a)

Suppose the ith customer arrives immediately after the $(i-1)$th customer. Then the waiting time for customer i is the waiting time for customer $i-1$ plus the service time for customer $i-1$, that is $w_{i-1} + s_{i-1}$. However, if there is a delay of a_i between these two arrivals this time will be reduced by a_i, unless the result $w_{i-1} - a_i + s_{i-1}$ is negative, in which case customer $i-1$ will have completed his service before customer i arrives, leading to a waiting time of zero for customer i.

(b)

```
> restart;

> wait:=proc(lambda,mu,w0,n) local
    a,b,j,r1,r2,c,d,w,u:
  w[0]:=w0;a:=evalf(1/lambda);b:=evalf(1/mu);
  for j from 1 to n do;
      r1:=evalf(rand()/10^12);
      r2:=evalf(rand()/10^12);
      c:=-a*ln(r1):d:=-b*ln(r2);
      w[j]:=max(w[j-1]-c+d,0);
      u[j]:=[j,w[j]];
  end do:
  [seq(u[j],j=1..n)]:
  end proc:
```

(c)

```
> with(stats):with(plots):
Warning, the name changecoords has been redefined.
```

 (i) Low traffic intensity, $\lambda/\mu = 0.5$

```
> randomize(89347);w0:=0;lambda:=0.5;mu:=1;
  n:=1000;v:=wait
```

```
(lambda,mu,w0,n):#rates are in customers per
minute
```

$$89347$$
$$w0 := 0$$
$$\lambda := 0.5$$
$$\mu := 1$$
$$n := 1000$$

```
> PLOT(CURVES(v),TITLE("waiting times"),AXESLABELS
("customer number","mins."),AXESSTYLE(BOX));
```

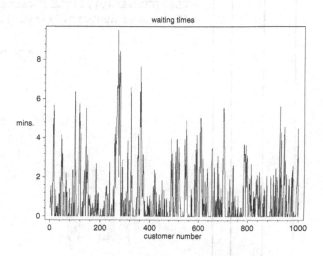

The plot suggests that stationarity is achieved almost immediately. This is as expected for a queue with low traffic intensity, λ/μ. Therefore the entire sample record is used to estimate the mean waiting time under stationarity.

```
> waittimes:=seq(op(2,v[i]),i=1..1000):
  mean_waiting_time=evalf(describe[mean]
  ([waittimes]));
  standard_deviation_of_waiting_time=evalf
  (describe[standard.
  deviation[1]]([waittimes]));
```

$$mean\_waiting\_time = 0.9375110082$$
$$standard\_deviation\_of\_waiting\_time = 1.478717653$$

The theoretical stationary mean waiting time is 1. The difference from the estimate is due to sampling variation.

(ii) High traffic intensity, $\lambda/\mu = 0.99$

```
> randomize(89347);w0:=0;lambda:=0.99;mu:=1;
  n:=1000; v:=wait (lambda,mu,w0,n):
```

$$89347$$
$$w0 := 0$$
$$\lambda := 0.99$$
$$\mu := 1$$
$$n := 1000$$

```
> PLOT(CURVES(v),TITLE("waiting times"),AXESLABELS
  ("customer number","mins."),AXESSTYLE(BOX));
```

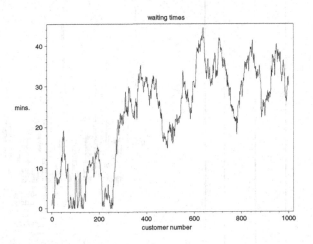

The plot suggests that perhaps 1000 or so customers have to arrive before stationarity is achieved. We will need to increase the number of customers to, say, 10 000, with a burn-in time of at least, say, 1000 customers, to obtain an estimate of stationary mean waiting times.

```
> randomize(89347);w0:=0;lambda:=0.99;mu:=1;
  n:=10000;v:=wait (lambda,mu,w0,n):
```

$$89347$$
$$w0 := 0$$
$$\lambda := 0.99$$
$$\mu := 1$$
$$n := 10000$$

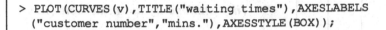

```
> PLOT(CURVES(v),TITLE("waiting times"),AXESLABELS
  ("customer number","mins."),AXESSTYLE(BOX));
```

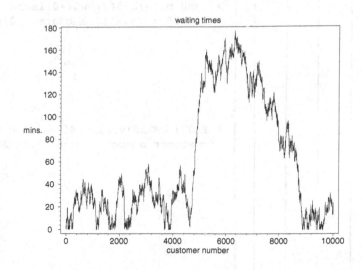

```
> waittimes:=seq(op(2,v[i]),i=1000..10000):
  mean_waiting_time=evalf(describe[mean]
  ([waittimes]));
  standard_deviation_of_waiting_time=evalf
  (describe[standard
  deviation[1]]([waittimes]));
```

$$mean\_waiting\_time = 65.42348523$$
$$standard\_deviation\_of\_waiting\_time = 54.91974974$$

The theoretical mean waiting time under stationarity is 99. Comparison with the estimate indicates that a large sampling variation is present, even with this large sample size of 9000. This is due to the large positive autocorrelation between the individual waiting times. It also seems likely that we have underestimated the burn-in time that is required. Note that even if this sample represents stationary behaviour we cannot say that the standard error of the estimate of mean waiting time is $54.92/\sqrt{9000}$ because of the correlation. It is much larger than this.

— (iii) Stationarity not achievable $(\lambda/\mu = 1.2)$

```
> randomize(89347);w0:=0;lambda:=1.2;mu:=1;
  n:=1000;v:=wait(lambda,mu,w0,n):
```

$$89347$$
$$w0 := 0$$
$$\lambda := 1.2$$
$$\mu := 1$$
$$n := 1000$$

```
> PLOT(CURVES(v),TITLE("waiting times"),AXESLABELS
  ("customer number","mins."),AXESSTYLE(BOX));
```

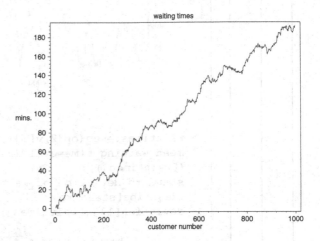

Note how the trend is for queue length to increase progressively, since the arrival rate exceeds the service rate.

Appendix 2: Random number generators

In addition to the in-built Maple generator 'rand', the following generators described in Chapter 2 can also be used.

```
> r1:= proc() global seed;
  seed:=(906185749 * seed+1) mod 2^31;
  evalf(seed/2^31);
  end proc:
```

```
> r2:= proc() global seed;
  seed:=(2862933555777941757 * seed+1) mod 2^64;
  evalf(seed/2^64);
  end proc:
```

```
> schrage:= proc() local s,r; global seed;
  s:= seed mod 62183;r:=(seed-s)/62183;
  seed:=-49669 * r+69069 * s+1;
  if seed<0 then seed:= seed + 2^32 end if;
  evalf(seed/2^32);
  end proc:
```

```
> r3:= proc() global seed;
  seed:=(seed * 630360016) mod (2^31-1);
  evalf(seed/(2^31-1));
  end proc:
```

Note that 'seed' is declared as a global variable in each of these. Therefore, any procedures calling these generators will likewise have to declare 'seed' as a global variable. The procedure 'expon' below generates a variate from the density $f(x) = e^{-x}$ on support $[0, \infty)$. It calls the random number generator 'schrage'. The final execution group sets an (initial) seed and calls 'expon' three times.

```
> expon:= proc() global seed;
  -ln(schrage());
  end proc:
```

```
> seed:= 47139;
  for j from 1 to 3 do;
  expon();
  end do;
```

$$seed := 47139$$
$$0.2769925308$$
$$0.7827805984$$
$$0.3164070444$$

Appendix 3: Computations of acceptance probabilities

3.1 Computation of acceptance probabilities for arbitrary envelope rejection generators

The procedure 'envelopeaccep' is used to compute the acceptance probability for generating variates from a p.d.f. proportional to $h(x)$ using an envelope (and therefore proposal density) proportional to $r(x)$. The parameters are:

h = a procedure supplied by the user,

r = a procedure supplied by the user,

$[x_1, x_2]$ = the connected support of h,

xinit = a suggested start point for the numerical maximization of $h(x)/r(x)$ over $[x_1, x_2]$.

```
> restart;with(Optimization):
```

```
> envelopeaccep:=proc(h::procedure,r::procedure,x1,x2,xinit)
  local u,K,acc;
  u:=NLPSolve(h(x)/r(x),
  x=x1..x2,initialpoint={x=xinit},maximize);
  K:=op(1,u);
  print("envelope is",K*r(x));
  acc:=int(h(x),x=x1..x2)/int(r(x),x=x1..x2)/K;
  end proc:
```

Now the acceptance probability will be found when generating from a generalized inverse Gaussian distribution having a p.d.f. proportional to

$x^{(\lambda-1)}e^{-\beta(x+1/x)/2}$ on support $(0,\infty)$ using a proposal density proportional to $x^{(\lambda-1)}e^{-\gamma x/2}$ for selected values of λ and β. The envelope has been optimized by setting

$$\gamma = \frac{2\lambda^2\left(\sqrt{1+\beta^2/\lambda^2}-1\right)}{\beta}$$

(see Section 4.8 of the text).

```
> lambda:=5;
  beta:=10;
  gam:=2*lambda^2*(sqrt(1+beta^2/lambda^2)-1)/beta;
```
$$\lambda := 5$$
$$\beta := 10$$
$$gam := 5\sqrt{5} - 5$$

```
> ha:=proc(x) global alpha,beta;
  x^(lambda-1)*exp(-beta*(x+1/x)/2);
  end proc;
```
$$ha := \mathbf{proc}(x)\ \mathbf{global}\ \alpha,\beta;x^{\wedge}(\lambda-1)*\exp(-1/2*\beta*(x+1/x))\ \mathbf{end\ proc}$$

```
> ra:=proc(x) global alpha,gam;
  x^(lambda-1)*exp(-gam^*x/2);
  end proc;
```
$$ra := \mathbf{proc}(x)\ \mathbf{global}\ \alpha,\ gam;x^{\wedge}(\lambda-1)*\exp(-1/2*gam*x)\ \mathbf{end\ proc}$$

```
> acceptance_prob:=evalf(envelopeaccep(ha, ra, 0, infinity,
  1));
```
$$\text{'envelope is'},\ 0.00206972426167299210\ x^4 e^{-(5\sqrt{5}-5)x/2}$$
$$acceptance\_prob := 0.6528327067$$

3.2 Computation of acceptance probabilities for arbitrary ratio of uniforms generators

The procedure 'ratioaccep' is used to compute the acceptance probability for a ratio of uniforms generator for a p.d.f. proportional to $h(x)$. The parameters are:

$h =$ a procedure supplied by the user,

$[x_1, x_2] =$ the connected support of h,

$xinit =$ a suggested start point for the numerical maximization of $h(x)$ over $[x_1, x_2]$, *xinit1* = a suggested start point for the numerical maximization of $x^2h(x)$ over $[x_1, 0]$,

xinit2 = a suggested start point for the numerical maximization of $x^2h(x)$ over $[0, x_2]$.

```
> restart;with(Optimization):
```

```
>
  ratioaccep:=proc(h::procedure,x1,x2,xinit,xinit1,xinit2)
  local u,u1,v1,v11,v2,v22,acc;
  u:=NLPSolve(h(x),
  x=x1..x2,initialpoint={x=xinit},maximize);u1:=sqrt(op(1,u));
  if x1<0 then
     v1:=NLPSolve(x*x*h(x),
  x=x1..0,initialpoint={x=xinit1},maximize);
     v11:=-sqrt(op(1,v1));
     else v11:=0
  end if;
  if x2>0 then
     v2:=NLPSolve(x*x*h(x),
  x=0..x2,initialpoint={x=xinit2},maximize);
     v22:=sqrt(op(1,v2));
     else v22:=0;
  end if;
  print("u+"=u1);
  print("v-"=v11);
  print("v+"=v22);
  acc:=int(h(x),x=x1..x2)/u1/(v22-v11)/2;
  acc;
  end proc:
```

Use 'ratioaccep' to compute the acceptance probabilities and $u+, \nu+, \nu-$ for a density proportional to $1/(1 + x^2/3)^2$ on support $(-\infty, \infty)$.

```
> h1:=proc(x);
  1/(1+x*x/3)^2;
  end proc;
              h1:=proc(x)1/(1+x*x*1/3)^2 end proc
```

```
> acceptance_prob:=ratioaccep(h1,-infinity,infinity,0,-2,2);
```

```
Warning, no iterations performed.
                'u+'=1.
            'v-'=-0.8660254038
            'v+'=0.8660254038
       acceptance_prob:=0.1443375672\pi\sqrt{3}
```

The warning arises because the initial point chosen for the maximization of h over the entire support happens to be the maximizing point.

Now repeat for a density proportional to $\left(\frac{1}{4} - x^2\right)^9$ on support $\left(-\frac{1}{2}, \frac{1}{2}\right)$.

```
> h2:=proc(x);
  (1/4-x*x)^9;
  end proc;
```

$$h2 := \mathbf{proc}(x)(1/4 - x*x)^\wedge 9 \ \mathbf{end \ proc}$$

```
> acceptance_prob:=ratioaccep(h2,-1/2,1/2,0,-1/4,1/4);
```

$$'u+' = 0.001953125000$$
$$'v-' = -0.0001922167969$$
$$'v+' = 0.0001922167969$$
$$acceptance\_prob := 0.7208585900$$

Appendix 4:
Random variate generators
(standard distributions)

The following variate generation procedures appear to be faster than the proprietary Maple functions and are therefore recommended. Note that the proprietary Maple beta and negative binomial generators do not accept noninteger shape parameters, whereas the corresponding procedures below do so. In all cases faster generators than the ones listed below could be devised. However, these ones are selected for their brevity and reasonable efficiency. No attempt has been made to save parameter values between successive calls of the same generator. Doing so would improve the efficiency when the parameter(s) of the distribution remains unchanged between successive calls.

 4.1   Standard normal generator

This is a polar implementation of the Box–Müller generator, as described in Section 4.1. Note the warning to set i and X_2 to global variables and to set i to 'false' on the first call.

```
> STDNORM:=proc( ) local ij,U1,U2,S,B,X1;global i,X2;

  #
  #      PROCEDURE GENERATES A STANDARD RANDOM NORMAL
  #      DEVIATE, USING THE POLAR BOX MULLER METHOD.
  #      SET i to 'false' ON FIRST CALL. Note that i and
```

```
#       X2 are global variables and have to be declared as
#       such in any procedure that calls STDNORM.
if type(i,boolean) then
if (i) then
    i:=not(i);X2;
else
    for ij from 1 to infinity do
        U1:=evalf(2.0*rand()/10^12)-1.0;
        U2:=evalf(2.0*rand()/10^12)-1.0;
        S:=U1*U1+U2*U2;
        if(S>=1.0 or S<=0.0) then next end if;
        B:=sqrt(-2.0*ln(S)/S);
        X1:=B*U1;
        X2:=B*U2;
        break;
    end do;
    i:=not(i);X1;
end if;
else ERROR("i should be boolean") end if;
end proc:
```

4.2 Beta generator

This is an implementation of Cheng's log-logistic generator as described in Section 4.5.1. It generates a beta(a,b) variate.

```
> beta:=proc(a,b) local z,r1,r2,w,rho,m,la;
  if a>0 and b>0 then
    rho:=a/b;m:=min(a,b);if m<=1 then la:=m else
  la:=sqrt((2*a*b-a-b)/(a+b-2)) end if;
    do;
    r1:=rand()/10^12;
    r2:=rand()/10^12;
    z:=(1/r1-1)^(1/la);
    if
  evalf(4*r1*r1*r2-evalf(z^(a-la)*((1+rho)/(1+rho*z))^
  (a+b)))<0 then break end if;
    end do;
    evalf(rho*z/(1+rho*z));
  else ERROR("a and b should be positive") end if;
  end proc:
```

4.3   Student's t distribution

This procedure generates a Student's *t* variate with *n* degrees of freedom using the method described in Section 4.7. If $n = 1$ then a Cauchy variate is delivered.

```
> tdistn:=proc(n) local r1,r2,c,x;
  if type(n,posint)then;
     r1:=evalf(rand()/10^12);
     if n>1 then
        r2:=evalf(rand()/10^12);
        c:=cos(6.2831853071795864770*r2);
        sqrt(n/(1/(1-r1^(2/(n-1)))-c^2))*c;
     else
        tan(1.5707963267948966192*(r1+r1-1));
     end if;
  else
  ERROR("n should be positive integer") end if;
  end proc:
```

4.4   Generalized inverse Gaussian

This procedure generates a variate from a p.d.f. that is proportional to $x^{\lambda-1}e^{-\beta(x+1/x)/2}$ on support $(0, \infty)$ where $0 < \lambda$ and $0 < \beta$. The method is described in Section 4.8.

```
> geninvgaussian:=proc(lambda,beta) local
  gam,a1,a2,a3,a4,r,j,x;
  if lambda>0 then
  if beta>0 then
    gam:=2*lambda^2*(sqrt(1+(beta/lambda)^2)-1)/beta;
    a1:=0.5*(beta-gam);
    a2:=0.5*beta;
    a3:=sqrt(beta*(beta-gam));
    a4:=2.0/gam;
    for j from 1 to infinity do;
      r:=evalf (rand()/10^12);
      x:=random[gamma[lambda,a4]] (1);
      if -ln(r)>x*a1+a2/x-a3 then break end if;
    end do;
  else ERROR("beta must be positive") end if;
  else ERROR("lambda must be positive") end if;
  end proc:
```

4.5   Poisson generator

This procedure uses unstored inversion of the c.d.f. as described in Section 4.9. It generates a Poisson (μ) variate.

```
> ipois1:=proc(mu) local y,x,r;
  if mu>=0 then
    y:=evalf(exp(-mu));
    x:=0;
    r:=evalf(rand()/10^12);
    while r>y do;
        x:=x+1;
        r:=r-y;
        y:=mu*y/x;
    end do;
    x;
  else ERROR("mu should be non-negative") end if;
  end proc:
>
```

4.6   Binomial

This procedure uses unstored inversion of the c.d.f. as described in Section 4.10. It generates a binomial (n, p) variate.

```
> ibinom:=proc(n,p) local ps,q,r,y,x,qs;
  if type(n,posint) then
  if p>=0 and p<=1 then
    if p>0.5 then ps:=1-p else ps:=p; end if;
    q:=1-ps;
    qs:=ps/q;
    y:=evalf(q^n);
    r:=evalf(rand()/10^12);
    x:=0;
    while r>y do;
        x:=x+1;
        r:=r-y;
        y:=evalf(y*(n-x+1)*qs/x);
    end do;
    if p>0.5 then n-x else x end if;
  else ERROR("p should belong to[0,1]") end if;
  else ERROR("n should be a positive integer") end if;
  end proc:
```

4.7   Negative binomial

This procedure generates a variate from a p.m.f. where

$$f(x) = \frac{\Gamma(x+k)\, p^x\, q^k}{\Gamma(k)\, \Gamma(x+1)} \quad (x = 0, 1, 2, \ldots)$$

k is a positive real. It uses unstored inversion as mentioned in Section 4.11. It is fast providing that kp/q is not too large. In that case the Poisson gamma method described in Section 4.11 is better.

```
> negbinom:=proc(k,p) local q,r,y,x;
  if k>0 then
  if p>=0 and p<=1 then
    q:=1-p;
    y:=evalf(q^k);
    r:=evalf(rand()/10^12);
    x:=0;
    while r>y do;
        x:=x+1;
        r:=r-y;
        y:=evalf(y*p*(x+k-1)/x);
    end do;
    x;
  else ERROR("p should belong to[0,1]") end if;
  else ERROR("k should be positive") end if;
  end proc:
```

Appendix 5: Variance reduction

 5.1 Antithetic variates and the estimation of $\Gamma(1.9)$

The procedure 'theta1_2' samples m unit negative exponential variates, and raises each to the power of 0.9. The sample mean of the latter is an estimate of $\Gamma(1.9)$.

```
> restart;with(stats):
```

```
> theta1_2:=proc(m) local j,r,x,y,u;
    for j from 1 to m do;
        r:=evalf(rand()/10^12);
        x:=-ln(r);
        y[j]:=x^0.9;
    end do;
    u:=[seq(y[j],j=1..m)];
    [describe[mean](u),describe[standarddeviation[1]](u)^2/m,u];
    print("theta1_hat"=describe[mean](u));
    print("standard
    error"=evalf(describe[standarddeviation[1]](u)/sqrt(m)));
    u;
    end proc:
```

Compute the estimate and estimated standard error for a sample size of 1000.

```
> seed:=randomize(341):
    print("seed"=seed);
    res1:=theta1_2(1000):
```

Simulation and Monte Carlo: With applications in finance and MCMC   J. S. Dagpunar
© 2007 John Wiley & Sons, Ltd

'seed' $= 341$
'thetal_hat' $= 0.9259670987$
'standard error' $= 0.02797125535$

Now replace r by $1 - r$ and *thetal_hat* by *theta2_hat* in the print statement of procedure 'theta1_2' and run the simulation again with the *same seed*.

```
> seed:=randomize(341):
  print("seed"=seed);
  res2:=theta1_2(1000):
```

'seed' $= 341$
'theta2_hat' $= 1.009153865$
'standard error' $= 0.02773074613$

The procedure 'theta_combined' does the two jobs above in one run, giving an estimate and estimated standard error using primary and antithetic variates.

```
> theta_combined:=proc(m) local j,r,x,y,u,r1,z;
  for j from 1 to m do;
    r:=evalf(rand()/10^12);r1:=1-r;
    x:=-ln(r);z:=-ln(r1);
    y[j]:=(x^0.9+z^0.9)/2;
  end do;
  u:=[seq(y[j],j=1..m)];
  print("mean"=describe[mean](u));
  print("standard
  error"=evalf(describe[standarddeviation[1]](u)/sqrt(m)));
  u;
  end proc:
```

```
> seed:=randomize(341):
  print("seed"=seed);
  res3:=theta_combined(1000):
```

'seed' $= 341$
'mean' $= 0.9675604815$
'standard error' $= 0.01072362734$

Calculate the sample correlation between theta1 and theta2.

```
> rho:=describe[linearcorrelation](res1,res2);
```

$$\rho := -0.7035269031$$

Calculate an estimate of the variance reduction ratio.

```
> 1/(1+rho);
```

$$3.372987332$$

Calculate an alternative estimate of the v.r.r. that avoids the need to find the correlation.

```
> (.2797125535e-1^2+.2773074613e-1^2)/4/.1072362734e-1^2;
```

$$3.372688907$$

Calculate the exact correlation and v.r.r. using numerical integration.

```
> e3:=evalf(int((ln(R)*ln(1-R))^0.9,R=0..1));
```

$$e3 := 0.3909740620$$

```
> e1:=evalf(int((-ln(R))^0.9,R=0..1));
```

$$e1: = 0.9617658319$$

```
> e2:=evalf(int((-ln(R))^1.8,R=0..1));
```

$$e2: = 1.676490788$$

```
> rho:=(e3-e1^2)/(e2-e1^2);
```

$$\rho := -0.7106073074$$

Calculate the exact variance reduction factor.

```
> 1/(1+rho);
```

$$3.455512270$$

5.2 Exceedance probabilities for the sum of i.i.d. random variables

The procedure 'impbeta' estimates $P\{a < \sum_{i=1}^{n} X_i\}$ where $\{X_i\}$ are i.i.d. beta distributed with shape parameters α and β, both greater than one. The importance sampling density is $g(x) = \prod_{i=1}^{n} \gamma x_i^{(\gamma - 1)}$ on support $(0,1)^n$ where $\alpha < \gamma$. A good choice is $\gamma = 1/\log(n/a)$ and a should satisfy $n\, e^{-1/\alpha} \leq a$.

```
> with(stats):

   Warning, these names have been redefined: anova, describe, fit, import-
   data, random, statevalf, statplots, transform

>

> impbeta:=proc(n,m,a,alpha,beta) local
  gam,gam1,s1,s2,j,s,p,k,x,t,prob,stderr,vrf,b,z,mean,std_dev;
  #
  #Procedure finds probability that sum of n i.i.d. beta
  variates (shape parameters alpha>=1,beta>=1) exceeds a.
  #Importance sampling distribution is product of n identical
  distributions having density gam*x^(gam-1) where gam>1.
  #gam(=1/log(n/a)) is chosen to minimize an easily computed
  upper bound on variance.
  #if a<=n*exp(-1/alpha), i.e. if this is not a small right
  tail probability, use naive Monte Carlo.
  #sample size =m
  #computes central limit approximation for comparison
  #
  mean:=n*alpha/(alpha+beta);
  std_dev:=sqrt(n*alpha*beta/(alpha+beta+1)/(alpha+beta)^2);
  print("alpha"=alpha,"beta"=beta,"n"=n,"mean "=mean,"std
  dev"=std_dev);
  if a<evalf(n*exp(-1/alpha)) then print("STOP, use naive monte
  carlo" )
  return end if;
  gam:=evalf(1/ln(n/a));
  gam1:=1/gam;
  b:=(GAMMA(alpha+beta)/GAMMA(alpha)/GAMMA(beta)/gam)^n;
  s1:=0;s2:=0;
  for j from 1 to m do;
      s:=0;p:=1;
      for k from 1 to n do;
         x:=(evalf(rand()/10^12))^gam1;
         s:=s+x;
         p:=p*x^(alpha-gam)*(1-x)^(beta-1);
      end do;
      if s>a then
         t:=p;
         s1:=s1+t;s2:=s2+t^2;
      end if;
  end do;
  prob:=s1*b/m;
  stderr:=b*evalf(sqrt((s2-s1^2/m)/m/(m-1)));
```

```
    vrf:=1/(m*stderr^2/prob/(1-prob));
    print("estimate of probability that sum of",n,"variates
    exceeds",a,"is",prob);
    print("standard error"=stderr);
    print("variance reduction ratio"=vrf);
    z:=(mean-a)/std_dev;
    print("central limit approx"=statevalf[cdf,normald](z));
    end proc:
```

```
> seed:=randomize(99356):
  print("seed"=seed);
  impbeta(12,5000,9,1.5,2.5);
```
$$\text{`seed'} = 99356$$
$$\text{`alpha'} = 1.5, \text{`beta'} = 2.5, \text{`n'} = 12, \text{`mean'} = 4.500000000,$$
$$\text{`std dev'} = 0.7500000000$$
$$\text{`estimate of probability that sum of', 12, `variates exceeds', 9, `is', } 1.146003559 \times 10^{-9}$$
$$\text{`standard error'} = 1.067644898 \times 10^{-10}$$
$$\text{`variance reduction ratio'} = 2.010769567 \times 10^{7}$$
$$\text{`central limit approx'} = 9.865876451 \times 10^{-10}$$

```
> seed:=randomize(99356):
  print("seed"=seed);
  impbeta(12,5000,9,2.5,1.5);
```
$$\text{`seed'} = 99356$$
$$\text{`alpha'} = 2.5, \text{`beta'} = 1.5, \text{`n'} = 12, \text{`mean'} = 7.500000000,$$
$$\text{`std dev'} = 0.7500000000$$
$$\text{`estimate of probability that sum of', 12, `variates exceeds', 9, `is', } 0.01936693724$$
$$\text{`standard error'} = 0.0005737629220$$
$$\text{`variance reduction ratio'} = 11.53805257$$
$$\text{`central limit approx'} = 0.02275013195$$

```
> seed:=randomize(6811357):
  print("seed"=seed);
  impbeta(12,5000,9,2.5,1.5);
```
$$\text{`seed'} = 6811357$$
$$\text{`alpha'} = 2.5, \text{`beta'} = 1.5, \text{`n'} = 12, \text{`mean'} = 7.500000000, \text{`std dev'} = 0.7500000000$$
$$\text{`estimate of probability that sum of', 12, `variates exceeds', 9, `is', } 0.01978833352$$
$$\text{`standard error'} = 0.0005813529138$$
$$\text{`variance reduction ratio'} = 11.47834773$$
$$\text{`central limit approx'} = 0.02275013195$$

```
> seed:=randomize(6811357):
  print("seed"=seed);
  impbeta(24,5000,18,2.5,1.5);
```

'seed' = 6811357
'alpha' = 2.5, 'beta' = 1.5, 'n' = 24, 'mean' = 15.00000000, 'std dev' = 1.060660172
'estimate of probability that sum of', 24, 'variates exceeds', 18, 'is', 0.001761757803
'standard error' = 0.00008414497712
'variance reduction ratio' = 49.67684543
'central limit approx' = 0.002338867496

5.3   Stratified sampling

5.3.1   Estimating

$$\int_0^1 \int_0^1 \left([-\ln (r_1)]^{2/3} + [-\ln(r_2)]^{2/3} \right)^{5/4} dr_1 \ dr_2$$

using naive Monte Carlo

The procedure 'weibullnostrat' gives the estimate and estimated standard
error for a sample of size k.

```
> weibullnostrat:=proc(k) local j,r1,r2,w,v,y,sum1,sum2,m;
  sum1:=0;sum2:=0;
  for m from 1 to k do;
      r1:=evalf(rand()/10^12);
      r2:=evalf(rand()/10^12);
      w:=(-ln(r2))^(2/3);
      v:=(-ln(r1))^(2/3);
      y:=(v+w)^(5/4);
      sum1:=sum1+y;
      sum2:=sum2+y^2;
  end do;
  print("mean"=sum1/k, "std error"=sqrt((sum2-sum1^2/k)/
  (k-1)/k));
  end proc:
```

```
> t1:=time():seed:=randomize(639156):
  weibullnostrat(20000):
  t3:=time()-t1;
```

'mean' = 2.158428985, 'std error' = 0.009127721506
$t3$: = 20.500

5.3.2 Stratified version using a single stratification variable

First plot

$$Y = \left([-\ln (r_1)]^{2/3} + [-\ln(r_2)]^{2/3} \right)^{5/4}$$

against stratified variable $X = r_1 r_2$. This confirms that much of the variation in Y is accounted for by variation in the conditional expectation of Y given X. Therefore, stratification on X will be effective.

```
> with(stats);
Warning, these names have been redefined: anova, describe,
fit, importdata, random, statevalf, statplots, transform
```

```
> strat:=proc(n) local j,r1,r2,x,a1,a2,y,z;
    for j from 1 to n do;
        r1:=evalf(rand()/10^12);
        r2:=evalf(rand()/10^12);
        x[j]:=(r1*r2);
        a1:=(-ln(r1))^(2/3);
        a2:=(-ln(r2))^(2/3);
        y[j]:=(a1+a2)^(1.25);
        z[j]:=[x[j],y[j]];
    end do;
    seq(z[j],j=1..n);
    end proc:
```

```
> seed:=randomize(5903):print("seed"=seed);
    v:=strat(500):
    PLOT(POINTS(v),SYMBOL(POINT),AXESLABELS('X','Y'));
    vx:=seq((op(1,v[j]),j=1..500)):vy:=seq((op(2,v[j]),
    j=1..500)):
    correlation:=describe[linearcorrelation]([vx],[vy]);
    CV_VRF:=1-correlation^2;
```

$$\text{'seed'} = 5903$$

$correlation := -0.8369837074$

$CV\_VRF := 0.2994582735$

The following procedure, 'weibullstrat', performs k independent realizations. Each realization comprises n observations over n equiprobable strata, with exactly one observation per stratum.

```
> weibullstrat:=proc(n,k) local
  j,u1,t,x,u2,r2,r1,y,w,v,tprev,sum1,sum2,mean,s1,m;
  sum1:=0;sum2:=0;
  for m from 1 to k do;
    s1:=0;tprev:=10^(-10);
    for j from 1 to n do;
      u1:=evalf(rand()/10^12); u2:=evalf(rand()/
      10^12);
      t:=fsolve(x-x*ln(x)-(j-1+u1)/n,x=tprev,
      tprev..1);
      tprev:=t;
      r2:=t^u2;w:=(-ln(r2))^(2/3);
      r1:=t/r2;v:=(-ln(r1))^(2/3);
      y:=(v+w)^(5/4);
      s1:=s1+y;
    end do;
    mean:=s1/n;
    sum1:=sum1+mean;sum2:=sum2+mean^2;
  end do;
  print("mean"=sum1/k,"std error"=sqrt((sum2-sum1^2/k)/
  (k-1)/k));
  end proc:
```

```
> t1:=time();seed:=randomize(639156);weibullstrat(100,200);
  t2:=time()-t1;
```

$$t1: = 109.077$$
$$seed: = 639156$$
$$\text{'mean'} = 2.166441095, \text{ 'std error'} = 0.001321419976$$
$$t2: = 110.250$$

```
estimated_vrr:=(.1321419976e-2/.9127721506e-2)^(-2);
estimated_efficiency:=t3*estimated_vrr/t2;
```

$$estimated\_vrr: = 47.71369240$$
$$estimated\_efficiency: = 8.871933734$$

5.3.3 Stratification on two variables

The procedure 'grid' estimates the same integral using k replications where there are now two stratification variables, r_1 and r_2. Each replication comprises n^2 equiprobable strata on $[0, 1]^2$.

```
> grid:=proc(n,k) local j,r1,r2,w,v,y,m,sum1,sum2,k1,s1,
  mean;
  sum1:=0;sum2:=0;
  for k1 from 1 to k do;
      s1:=0;
      for j from 1 to n do;
          for m from 1 to n do;
              r1:=(j-1+evalf(rand()/10^12))/n;
              r2:=(m-1+evalf(rand()/10^12))/n;
              w:=(-ln(r2))^(2/3);
              v:=(-ln(r1))^(2/3);
              y:=(v+w)^(5/4);
              s1:=s1+y;
          end do;
      end do;
  mean:=s1/n/n;
  sum1:=sum1+mean;
  sum2:=sum2+mean^2;
  end do;
  print("mean"=sum1/k,"standard error"=sqrt((sum2-
  sum1^2/k)/k/(k-1)));
  end proc:
> t1:=time();
  seed:=randomize(639156):
  print("seed"=seed);
  grid(10,200);
  t4:=time()-t1;
```

$t1: = 152.688$
'seed' $= 639156$
'mean' $= 2.167096762$, 'standard error' $= 0.002507078421$
$t4: = 21.531$

```
> estimated_vrr:=(.2507078421e-2/.9127721506e-2)^(-2);
  estimated_efficiency:=t3*estimated_vrr/t4;
```

estimated_vrr:=13.25528056

estimated_efficiency:=12.62055880

Appendix 6: Simulation and finance

6.1 Brownian motion

'STDNORM' is a standard normal generator, using the polar Box–Müller method.

```
>       STDNORM:=proc( ) local ij,U1,U2,S,B,X1;global i,X2;
 #
 #      PROCEDURE GENERATES A STANDARD RANDOM NORMAL
 #      DEVIATE, USING THE POLAR BOX MULLER METHOD.
 #      SET i to 'false' ON FIRST CALL. Note that i and
 #      X2 are global variables and have to be declared as
 #      such in any procedure that calls this STDNORM.
        if type(i,boolean)then
        if (i) then
            i:=not(i);X2;
        else
            for ij from 1 to infinity do
                U1:=evalf(2.0*rand()/10^12)-1.0;
                U2:=evalf(2.0*rand()/10^12)-1.0;
                S:=U1*U1+U2*U2;
                if(S>=1.0 or S<=0.0) then next end if;
                B:=sqrt(-2.0*ln(S)/S);
                X1:=B*U1;
                X2:=B*U2;
                break;
            end do;
            i:=not(i);X1;
```

```
            end if;
          else ERROR("i should be boolean") end if;
          end proc:
```

The procedure Brownian generates the position of a standard Brownian motion $\{B(t)\}$ at times $0, h, 2h, \ldots, nh$ where $nh = T$.

```
> Brownian:=proc(T,n,seed) local h,sh,B,P,j,z;
  #
  # Simulation of standard Brownian motion over [0,T] using n
  subintervals each of length h.
  # Procedure generates a list, [{[j*h,B(j*h)],j=0..n}] where
  B(j*h) is the position of the Brownian motion at time j*h.
  #
  global i,X2;
  i:=false;
  randomize(seed);
  h:=T/n;
  sh:=sqrt(h);
  B:=0;
  P[0]:=[0,0];
  for j from 1 to n do;
      z:=STDNORM();
      B:=B+evalf(sh*z);
      P[j]:=[evalf(j*h),B];
  end do;
  [seq(P[j],j=0..n)];
  end proc:
```

```
> u1:=Brownian(100,10000,3671):
  u2:=Brownian(100,10000,4791023):
  u3:=Brownian(100,10000,4591038):
  PLOT(CURVES(u1,u2,u3,COLOR(RGB, 1, 0, 0, 0, 1, 0,0,0,1)),
  TITLE("Three realizations of a standard Brownian
  motion"),AXESLABELS("t","B(t)"),AXESSTYLE(NORMAL));
```

Three realizations of a standard Brownian motion

6.2 Geometric Brownian motion

Let $\{B(t)\}$ denote a standard Brownian motion. The procedure 'Geometric-Brownian' generates the position, $X(t)$, of a Brownian motion

$$\frac{\mathrm{d}X}{X} = \mu\,\mathrm{d}t + \sigma\,\mathrm{d}\beta$$

at times 0, h, $2h$,...,nh where $nh = T$. The solution to the stochastic differential equation is $X(t) = X(s)\,e^{(\mu - \sigma^2/2)(t-s) + \sigma B(t-s)}$. Suppose $X(s) = x(s)$. Since $E(X(t)) = x(s)e^{\mu(t-s)}$, μ is interpreted as the expected growth rate.

```
> GeometricBrownian:=proc(T,n,seed,mu,sigma,x0) local
  h,sh,X,P,j,z,mh;
  #
  # Procedure generates a list [{[j*h,X(j*h)],j=0..n}] where
  X(j*h) is the position at time jh of the geometric Brownian
  motion(with expected growth rate, mu, and volatility, sigma).
  #
  global i,X2;
  i:=false;
  randomize(seed);
  h:=T/n; r:=mu-sigma^2/2;
  sh:=sqrt(h);
  mh:=r*h;
  X:=x0;
  P[0]:=[0,x0];
  for j from 1 to n do;
      z:=STDNORM();
      X:=X*exp(evalf(mh+sh*sigma*z));
      P[j]:=[evalf(j*h),X];
  end do;
  [seq(P[j],j=0..n)];
  end proc:
```

```
> v1:=GeometricBrownian(10,2000,1845,0.1,0.3,100):
  v2:=GeometricBrownian(10,2000,35915,0.1,0.3,100):
  v3:=GeometricBrownian(10,2000,159284,0.1,0.3,100):PLOT
  (CURVES(v1,v2,v3,COLOR(RGB, 1, 0, 0, 0, 1, 0,0,0,1)),
  TITLE("Three independent realizations of a geometric
  Brownian motion with expected \n growth rate of 0.1,
  volatility 0.3, initial price
  =100"),AXESLABELS("t","X(t)"),AXESSTYLE(NORMAL));
```

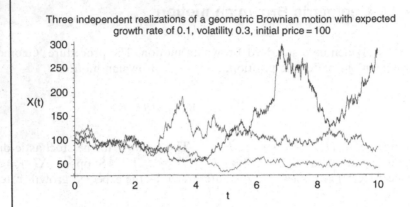

Three independent realizations of a geometric Brownian motion with expected growth rate of 0.1, volatility 0.3, initial price = 100

```
> v1:=GeometricBrownian(10,2000,1845,0.15,0.02,100):
  v2:=GeometricBrownian(10,2000,35915,0.15,0.04,100):
  v3:=GeometricBrownian(10,2000,159284,0.15,0.08,100):
  PLOT(CURVES(v1,v2,v3,COLOR(RGB, 1, 0, 0, 0, 1,
  0,0,0,1)),TITLE("Independent Geometric Brownian motions
  over 10 years with expected growth rate \n of 0.15 p.a.,
  volatility 0.02, 0.04, 0.08 p.a., \n initial price
  =100pence"),AXESLABELS("t","X(t)"),AXESSTYLE(NORMAL));
```

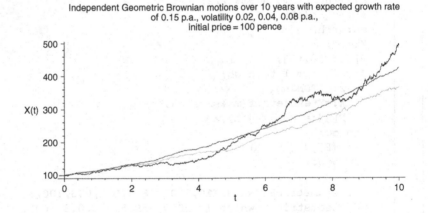

Independent Geometric Brownian motions over 10 years with expected growth rate of 0.15 p.a., volatility 0.02, 0.04, 0.08 p.a., initial price = 100 pence

6.3 Black–Scholes formula for a European call

Using the built-in 'blackscholes' procedure (part of Maple's finance package), find the price of a European call option on a share that is currently priced at £100, has volatility 20 % per annum (this means that the standard deviation on the return in 1 year is 0.2), and where the option has 103 trading days till expiry (252 trading days in a year). The strike price is £97. The risk-free interest rate is 5 %. Assume no dividends.

```
> with(finance):
```

```
> B:=blackscholes(100, 97, 0.05, 103/252, 0.2);
  evalf(B);
```

$$B := 50\,\mathrm{erf}\left(0.02427184466\left(\ln\left(\frac{100}{97}\right) + 0.02861111111\right)\sqrt{103}\sqrt{252}\sqrt{2}\right)$$
$$+\, 2.48111124 - 47.51888876\,\mathrm{erf}\left(\frac{1}{2}\left(0.04854368932\left(\ln\left(\frac{100}{97}\right)\right.\right.\right.$$
$$\left.\left.\left. +0.02861111111\right)\sqrt{103}\sqrt{252} - 0.0007936507936\sqrt{103}\sqrt{252}\right)\sqrt{2}\right)$$

$$7.84025659$$

6.4 Monte Carlo estimation of the price of a European call option

The procedure 'BS' below simulates m independent payoffs together with their antithetic counterparts; x_0 is the current asset price at time t, and t_e is the exercise time of the option.

```
> BS:=proc(m,t,te,sigma,r,k,x0)local
  m1,m2,s1,s2,j,payoff_avg,disc,payoffa,payoffb,xa,xb,z,
  call,std;global i,X2;
  i:=false;
  m1:=(r-0.5*sigma^2)*(te-t);
  m2:=sigma*sqrt(te-t);
  disc:=exp(-r*(te-t));
  s1:=0;s2:=0;
  for j from 1 to m do;
    z:=STDNORM();
    xa:=evalf(x0*exp(m1+m2*z)); xb:=evalf(x0*exp(m1-m2*z));
    payoffa:=max(xa-k,0);payoffb:=max(xb-k,0);
    payoff_avg:=(payoffa+payoffb)/2;
    s1:=s1+payoff_avg;
    s2:=s2+payoff_avg^2;
  end do;
```

```
        call:=disc*s1/m;
        std:=disc*sqrt((s2-s1^2/m)/m/(m-1));
        print("call_price"=call,"std_error"=std);
        print("confidence interval (large m)
        is",call-1.96*std,call+1.96*std);
        end proc:
```

Now call this procedure, using 10 000 payoffs, to estimate the price that was calculated exactly using the Black–Scholes formula in Appendix 6.3 above.

```
> seed:=randomize(87125635);BS(10000,23/252,126/
  252,0.2,0.05,97,100);
```
$$seed:=87125635$$
'call_price' = 7.835472640, 'std_error' = 0.04129181009
'confidence interval (large m) is', 7.754540692, 7.916404588

Now reduce the remaining life (from 103 to 50 trading days) of the option and make the option more 'in the money' by reducing the strike price from £97 to £85. Note how the standard error is now smaller, as a result of the antithetic design being more effective (see text).

```
> seed:=randomize(87125635);BS(10000,76/252,126/
  252,0.2,0.05,85,100);
```
$$seed:=87125635$$
'call_price' = 15.91630722, 'std_error' = 0.009405292445
'confidence interval (large m) is', 15.89787285, 15.93474159

6.5 Delta hedging for a currency

The procedure 'hedge' returns the cost of writing and hedging a call option, where the hedging is performed $nn + 1$ times. The asset earns interest at rate r_f. The risk-free interest rate is r and the (unknown) expected growth rate is μ. Print statements are currently suppressed using '#'. Removing '#' will show the price (x) of the asset, the Δ, the change in Δ since the last hedge, and the cumulative borrowings to finance the hedge, all at each hedging instant. Remember to load the procedure 'STDNORM' in Appendix 6.1.

```
> hedge:=proc(K,r,rf,sigma,T,n,x0,mu) local
  a1,a2,a3,a4,a5,a6,h,x,d,delta,delta_prev,c,j,z,cost,xprev;
  global i,X2;
```

```
    i:=false;
    h:=T/n;
    a1:=exp(h*(mu-0.5*sigma^2));
    a2:=sigma*sqrt(h);
    a3:=(r-rf+0.5*sigma^2)*h;
    a4:=exp(r*h);
    a5:=exp(-r*T);
    a6:=exp(rf*h);
    x:=x0;
    d:=(ln(x/K)+a3*n)/a2/sqrt(n);
    delta:=exp(-rf*T)*statevalf[cdf,normald](d);
    c:=delta*x;
    #print("x"=x, "delta"=delta, "c"=c);
    for j from 1 to n do;
        xprev:=x;
        z:=STDNORM();
        x:=x*a1*exp(a2*z);
        delta_prev:=delta*a6;# The holding in assets changes due
    to the interest earned on them at rate rf
        if j=n then;
            if x>K then delta:=1 else delta:=0 end if;
        else;
            d:=(ln(x/K)+a3*(n-j))/a2/sqrt(n-j);
            delta:=exp(-rf*(n-j)*h)*statevalf[cdf,normald](d);
        end if;
        c:=c*a4+x*(delta-delta_prev);
    end do;
    #print("x"=x, "delta"=delta, "dif"=delta-delta_prev,
    "c"=c);
    if delta>0 then cost:=a5*(c-K) else cost:=a5*c end if;
    cost:
    end proc:

> with(stats);

    [anova, describe, fit, importdata, random, statevalf, statplots, transform]
```

The procedure 'effic' calls 'hedge' *replic* times. It stores the costs in the list p and then prints the resulting mean and standard deviation of hedging cost together with a histogram. If the option were continuously hedged, then the standard deviation would be zero, and the hedging cost would always equal the Black–Scholes price. The current price of the asset is £680 and the strike price is £700. The risk-free interest rate is 5% per annum and the asset earns interest continuously at the rate of 3% per annum. The volatility is 0.1 per annum and the exercise time is 0.5 years from now.

```
> effic:=proc(nn,replic)local m,u,p,e2,e1;
  randomize(3640651);
  for m from 1 to replic do:
     u[m]:=hedge(700,0.05,0.03,0.1,0.5,nn,680,0.15):
  end do:
  p:=[seq(u[m],m=1..replic)]:
  e2:=describe[standarddeviation[1]](p);
  e1:=describe[mean](p);
  print("mean cost of hedging"=e1, "std dev of cost"=e2,
  "number of hedges"=nn+1, "number of contracts"=replic);
  statplots[histogram](p);
  end proc:
```

Now perform 10 000 replications for contracts that are hedged 2, 3, 4, 13, and 127 times respectively. It is supposed that the unknown expected growth rate of the underlying asset (euros) is 15 % per annum. The last example (127 hedges) takes a considerable amount of time ($\sim$10 minutes on a Pentium M 730, 1.6 GHz processor).

```
> effic(1,10000);
```

'mean cost of hedging' $= 21.53800165$, 'std dev of cost' $= 22.17558076$, 'number of hedges' $= 2$, 'number of contracts' $= 10000$

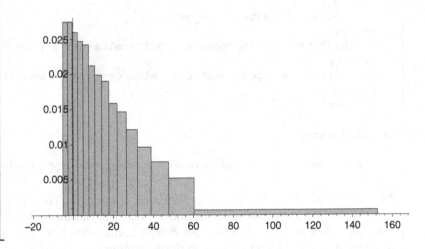

```
> effic(2,10000);
```

'mean cost of hedging' $= 17.26850774$, 'std dev of cost' $= 13.21884756$, 'number of hedges' $= 3$, 'number of contracts' $= 10000$

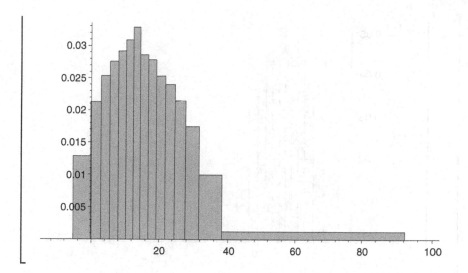

```
> effic(3,10000);
```

'mean cost of hedging' = 16.02928370, 'std dev of cost' = 10.25258825,
'number of hedges' = 4, 'number of contracts' = 10000

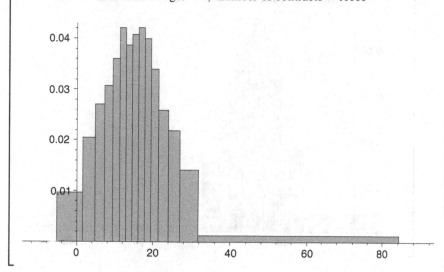

```
> effic(4,10000);
```

'mean cost of hedging' = 15.30498281, 'std dev of cost' = 8.665076040,
'number of hedges' = 5, 'number of contracts' = 10000

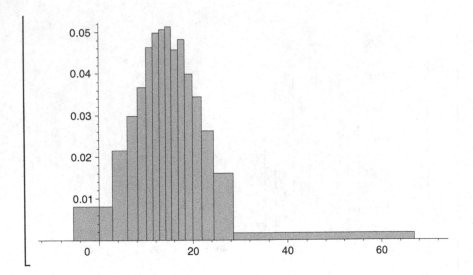

```
> effic(12,10000);
```

'mean cost of hedging' = 13.99661047, 'std dev of cost' = 4.750293651,
'number of hedges' = 13, 'number of contracts' = 10000

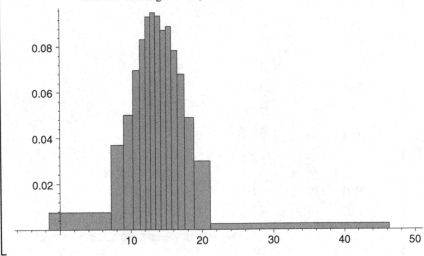

```
> effic(126,10000);
```

'mean cost of hedging' = 13.40525657, 'std dev of cost' = 1.452085739,
'number of hedges' = 127, 'number of contracts' = 10000

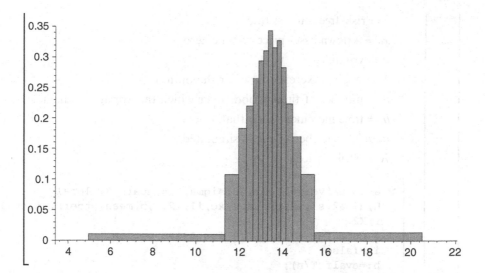

The cost of continuous hedging is the Black–Scholes cost, which is computed in the procedure 'bscurrency'.

```
> bscurrency:=proc(r,rf,K,x0,sigma,T,t) local d,price;
  d:=(ln(x0/K)+(r-rf+0.5*sigma^2)*(T-t))/sigma/sqrt(T-t);
  price:=x0*exp(-rf*(T-t))*statevalf[cdf,normald](d)-K*exp(-r*
  (T-t))*statevalf[cdf,normald](d-sqrt(T-t)*sigma);
  end proc:
```

```
> bscurrency(0.05,0.03,700,680,0.1,0.5,0);
                    13.3368447
```

Now, as suggested in the text, perform your own experiments to verify that (subject to sampling error) the *expected* cost of writing and discrete hedging the option is the Black–Scholes price, when the expected growth rate of the underlying asset happens to be the same as the risk-free interest rate. Of course, there is still variation in this cost, reflecting the risk from not hedging continuously.

 ## 6.6 Asian option

 ### 6.6.1 Naive Monte Carlo

The procedure 'asiannaive' computes the price of an Asian call option (arithmetic average) using standard Monte Carlo with no variance reducation. The parameters are:

r = risk-free interest rate

x_0 = known asset price at time zero

σ = volatility

T = expiry (exercise) time for the option

n = number of time periods over which the average is taken

h = time increment such that $T = nh$

$npath$ = number of paths simulated

K = strike price

```
> asiannaive:=proc(r,x0,sigma,T,n,npath,K) local
  R,a1,a2,s1,s2,theta,x,xc,i1,i2,z,h,mean,stderr; global
  i,X2;
  # Computes call price for Asian option.
  i:=false;
  h:=evalf(T/n);
  a1:=exp((r-0.5*sigma^2)*h);
  a2:=evalf(sigma*sqrt(h));
  s1:=0;s2:=0;
  for i2 from 1 to npath do;
      x:=x0;
      xc:=0;
      for i1 from 1 to n do;
          z:=STDNORM();
          x:=x*a1*exp(a2*z);
          xc:=xc+x;
      end do;
      R:=max(xc/n-K,0);
      s1:=s1+R;
      s2:=s2+R^2;
  end do;
  theta:=s1/npath;
  stderr:=sqrt((s2-s1^2/npath)/npath/(npath-1));
  print("K"=K,"sigma"=sigma,"n"=n);
  print("# paths"=npath);
  print("point estimate of price"=exp(-r*T)*theta);
  print("estimated standard error"=exp(-r*T)*stderr);
  end proc:
```

Now call 'asiannaive' for the parameter values given below, remembering first to load the procedure STDNORM (Section 6.1). These values are used in the paper by P. Glasserman, P. Heidelberger, and P. Shahabuddin (1999), Asymptotically optimal importance sampling and stratification for pricing path-dependent options, *Mathematical Finance*, **9**, 117–52. The present results can be compared with those appearing in that paper.

```
> randomize(13753);asiannaive(0.05,50,0.3,1,16,25000,55);
```
                        13753
           'K' = 55, 'sigma' = 0.3,'n' = 16
                   '#paths' = 25000
           'point estimate of price' = 2.214911961
           'estimated standard error' = 0.03004765102

```
> randomize(13753);asiannaive(0.05,50,0.3,1,16,25000,50);
  randomize(13753);asiannaive(0.05,50,0.3,1,16,25000,45);
  randomize(13753);asiannaive(0.05,50,0.1,1,16,25000,55);
  randomize(13753);asiannaive(0.05,50,0.1,1,16,25000,50);
  randomize(13753);asiannaive(0.05,50,0.1,1,16,25000,45);
```
                        13753
           'K' = 50, 'sigma' = 0.3, 'n' = 16
                   '# paths' = 25000
           'point estimate of price' = 4.166563741
           'estimated standard error' = 0.03986122915
                        13753
           'K' = 45, 'sigma' = 0.3, 'n' = 16
                   '# paths' = 25000
           'point estimate of price' = 7.145200333
           'estimated standard error' = 0.04857962659
                        13753
           'K' = 55, 'sigma' = 0.1, 'n' = 16
                   '# paths' = 25000
           'point estimate of price' = 0.2012671276
           'estimated standard error' = 0.004621266108
                        13753
           'K' = 50, 'sigma' = 0.1, 'n' = 16
                   '# paths' = 25000
           'point estimate of price' = 1.917809699
           'estimated standard error' = 0.01401995095
                        13753
           'K' = 45, 'sigma' = 0.1, 'n' = 16
                   '# paths' = 25000
           'point estimate of price' = 6.048215279
           'estimated standard error' = 0.01862584391

6.6.2 Monte Carlo with importance and stratified sampling

The procedure 'asianimpoststrat' computes the price of an Asian average price call option (*arithmetic* average) using importance sampling and *post* stratified sampling. Let n be the number of time points over which the average is calculated. The drift is set to $\boldsymbol{\beta}$ where $\beta_i = (\partial/\partial Z_i)\ln$ (payoff) at $z_i = \beta_i$, $i = 1,\ldots,n$, and where *payoff* is based on the *geometric* average. This gives $\beta_i = \lambda(n - i + 1)$ where

$$\lambda = \frac{x_0 \exp\left((r - 0.5\sigma^2)(n+1)h/2 + \sigma\lambda(n+1)(2n+1)\sqrt{h}/6\right)\sigma\sqrt{h}}{n\left(x_0 \exp\left((r - 0.5\sigma^2)(n+1)h/2 + \sigma\lambda(n+1)(2n+1)\sqrt{h}/6\right) - K\right)}$$

A lower bound on the search for λ is the value that makes the denominator zero. With the resulting importance sampling distribution, $z_i \sim N(\beta_i, 1)$ for $i = 1, \ldots, n$. The stratification variable is

$$X = \frac{\sum_{i=1}^{n} \beta_i(z_i - \beta_i)}{\sqrt{\sum_{i=1}^{n} \beta_i{}^2}} = \frac{\left(\sum_{i=1}^{n}(n-i+1)z_i\right) - \lambda n(n+1)(2n+1)/6}{\sqrt{n(n+1)(2n+1)/6}} \sim N(0,1).$$

Remember to load the standard normal generator STDNORM() in Appendix 6.1.

```
> with(stats):
Warning, these names have been redefined: anova, describe,
fit, importdata, random, statevalf, statplots, transform
```

```
> asianimppoststrat:=proc(r,x0,sigma,T,t,n,m,npath,K,
  p,upper) local
  R,a1,a2,a3,a4,b3,theta,j,s,f,i2,x,xc,x1,i1,z,h,st,
  xbar,v,c1,c2,mean,stderr,jj,lambda; global i,X2;
  #
  # Computes price of an Asian average price call option at
  time t, with expiry time T, and strike K, using importance
  sampling combined with post stratification.
  #
  # r=risk-free interest rate
  # x0=asset price at time t
  # sigma=volatility
  # T=exercise time
  # T=n*h where the average is taken over times h,2h,...,nh
  # m=number of strata
  # npath=number of paths in one replication. It should be
  at least 20*m for post stratification to be efficient
  # K=strike price
  # p=number of replications
  # upper=an upper bound for lambda
  #
  i:=false;
  h:=evalf((T-t)/n);
  a1:=exp((r-0.5*sigma^2)*h);
  a2:=evalf(sigma*sqrt(h));
  a3:=evalf(n*(2*n+1)*(n+1)/6);
  b3:=evalf(sqrt(a3));
```

```
a4:=max(0,6*(ln(K/x0)-(r-sigma^2)*h*(n+1)/2)/a2/(n+1)/
(2*n+1));
#
# solve for lambda
#
xbar:=x0*a1^(0.5*n+0.5)*exp(a3*u*a2/n);
lambda:=fsolve(u-xbar*a2/n/(xbar-K),u=a4..upper);
#lambda:=0; #exchange with previous line for no importance
sampling;
print('lambda'=lambda);
#
# end of solve for lambda
#
c1:=0;c2:=0;
for jj from 1 to p do:
  theta:=0;
  for j from 1 to m do s[j]:=0;f[j]:=0 end do;
  for i2 from 1 to npath do;
    x:=x0;
    xc:=0;
    x1:=0;
    for i1 from 1 to n do;
      z:=lambda*(n-i1+1)+STDNORM();
      x:=x*a1*exp(a2*z);
      xc:=xc+x;
      x1:=x1+(n-i1+1)*z;
    end do;
    R:=max(xc/n-K,0)*exp(-lambda*x1);
    st:=(x1-lambda*a3)/b3;
    j:=1+floor(m*statevalf[cdf,normald](st)); #j is the
stratum number;
    f[j]:=f[j]+1; # increment frequency, this stratum;
    s[j]:=s[j]+R; # increment sum, this stratum;
  end do;
  for j from 1 to m do;
    theta:=theta+s[j]/f[j];
  end do;
  theta:=theta/m;
theta:=theta*exp(-r*(T-t))*exp(0.5*lambda^2*a3);
c1:=c1+theta;
c2:=c2+theta^2;
end do;
mean:=c1/p;
stderr:=sqrt((c2-c1^2/p)/p/(p-1));
print("K"=K,"sigma"=sigma,"n"=n);
print("# replications"=p,"paths per
replication"=npath,"strata"=m);
```

```
print("point estimate of price"=mean);
print("estimated standard error"=stderr);
end proc:
```

Now call 'asianimppoststrat' where the risk-free interest rate is 5 % per annum and the exercise time is 1 year from now. The average is taken over 16 equally spaced time periods over the year. The volatility is either 10 % or 30 % per annum, and the strike prices are either £55, £50, or £45, as shown in the six sample runs below. Each run took approximately 15 minutes on a Pentium M730 1.6 GHz processor. This data set is part of one appearing in the paper by P. Glasserman, P. Heidelberger, and P. Shahabuddin (1999), Asymptotically optimal importance sampling and stratification for pricing path-dependent options, *Mathematical Finance*, **9**, 117–52.

```
> randomize(13753);asianimppoststrat(0.05,50,0.3,1,0,16,
  100, 2500,55,100,0.5);
  randomize(13753);asianimppoststrat(0.05,50,0.3,1,0,16,
  100,2500,50,100,0.5);
  randomize(13753);asianimppoststrat(0.05,50,0.3,1,0,16,
  100, 2500,45,100,0.5);
  randomize(13753);asianimppoststrat(0.05,50,0.1,1,0,16,
  100,2500,55,100,0.5);
  randomize(13753);asianimppoststrat(0.05,50,0.1,1,0,16,
  100,2500,50,100,0.5);
  randomize(13753);asianimppoststrat(0.05,50,0.1,1,0,16,
  100,2500,45,100,0.5);
```

$$13753$$
$$\lambda = 0.03422043734$$
'K' = 55, 'sigma' = 0.3, 'n' = 16
'# replications' = 100, 'paths per replication' = 2500, 'strata' = 100
'point estimate of price' = 2.211606984
'estimated standard error' = 0.0003131946824
$$13753$$
$$\lambda = 0.02691202349$$
'K' = 50, 'sigma' = 0.3, 'n' = 16
'# replications' = 100, 'paths per replication' = 2500, 'strata' = 100
'point estimate of price' = 4.170829200
'estimated standard error' = 0.0003747052714
$$13753$$
$$\lambda = 0.02087070210$$
'K' = 45, 'sigma' = 0.3, 'n' = 16
'# replications' = 100, 'paths per replication' = 2500, 'strata' = 100
'point estimate of price' = 7.152065560
'estimated standard error' = 0.0004827321234
$$13753$$
$$\lambda = 0.04549738209$$

'K' = 55, 'sigma' = 0.1, 'n' = 16
'# replications' = 100, 'paths per replication' = 2500, 'strata' = 100
'point estimate of price' = 0.2023819565
'estimated standard error' = 0.00002354664402
13753
$\lambda = 0.02171666756$
'K' = 50, 'sigma' = 0.1, 'n' = 16
'# replications' = 100, 'paths per replication' = 2500, 'strata' = 100
'point estimate of price' = 1.919506766
'estimated standard error' = 0.00006567443424
13753
$\lambda = 0.01089879722$
'K' = 45, 'sigma' = 0.1, 'n' = 16
'# replications' = 100, 'paths per replication' = 2500, 'strata' = 100
'point estimate of price' = 6.055282128
'estimated standard error' = 0.0001914854216

6.7 Basket options

Consider a basket (or portfolio) consisting of n assets. The basket contains a
quantity q_i of asset i where $i = 1, \ldots, n$. Let $r, \sigma_i, \{X_i(u), u \leq T, 0 \leq u\}$
denote the risk-free interest rate, volatilty, and prices in $[0,T]$ of one unit of
the ith asset. At time t the spot price is $\sum_{i=1}^{n} q_i x_i(t)$. Let ρ denote the
correlation between the returns on the assets and let the Cholesky decomposi-
tion of this be $\boldsymbol{b} \, \boldsymbol{b}^T = \boldsymbol{\rho}$. Then the price of a European call option at time t
with strike price K and exercise time T is the discounted expected payoff in a
risk-neutral world, that is

$$c = \mathrm{e}^{-r(T-t)} \max\left(0, \left(\sum_{i=1}^{n} q_i x_i \exp\left[(r - 0.5\,\sigma_i^2)(T-t) + \sigma_i\sqrt{T-t}\,W_i\right]\right) - K\right)$$

where $\mathbf{W} \sim N(\mathbf{0}, \boldsymbol{\rho})$. Put $\mathbf{W} = \boldsymbol{b}\mathbf{Z}$ where $\mathbf{Z} \sim N(\mathbf{0}, \mathbf{I})$.

6.7.1 Naive Monte Carlo

The procedure 'basket' estimates c using naive Monte Carlo. It performs
npath replications of the payoff. Ensure that the procedure 'STDNORM'
in Appendix 6.1 and the Linear Algebra and statistics packages are loaded.

```
> with(LinearAlgebra):
  with(stats):
Warning, these names have been redefined: anova, describe,
fit, importdata, random, statevalf, statplots, transform

> basket:=proc(r,x,sigma,q,rho,T,t,n,npath,K) local
  b,c1,c2,i1,i2,mean,R,spot,stderr,theta,w,xav,xi,z; global
  i,X2;
#
# Computes call price for basket option using naive Monte Carlo;
# load STDNORM and Linear Algebra package;
#
# r=risk-free interest rate;
# x[i1]= i1th. asset price at time t;
# sigma[i1]=volatility of asset i1;
# q[i1]=quantity of asset i1 in basket;
# rho=correlation matrix for returns between assets;
# T=exercise time;
# npath=number of paths
# K=strike price;
#
spot:=Transpose(x).q;
b:= LUDecomposition(rho, method='Cholesky');
i:=false;
z:=Vector(n);
w:=Vector(n);
c1:=0;c2:=0;
for i2 from 1 to npath do;
   for i1 from 1 to n do;
      z[i1]:=STDNORM();
   end do;
   w:=b.z;
   xav:=0;
   for i1 from 1 to n do;
xi:=q[i1]*x[i1]*exp((r-0.5*sigma[i1]^2)*(T-t)+sigma[i1]*
sqrt(T-t)*w[i1]);
      xav:=xav+xi;
   end do;
   theta:=max(0,xav-K)*exp(-r*T-t));
   c1:=c1+theta;
   c2:=c2+theta^2;
end do;
mean:=c1/npath;
stderr:=sqrt((c2-c1^2/npath)/npath/(npath-1));
print("K"=K, "spot"=spot, "r"=r, "n"=n, "t"=t, "T"=T,
"x"=x, "q"=q, "sigma"=sigma, "rho"=rho);
print("# paths"=npath);
print("point estimate of price"=mean);
print("estimated standard error"=stderr);
end proc:
```

[Set up the input data.

```
> r:=0.04:
  x:=Vector([5,2.5,4,3]):
  sigma:=Vector([0.3,0.2,0.3,0.4]):
  q:=Vector([20,80,60,40]):
  rho:=Matrix([[1,0.7,0.5,0.3], [0.7,1,0.6,0.2],
  [0.5,0.6,1,0.4], [0.3,0.2,0.4,1]]):
  T:=0.5:t:=0:n:=4:
  npath:=10000:
  spot:=Transpose(x).q:
```

[Now price the option with the following strike prices, K.

```
> K:=660;seed:=randomize(9624651);basket(r,x,sigma,q,
  rho,T,t,n,npath,K);
```

$$K := 660$$
$$seed := 9624651$$
$$\text{`K'} = 660, \text{`spot'} = 660., \text{`r'} = 0.04, \text{`n'} = 4, \text{`t'} = 0, \text{`T'} = 0.5,$$

$$\text{`x'} = \begin{bmatrix} 5 \\ 2.5 \\ 4 \\ 3 \end{bmatrix}, \text{`q'} = \begin{bmatrix} 20 \\ 80 \\ 60 \\ 40 \end{bmatrix},$$

$$\text{`sigma'} = \begin{bmatrix} 0.3 \\ 0.2 \\ 0.3 \\ 0.4 \end{bmatrix}, \text{`rho'} = \begin{bmatrix} 1 & 0.7 & 0.5 & 0.3 \\ 0.7 & 1 & 0.6 & 0.2 \\ 0.5 & 0.6 & 1 & 0.4 \\ 0.3 & 0.2 & 0.4 & 1 \end{bmatrix}$$

$$\text{`\# paths'} = 10000$$
$$\text{`point estimate of price'} = 47.20505098$$
$$\text{`estimated standard error'} = 0.7072233067$$

```
> K:=600;seed:=randomize(9624651);basket(r,x,sigma,q,
  rho,T,t,n,npath,K);
```

$$K := 600$$
$$seed := 9624651$$
$$\text{`K'} = 600, \text{`spot'} = 660., \text{`r'} = 0.04, \text{`n'} = 4, \text{`t'} = 0, \text{`T'} = 0.5,$$

$$\text{`x'} = \begin{bmatrix} 5 \\ 2.5 \\ 4 \\ 3 \end{bmatrix}, \text{`q'} = \begin{bmatrix} 20 \\ 80 \\ 60 \\ 40 \end{bmatrix},$$

$$\text{'sigma'} = \begin{bmatrix} 0.3 \\ 0.2 \\ 0.3 \\ 0.4 \end{bmatrix}, \text{'rho'} = \begin{bmatrix} 1 & 0.7 & 0.5 & 0.3 \\ 0.7 & 1 & 0.6 & 0.2 \\ 0.5 & 0.6 & 1 & 0.4 \\ 0.3 & 0.2 & 0.4 & 1 \end{bmatrix}$$

'# paths' = 10000
'point estimate of price' = 84.02729573
'estimated standard error' = 0.8807106819

```
> K:=720;seed:=randomize(9624651);basket(r,x,sigma,q,
  rho,T,t,n,npath,K);
```

$$K := 720$$
$$seed := 9624651$$

'K' = 720, 'spot' = 660., 'r' = 0.04, 'n' = 4, 't' = 0, 'T' = 0.5,

$$\text{'x'} = \begin{bmatrix} 5 \\ 2.5 \\ 4 \\ 3 \end{bmatrix}, \text{'q'} = \begin{bmatrix} 20 \\ 80 \\ 60 \\ 40 \end{bmatrix},$$

$$\text{'sigma'} = \begin{bmatrix} 0.3 \\ 0.2 \\ 0.3 \\ 0.4 \end{bmatrix}, \text{'rho'} = \begin{bmatrix} 1 & 0.7 & 0.5 & 0.3 \\ 0.7 & 1 & 0.6 & 0.2 \\ 0.5 & 0.6 & 1 & 0.4 \\ 0.3 & 0.2 & 0.4 & 1 \end{bmatrix}$$

'# paths' = 10000
'point estimate of price' = 23.48827444
'estimated standard error' = 0.5139235861

[Now change the vector of volatilities.

```
> sigma:=Vector([0.05,0.1,0.15,0.05]);
```

$$\sigma := \begin{bmatrix} 0.05 \\ 0.1 \\ 0.15 \\ 0.05 \end{bmatrix}$$

```
> K:=660;seed:=randomize(9624651);basket(r,x,sigma,q,
  rho,T,t,n,npath,K);
  K:=600;seed:=randomize(9624651);basket(r,x,sigma,q,
  rho,T,t,n,npath,K);
  K:=720;seed:=randomize(9624651);basket(r,x,sigma,q,
  rho,T,t,n,npath,K);
```

$$K := 660$$
$$seed := 9624651$$

'K' = 660, 'spot' = 660., 'r' = 0.04, 'n' = 4, 't' = 0, 'T' = 0.5,

$$\text{'x'} = \begin{bmatrix} 5 \\ 2.5 \\ 4 \\ 3 \end{bmatrix}, \text{'q'} = \begin{bmatrix} 20 \\ 80 \\ 60 \\ 40 \end{bmatrix},$$

$$\text{'sigma'} = \begin{bmatrix} 0.05 \\ 0.1 \\ 0.15 \\ 0.05 \end{bmatrix}, \text{'rho'} = \begin{bmatrix} 1 & 0.7 & 0.5 & 0.3 \\ 0.7 & 1 & 0.6 & 0.2 \\ 0.5 & 0.6 & 1 & 0.4 \\ 0.3 & 0.2 & 0.4 & 1 \end{bmatrix}$$

'# paths' = 10000
'point estimate of price' = 22.73700052
'estimated standard error' = 0.2834997907
$$K := 600$$
$$seed := 9624651$$

'K' = 600, 'spot' = 660., 'r' = 0.04, 'n' = 4, 't' = 0, 'T' = 0.5,

$$\text{'x'} = \begin{bmatrix} 5 \\ 2.5 \\ 4 \\ 3 \end{bmatrix}, \text{'q'} = \begin{bmatrix} 20 \\ 80 \\ 60 \\ 40 \end{bmatrix},$$

$$\text{'sigma'} = \begin{bmatrix} 0.05 \\ 0.1 \\ 0.15 \\ 0.05 \end{bmatrix}, \text{'rho'} = \begin{bmatrix} 1 & 0.7 & 0.5 & 0.3 \\ 0.7 & 1 & 0.6 & 0.2 \\ 0.5 & 0.6 & 1 & 0.4 \\ 0.3 & 0.2 & 0.4 & 1 \end{bmatrix}$$

'# paths' = 10000
'point estimate of price' = 71.67610118
'estimated standard error' = 0.3903638189
$$K := 720$$
$$seed := 9624651$$

'K' = 720, 'spot' = 660., 'r' = 0.04, 'n' = 4, 't' = 0, 'T' = 0.5,

$$\text{'x'} = \begin{bmatrix} 5 \\ 2.5 \\ 4 \\ 3 \end{bmatrix}, \text{'q'} = \begin{bmatrix} 20 \\ 80 \\ 60 \\ 40 \end{bmatrix},$$

$$\text{'sigma'} = \begin{bmatrix} 0.05 \\ 0.1 \\ 0.15 \\ 0.05 \end{bmatrix}, \text{'rho'} = \begin{bmatrix} 1 & 0.7 & 0.5 & 0.3 \\ 0.7 & 1 & 0.6 & 0.2 \\ 0.5 & 0.6 & 1 & 0.4 \\ 0.3 & 0.2 & 0.4 & 1 \end{bmatrix}$$

'# paths' = 10000
'point estimate of price' = 2.744216261
'estimated standard error' = 0.1003118091

 ### 6.7.2 Monte Carlo with importance and stratified sampling

Now define

$$x_0 = \sum_{i=1}^{n} q_i x_i \exp\left[(r - .5\sigma_i^2)(T - t)\right]$$

and

$$w_i = \frac{q_i x_i \exp\left[(r - 0.5\sigma_i^2)(T - t)\right]}{x_0}$$

Changing the measure, as described in the text, gives

$$c = x_0 \exp[-r(T - t)] \exp(0.5\,\boldsymbol{\beta}'\boldsymbol{\beta}) \times$$

$$E\left(\max\left(0, \left(\sum_{i=1}^{n} w_i \exp\left[\sqrt{T - t}\, \sigma_i \left(\sum_{j=1}^{n} b_{i,j} z_j \right) \right] \right) - \frac{K}{x_0} \right) \exp(\boldsymbol{\beta}'\mathbf{z}) \right)$$

where $\mathbf{Z} \sim N(\boldsymbol{\beta}, \mathbf{I})$ and $\boldsymbol{\beta}$ is chosen as described in the text. The stratification variable is

$$X = \frac{(\boldsymbol{\beta}'\mathbf{Z}) - (\boldsymbol{\beta}'\boldsymbol{\beta})}{\sqrt{\boldsymbol{\beta}'\boldsymbol{\beta}}}$$

The procedure 'basketimppoststratv2' below implements these two variance reduction devices. See Table 6.3 in the text for the variance reduction ratios achieved.

```
> with(LinearAlgebra):
  with(stats):

Warning, these names have been redefined: anova, describe,
fit,importdata, random, statevalf, statplots, transform
```

```
> basketimppoststratv2:=proc(r,x,sigma,q,rho,T,t,n,m,
  npath,K,p,upper) local
  a0,a4,b,beta,betasq,beta2,c,c1,c2,f,i1,i2,j,jj,K
  d,lambda,
  mean,stderr,R,T1,Tbeta,s,spot,theta,v,w,x0,xs,xi,xav,
  xstrat,z,zi1; global i,X2,d1,d2;
  #
  # Computes call price for basket option using importance
  sampling with post stratification.
  # load STDNORM, Linear Algebra package, and Statistics
  package
  #
  # r=risk-free interest rate;
  # x[i1]= i1th. asset price at time t;
  # sigma[i1]=volatility of asset i1;
  # q[i1]:=quantity of asset i1;
  # rho[i1, j]:=correlation between returns on assets i1 and j;
  # T=exercise time;
  # m=number of strata
  # npath=number of paths in one replication; should be at
  least 20*m for post stratification to be efficient;
  # K=strike price;
  # p=number of replications;
  # upper=an upper bound for lambda;
  # f[j]=number of paths falling in stratum j in one
  replication
  #
  spot:=Transpose(x).q;
  b:= LUDecomposition(rho, method='Cholesky' );
  i:=false;
  c:=Vector(n);
  z:=Vector(n);
  w:=Vector(n);
  beta:=Vector(n);
  v:=Vector(n);
  #
  x0:=0;T1:=sqrt(T-t);
  for i1 from 1 to n do;
    x0:=x0+q[i1]*x[i1]*exp((r-0.5*sigma[i1]^2)*(T-t));
  end do;
  for i1 from 1 to n do;
    w[i1]:=q[i1]*x[i1]*exp((r-0.5*sigma[i1]^2)*(T-t))/x0;
    c[i1]:=w[i1]*sigma[i1];
  end do;
  Kd:=K/x0;
  a0:=Transpose(c).rho.c;
  a4:=ln(Kd)/T1/a0;
```

```
lambda:=fsolve(u-T1*exp(T1*u*a0)/(exp(T1*u*a0)-
Kd),u=a4..upper);
beta:=ScalarMultiply(Transpose(b).c,lambda);
print("lambda"=lambda,"beta"=beta);
Tbeta:=Transpose(beta);
betasq:=Tbeta.beta;
beta2:=sqrt(betasq);
#
c1:=0;c2:=0;
for jj from 1 to p do:
  theta:=0;
  for j from 1 to m do s[j]:=0;f[j]:=0 end do;
  for i2 from 1 to npath do;
    xs:=0;
    for i1 from 1 to n do;
      zi1:=beta[i1]+STDNORM();
      xs:=xs+beta[i1]*zi1;
      z[i1]:=zi1;
    end do;
# z is N(beta,I),v is N(b*beta,rho);
    v:=b.z;
    xav:=0;
    for i1 from 1 to n do;
      xi:=w[i1]*exp(sigma[i1]*v[i1]*T1);
      xav:=xav+xi;
    end do;
    xstrat:=(xs-betasq)/beta2;
    j:=1+floor(m*statevalf[cdf,normald](xstrat));
    R:=max(0,xav-Kd)*exp(-xs);
    f[j]:=f[j]+1;
    s[j]:=s[j]+R;
  end do;
  for j from 1 to m do;
    theta:=theta+s[j]/f[j];
  end do;
  theta:=theta/m;
  c1:=c1+theta;
  c2:=c2+theta^2;
end do;
mean:=exp(-r*(T-t))*x0*exp(0.5*betasq)*c1/p;
stderr:= exp(-r*(T-t))*x0*exp(0.5*betasq)*sqrt
((c2-c1^2/p)/p/(p-1));
print("K"=K,"spot"=spot,"r"=r,"n"=n,"t"=t,"T"=T,"x"=x,
"q"=q,"sigma"=sigma,"rho"=rho);
print("# replications"=p,"paths per
replication"=npath,"strata"=m);
print("point estimate of price"=mean);
print("estimated standard error"=stderr);
```

```
d1:=statevalf[icdf,chisquare[p-1]](0.025)/(p-1)/stderr^2;
d2:=statevalf[icdf,chisquare[p-1]](0.975)/(p-1)/stderr^2;
print("approximate 95% confidence interval for reciprocal
variance of error=",d1,"to",d2);
end proc:
```

[Set up the input parameters.

```
> r:=0.04:
x:=Vector([5,2.5,4,3]):
sigma:=Vector([0.3,0.2,0.3,0.4]):
q:=Vector([20,80,60,40]):
rho:=Matrix([[1,0.7,0.5,0.3],[0.7,1,0.6,0.2],
[0.5,0.6,1,0.4],[0.3,0.2,0.4,1]]):
T:=0.5:t:=0:n:=4:m:=20:npath:=400:upper:=200:p:=25:
spot:=Transpose(x).q:
```

```
> seed:=randomize(9624651);K:=660:basketimppoststratv
2(r,x,sigma,q,rho,T,t,n,m,npath,K,p,upper);
```
$$seed := 9624651$$

$$\text{`lambda'} = 4.700068356, \quad \text{`beta'} = \begin{bmatrix} 0.772172051123948266 \\ 0.380523654341933582 \\ 0.513898302957234265 \\ 0.301405189763077175 \end{bmatrix}$$

'K' = 660, 'spot' = 660., 'r' = 0.04, 'n' = 4, 't' = 0, 'T' = 0.5,

$$\text{`x'} = \begin{bmatrix} 5 \\ 2.5 \\ 4 \\ 3 \end{bmatrix}, \text{`q'} = \begin{bmatrix} 20 \\ 80 \\ 60 \\ 40 \end{bmatrix},$$

$$\text{`sigma'} = \begin{bmatrix} 0.3 \\ 0.2 \\ 0.3 \\ 0.4 \end{bmatrix}, \text{`rho'} = \begin{bmatrix} 1 & 0.7 & 0.5 & 0.3 \\ 0.7 & 1 & 0.6 & 0.2 \\ 0.5 & 0.6 & 1 & 0.4 \\ 0.3 & 0.2 & 0.4 & 1 \end{bmatrix}$$

'# replications' = 25, 'paths per replication' = 400, 'strata' = 20
'point estimate of price' = 48.03048912
'estimated standard error' = 0.04919665583
'approximate 95% confidence interval for reciprocal variance
of error =', 213.4909953, 'to', 677.6690736

```
> seed:=randomize(9624651);K:=600:basketimppoststratv
  2(r,x,sigma,q,rho,T,t,n,m,npath,K,p,upper);
```
$$seed := 9624651$$

$$\text{'lambda'} = 3.589486716, \text{ 'beta'} = \begin{bmatrix} 0.589715108385092890 \\ 0.290609518612742979 \\ 0.392469000899768161 \\ 0.230185997913606666 \end{bmatrix}$$

'K' = 600, 'spot' = 660., 'r' = 0.04, 'n' = 4, 't' = 0, 'T' = 0.5,

$$\text{'x'} = \begin{bmatrix} 5 \\ 2.5 \\ 4 \\ 3 \end{bmatrix}, \text{'q'} = \begin{bmatrix} 20 \\ 80 \\ 60 \\ 40 \end{bmatrix},$$

$$\text{'sigma'} = \begin{bmatrix} 0.3 \\ 0.2 \\ 0.3 \\ 0.4 \end{bmatrix}, \text{'rho'} = \begin{bmatrix} 1 & 0.7 & 0.5 & 0.3 \\ 0.7 & 1 & 0.6 & 0.2 \\ 0.5 & 0.6 & 1 & 0.4 \\ 0.3 & 0.2 & 0.4 & 1 \end{bmatrix}$$

'# replications' = 25, 'paths per replication' = 400, 'strata' = 20
'point estimate of price' = 85.18136772
'estimated standard error' = 0.06450488957
'approximate 95% confidence interval for reciprocal variance of error =',
124.1839457,
'to', 394.1881451

```
> seed:=randomize(9624651);K:=720:basketimppoststratv
  2(r,x,sigma,q,rho,T,t,n,m,npath,K,p,upper);
```
$$seed := 9624651$$

$$\text{'lambda'} = 6.065827969, \text{ 'beta'} = \begin{bmatrix} 0.996552064739320342 \\ 0.491097331047708086 \\ 0.663228375246980528 \\ 0.388988391569389748 \end{bmatrix}$$

'K' = 720, 'spot' = 660., 'r' = 0.04, 'n' = 4, 't' = 0, 'T' = 0.5,

$$\text{'x'} = \begin{bmatrix} 5 \\ 2.5 \\ 4 \\ 3 \end{bmatrix}, \text{'q'} = \begin{bmatrix} 20 \\ 80 \\ 60 \\ 40 \end{bmatrix},$$

$$\text{`sigma'} = \begin{bmatrix} 0.3 \\ 0.2 \\ 0.3 \\ 0.4 \end{bmatrix}, \text{`rho'} = \begin{bmatrix} 1 & 0.7 & 0.5 & 0.3 \\ 0.7 & 1 & 0.6 & 0.2 \\ 0.5 & 0.6 & 1 & 0.4 \\ 0.3 & 0.2 & 0.4 & 1 \end{bmatrix}$$

'# replications' = 25, 'paths per replication' = 400, 'strata' = 20
'point estimate of price' = 24.03898507
'estimated standard error' = 0.02821965008
'approximate 95% confidence interval for reciprocal variance of
error =', 648.8547348, 'to', 2059.612802

>

[Change the vector of volatilities.

```
> sigma:=Vector([0.05,0.1,0.15,0.05]);
```

$$\sigma := \begin{bmatrix} 0.05 \\ 0.1 \\ 0.15 \\ 0.05 \end{bmatrix}$$

```
> seed:=randomize(9624651);K:=660:basketimppoststratv2
(r,x,sigma,q,rho,T,t,n,m,npath,K,p,upper);
```
$$seed := 9624651$$

$$\text{`lambda'} = 10.38879243 \ \text{`beta'} = \begin{bmatrix} 0.610362601267784410 \\ 0.421462456610914860 \\ 0.478154378789704460 \\ 0.0850046460918798957 \end{bmatrix}$$

'K' = 660, 'spot' = 660., 'r' = 0.04, 'n' = 4, 't' = 0, 'T' = 0.5,

$$\text{`x'} = \begin{bmatrix} 5 \\ 2.5 \\ 4 \\ 3 \end{bmatrix}, \text{`q'} = \begin{bmatrix} 20 \\ 80 \\ 60 \\ 40 \end{bmatrix},$$

$$\text{`sigma'} = \begin{bmatrix} 0.05 \\ 0.1 \\ 0.15 \\ 0.05 \end{bmatrix}, \text{`rho'} = \begin{bmatrix} 1 & 0.7 & 0.5 & 0.3 \\ 0.7 & 1 & 0.6 & 0.2 \\ 0.5 & 0.6 & 1 & 0.4 \\ 0.3 & 0.2 & 0.4 & 1 \end{bmatrix}$$

'# replications' = 25, 'paths per replication' = 400, 'strata' = 20
'point estimate of price' = 23.09053636

"estimated standard error" = 0.009617279541

'approximate 95% confidence interval for reciprocal variance of error =',

5586.582894, "to", 17733.08729

```
> seed:=randomize(9624651);K:=600:basketimppoststratv2
(r,x,sigma,q,rho,T,t,n,m,npath,K,p,upper);
```

$$seed := 9624651$$

$$\text{'lambda'} = 5.407253416 \text{ 'beta'} = \begin{bmatrix} 0.317687092406742088 \\ 0.219366622596493726 \\ 0.248874151207387928 \\ 0.0442439933277393033 \end{bmatrix}$$

'K' = 600, 'spot' = 660., 'r' = 0.04, 'n' = 4, 't' = 0, 'T' = 0.5,

$$\text{'x'} = \begin{bmatrix} 5 \\ 2.5 \\ 4 \\ 3 \end{bmatrix}, \text{'q'} = \begin{bmatrix} 20 \\ 80 \\ 60 \\ 40 \end{bmatrix},$$

$$\text{'sigma'} = \begin{bmatrix} 0.05 \\ 0.1 \\ 0.15 \\ 0.05 \end{bmatrix}, \text{'rho'} = \begin{bmatrix} 1 & 0.7 & 0.5 & 0.3 \\ 0.7 & 1 & 0.6 & 0.2 \\ 0.5 & 0.6 & 1 & 0.4 \\ 0.3 & 0.2 & 0.4 & 1 \end{bmatrix}$$

'# replications' = 25, 'paths per replication' = 400, 'strata' = 20

'point estimate of price' = 72.24856764

'estimated standard error' = 0.03901100494

'approximate 95% confidence interval for reciprocal variance of error =',

339.5286687, 'to', 1077.741373

```
> seed:=randomize(9624651);K:=720:basketimppoststratv2
(r,x,sigma,q,rho,T,t,n,m,npath,K,p,upper);
```

$$seed := 9624651$$

$$\text{'lambda'} = 20.43485296, \text{'beta'} = \begin{bmatrix} 1.20058900909143329 \\ 0.829020638061244374 \\ 0.940534185140876167 \\ 0.167204942779322074 \end{bmatrix}$$

'K' = 720, 'spot' = 660., 'r' = 0.04, 'n' = 4, 't' = 0, 'T' = 0.5,

$$\text{'x'} = \begin{bmatrix} 5 \\ 2.5 \\ 4 \\ 3 \end{bmatrix}, \text{'q'} = \begin{bmatrix} 20 \\ 80 \\ 60 \\ 40 \end{bmatrix},$$

$$\text{'sigma'} = \begin{bmatrix} 0.05 \\ 0.1 \\ 0.15 \\ 0.05 \end{bmatrix}, \text{'rho'} = \begin{bmatrix} 1 & 0.7 & 0.5 & 0.3 \\ 0.7 & 1 & 0.6 & 0.2 \\ 0.5 & 0.6 & 1 & 0.4 \\ 0.3 & 0.2 & 0.4 & 1 \end{bmatrix}$$

'# replications' = 25, 'paths per replication' = 400, 'strata' = 20
'point estimate of price' = 2.865814172
'estimated standard error' = 0.002971217579
'approximate 95% confidence interval for reciprocal variance of error =',
58530.44349, 'to', 185788.9668

6.8 Stochastic volatility

The procedure 'meanreverting' generates a volatility process $\{\sigma(t)\}$ in $[0, T]$ at $t = 0, \ldots, nh$, where $\sigma(t) = e^{Y(t)}$ and $Y(t)$ is an Ornstein–Uhlenbeck process,

$$dY = \alpha(m - Y)dt + \beta \, dB,$$

where $0 < \alpha$, β and $\{B(t)\}$ is a standard Brownian motion. Remember to load 'STDNORM' in Appendix 6.1. Define $\nu^2 = \beta^2/(2\alpha)$. There is a closed-form solution

$$Y(t) = y(0)e^{-\alpha t} + (1 - e^{-\alpha t})m + \frac{\nu\sqrt{1 - e^{-2\alpha t}} \, B(t)}{t}$$

and so

$$E(\sigma(t)) = \exp[y(0)e^{-\alpha t} + (1 - e^{-\alpha t})m + 0.5 \, \nu^2(1 - e^{-2\alpha t})]$$

```
> with(stats);
Warning, these names have been redefined: anova, describe,
fit, importdata, random, statevalf, statplots, transform
```

```
> meanreverting:=proc(T,n,seed,alpha,m,nu,y0) local
  h,P,j,z,a,b,d,A,Y;
  #
  # Procedure generates a list [{[j*h,exp(Y(j*h))],j=0..n}]
  where Y(j*h) is the position of the OU process (with
  parameters alpha,m,nu) at time jh..
  #
  global i,X2;
  i:=false;
  randomize(seed);
  h:=T/n;
  Y:=y0;
  P[0]:=[0,evalf(exp(y0))];
  a:=evalf(exp(-alpha*h));
  b:=sqrt(1-a^2);
  d:=m*(1-a);
  for j from 1 to n do;
      z:=STDNORM();
      Y:=d+Y*a+nu*b*z;
      P[j]:=[evalf(j*h),exp(Y)];
  end do;
  [seq(P[j],j=0..n)];
  end proc:
```

Now plot a realization over $[0, T]$ with the parameter values shown below together with $E(\sigma(t))$.

```
> n:=1000:T:=5;alpha:=0.5;m:=-2;nu:=0.3;y0:=-2;
  v1:=meanreverting(T,n,14457154,alpha,m,nu,y0):
  h:=T/n:
  for j from 0 to n do:
      m1:=m+(y0-m)*exp(-alpha*j*h):
      m2:=nu^2*(1-exp(-2*alpha*j*h)):
      Q[j]:=[j*h,evalf(exp(m1+0.5*m2))]:
  end do:
  v2:=[seq(Q[j],j=0..n)]:
  PLOT(CURVES(v1,v2),TITLE("A slow mean reverting volatility
  process\n (alpha=0.5) with the expected
  volatility\n"),AXESLABELS("t","sigma(t)"),
  AXESSTYLE(NORMAL));
  g:=[seq(op(2,v1[j]),j=1..n)]:
  sample_mean:=describe[mean](g);
```

$$T := 5$$
$$\alpha := 0.5$$
$$m := -2$$
$$\nu := 0.3$$
$$y0 := -2$$

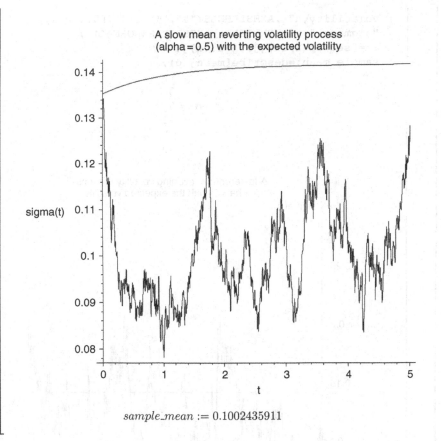

A slow mean reverting volatility process
(alpha = 0.5) with the expected volatility

$sample\_mean := 0.1002435911$

Notice in the realization above that the volatility is taking an exceptionally long time to reach a value equal to its asymptotic expectation. Now increase α from 0.5 to 5. Observe in the plot below that the reversion to the mean is far more rapid.

```
>  n:=1000:T:=5;alpha:=5;m:=-2;nu:=0.3;y0:=-2;
   v1:=meanreverting(T,n,14457154,alpha,m,nu,y0):
   h:=T/n:
   for j from 0 to n do:
   m1:=m+(y0-m)*exp(-alpha*j*h):
   m2:=nu^2*(1-exp(-2*alpha*j*h)):
   Q[j]:=[j*h,evalf(exp(m1+0.5*m2))]:
   end do:
   v2:=[seq(Q[j],j=0..n)]:
   PLOT(CURVES(v1,v2),TITLE("A faster mean reverting volatility
   process\n (alpha=5) with the expected
```

```
volatility\n"),AXESLABELS("t",""),TEXT([0.2,0.3],
"sigma(t)", ALIGNRIGHT),AXESSTYLE(NORMAL));
g:=[seq(op(2,v1[j]),j=1..n)]:
sample_mean:=describe[mean](g);
```

$$T := 5$$
$$\alpha := 5$$
$$m := -2$$
$$\nu := 0.3$$
$$y0 := -2$$

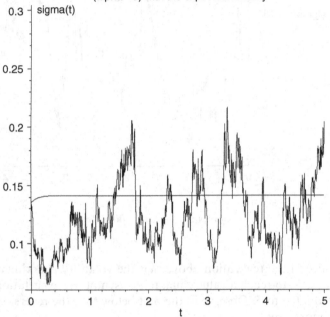

A faster mean reverting volatility process
(alpha = 5) with the expected volatility

$$sample\_mean := 0.1234165404$$

Finally, increase α from 5 to 50, giving very fast reversion to the mean.

```
> n:=1000:T:=5;alpha:=50;m:=-2;nu:=0.3;y0:=-2;
  v1:=meanreverting(T,n,14457154,alpha,m,nu,y0):
  h:=T/n:
  for j from 0 to n do:
  m1:=m+(y0-m)*exp(-alpha*j*h):
  m2:=nu^2*(1-exp(-2*alpha*j*h)):
  Q[j]:=[j*h,evalf(exp(m1+0.5*m2))]:
  end do:
```

```
v2:=[seq(Q[j],j=0..n)]:
PLOT(CURVES(v1,v2),TITLE("A fast mean reverting volatility
process \n (alpha=50) with the expected
volatility"),AXESLABELS("t","sigma(t)"),AXESSTYLE(NORMAL));
g:=[seq(op(2,v1[j]),j=1..n)]:
sample_mean:=describe[mean](g);
```

$$T := 5$$
$$\alpha := 50$$
$$m := -2$$
$$\nu := 0.3$$
$$y0 := -2$$

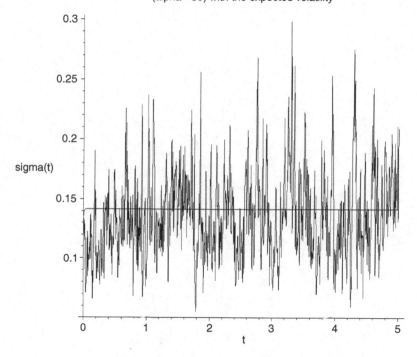

A fast mean reverting volatility process
(alpha = 50) with the expected volatility

$$sample\_mean := 0.1353869997$$

Appendix 7: Discrete event simulation

7.1 G/G/1 queue simulation using the regenerative technique

For procedure 'gg1', the aim is to estimate the long-run average line length. Line length is the number of customers in the queueing system. It includes any customer currently being served. (The queue length is the number of customers waiting for service.) The program assumes that interarrival times are independently Weibull distributed with $P(x < X) = \exp[-(\lambda x)^\gamma]$ for $0 < \gamma$, $0 < x$, and that service durations are independently distributed with complementary cumulative distribution $P(x < D) = \exp[-(\mu x)^\beta]$ for $0 < \beta$, $0 < x$. The expectation and standard deviation of these are printed out. For stationary behaviour, the mean interarrival time should be greater than the mean service duration. Other distributions can be used by suitable modifications to the program. The three-phase method for discrete event simulation is used. The bound state changes are (i) customer arrival and (ii) customer departure. The only conditional state change is (iii) start customer service. Separate streams are used for the two types of sampling activity. This has benefits when performing antithetic replications. The regeneration points are those instants at which the system enters the empty and idle state and n regeneration cycles are simulated.

```
> with(stats):

Warning, these names have been redefined: anova, describe,
fit, importdata, random, statevalf, statplots, transform.
```

```
> gg1:=proc(lambda,mu,gam,beta,n,seed1,seed2) local
  rho,seeda,seedb,m,clock,L,Lp,a,d,reward,k,tau,prev,delta,
  u,v,R,tau,TAU,b1,b2,a1,a2,a3,a4,a5,a6,f,stderr,a7,gam1,
  beta1,gam2,gam3,beta2,beta3,v1;
  gam1:=1/gam;beta1:=1/beta;
  gam2:=evalf(GAMMA(gam1+1));
  gam3:=evalf(GAMMA(2/gam+1));
  beta2:=evalf(GAMMA(beta1+1));
  beta3:=evalf(GAMMA(2/beta+1));
  print("Mean inter-arrival time"=gam2/lambda, "Std dev of
  inter-arrival time"=sqrt(gam3-gam2^2)/mu);
  print("Mean service duration"=beta2/mu,"Std dev of service
  duration"=sqrt(beta3-beta2^2)/mu);
  seeda:=seed1;seedb:=seed2;
  # SIMULATE n REGENERATIVE CYCLES - REGENERATIVE POINTS ARE
  WHEN SYSTEM ENTERS THE BUSY STATE FROM THE EMPTY & IDLE
  STATE
  for m from 1 to n do;
  # START OF NEW CYCLE
     tau:=0;
     L:=1;
     reward:=0;
     randomize(seeda);
     seeda:=rand();
     u:=evalf(seeda/10^12);
     a:=tau+(-ln(u))^gam1/lambda;
     randomize(seedb);
     seedb:rand();
     v:=evalf(seedb/10^12);
     d:=tau+(-ln(v))^beta1/mu;
     do
     tauprev:=tau;
     tau:=min(a,d);
     delta:=tau-tauprev;
     reward:=reward+delta*L;
  # CHECK FOR 'END OF CYCLE'
     if tau=a and L=0 then break end if;
  # CHECK FOR BOUND ACTIVITY 'ARRIVAL'
     if tau=a then
         L:=L+1;
         randomize(seeda);
         seeda:=rand();
         u:=evalf(seeda/10^12);
         a:=clock+(-ln(u))^gam1/lambda;
     end if;
  # CHECK FOR BOUND ACTIVITY 'DEPARTURE'
     if tau=d then;
         L:=L-1;
```

```
            d:=infinity
        end if;
# CHECK FOR CONDITIONAL ACTIVITY 'START SERVICE'
        if L>0 and d=infinity then;
            randomize(seedb);
            seedb:=rand();
            v:=evalf(seedb/10^12);
            d:=tau+(-ln(v))^beta1/mu;
            end if;
    end do;
    # END OF CYCLE
    # TAU[m]=LENGTH OF m th. REGENERATIVE CYCLE
    # R[m]=CYCLE 'REWARD' =int(L(t),t=0..TAU[m]), where L(t) IS
    NUMBER IN SYSTEM AT TIME t, AND t IS THE ELAPSED TIME SINCE
    START OF m th. CYCLE
    TAU[m]:=tau;
    R[m]:=reward;
    end do;
    # END OF n th CYCLE
    b1:=[seq(R[m],m=1..n)];
    b2:=[seq(TAU[m],m=1..n)];
    a1:=stats[describe,mean](b1);
    a2:=stats[describe,mean](b2);
    a3:=stats[describe,variance](b1);
    a4:=stats[describe,variance](b2);
    a5:=stats[describe,covariance](b1,b2);
    a6:=a1/a2;
    a7:=a6*(1+(a5/a1/a2-a4/a2/a2)/n); #a7 IS MODIFIED RATIO
    ESTIMATOR OF TIN, M.(1965),JASA, 60, 294-307
    f:=a3+a4*a7^2-2*a7*a5;
    stderr:=sqrt(f/n)/a2;
    print("mean cycle length"=a2);
    print("estimate of mean line length"=a7);
    print("95% Conf Interval for mean line
    length"=a7-1.96*stderr,"to",a7+1.96*stderr);
    end proc:
```

Now check that the program performs as expected for the M/M/1 system.

```
> gg1(1,2,1,1,10000,246978,71586);
  print("_____
  _____");
  gg1(1,1.5,1,1,10000,246978,71586);
  print("_____
  _____");
  gg1(1,1.1,1,1,10000,246978,71586);
```

'Mean interarrival time' = 1 'Std dev of interarrival time' = 0.5000000000
'Mean service duration' = 0.5000000000, 'Std dev of service duration' = 0.5000000000
'mean cycle length' = 2.003817572

'estimate of mean line length' = 0.9993680613
'95% Confidence interval for mean line length' = 0.9503482328,'to', 1.048387890

'Mean interarrival time' = 1., 'Std dev of interarrival time' = 0.6666666667
'Mean service duration' = 0.6666666667, 'Std dev of service duration' = 0.6666666667
'mean cycle length' = 3.014993555
'estimate of mean line length' = 2.030197245
'95%Confidence interval for mean line length' = 1.903465437, 'to', 2.156929053

'Mean interarrival time' = 1., 'Std dev of interarrival time' = 0.9090909091
'Mean service duration' = 0.9090909091, 'Std dev of service duration' = 0.9090909091
'mean cycle length' = 11.44855979
'estimate of mean line length' = 10.32733293
'95% Confidence interval for mean line length' = 9.087027523, 'to', 11.56763834

Note how the confidence interval becomes wider as the traffic intensity approaches 1. Replace $U(0,1)$ random numbers in 'gg1' by $1 - U(0,1)$ random numbers, i.e. perform antithetic replications.

```
> gg1(1,2,1,1,10000,246978,71586);
  print("_____
  _____");
  gg1(1,1.5,1,1,10000,246978,71586);
  print("_____
  _____");
  gg1(1,1.1,1,1,10000,246978,71586);
```
'Mean interarrival time' = 1., 'Std dev of interarrival time' = 0.5000000000
'Mean service duration' = 0.5000000000, 'Std dev of service duration' = 0.5000000000
'mean cycle length' = 1.980971270
'estimate of mean line length' = 1.006757945
'95% Confidence interval for mean line length' = 0.9547306296, 'to', 1.053785260

'Mean interarrival time' = 1., 'Std dev of interarrival time' = 0.6666666667
'Mean service duration' = 0.6666666667, 'Std dev of service duration' = 0.6666666667
'mean cycle length' = 2.958999958
'estimate of mean line length' = 1.952424988
'95% Confidence interval for mean line length' = 1.839236291, 'to', 2.065613685

'Mean interarrival time' = 1.,'Std dev of interarrival time' = 0.9090909091
'Mean service duration' = 0.9090909091, Std dev of service duration' = 0.9090909091
'mean cycle length' = 10.86151154
'estimate of mean line length' = 10.15972516
'95% Confidence interval for mean line length' = 8.897052772, 'to',11.42239555

The results are summarized in the text. Now experiment with the program for systems where no closed form results are available. Note that mean waiting times in the system (waiting in the line) and in the queue can then be obtained using Little's result.

7.2 Simulation of a hospital ward

7.2.1 Procedure hospital_ward

The procedure 'hospital_ward' simulates a ward of n beds. Arrivals form a Poisson process of rate λ and patient occupancy durations are Weibull distributed with the complementary cumulative distribution function, $P(x < X) = \exp[-(\mu x)^{\beta}]$ for $0 < \beta$, $0 < x$. If an arriving patient finds all beds occupied the following protocol applies. If the time until the next 'scheduled' departure is less than α, then that patient leaves now (early departure) and the arriving patient is admitted to the momentarily vacant bed. Otherwise, the arriving patient cannot be admitted and is referred elsewhere. Using the 'three-phase' terminology, the bound events are (i) an arrival and (ii) a normal (scheduled) departure. The conditional events are (iii) normal admission of patient, (iv) admission of patient following early departure, and (v) referral of the arriving patient. The following variables are used:

$simtim$ = duration of simulation
n = number of beds in unit
$t0[j]$ = ranked (ascending) scheduled departure times for bed j, at start of simulation, with $t0[j]$ = infinity indicating bed is currently unoccupied, $j = 1, \ldots, n$
α = threshold early departure parameter
$seeda$ = seed for random number stream generating arrivals
$seedb$ = seed for random number stream generating durations of time patient spends in unit
$clock$ = present time

clockprev = time of previous event

δ = time since last event

$t[j]$ = ranked (ascending) scheduled departure times for bed j, at current time, with $t[j]$ =infinity indicating bed is currently unoccupied, $j = 1, \dots, n$

a = time of next arrival

nocc = number of occupied beds

$q = 1, 0$ according to whether or not there is a patient momentarily requiring admission

na = cumulative number of arrivals

nout1 = cumulative number of normal departures

nout2 = cumulative number of early departures

na = cumulative number of admissions

nrefer = cumulative number of referrals

$cum[j]$ = cumulative time for which occupancy is j beds, $j = 1, \dots, n$

```
> restart;
```

```
> hospital_ward:=proc(simtim,n,t0,alpha,lambda,mu,beta,
  seeda,seedb) local
  q,d,t,clock,r1,a,clockprev,cum,na,nout1,nin,r2,nout2,
  nrefer,delta,nocc,seed1,seed2,j,beta1,f,s,util,r,beta2,
  beta3,short,ed;
  seed1:=seeda;seed2:=seedb;
  beta1:=1/beta;
  beta2:=evalf(GAMMA(beta1+1));
  beta3:=evalf(GAMMA(2/beta+1));
  print("_____INPUT
  DATA_____");
  print("Arrival and service seeds"=seeda,seedb);
  print("Run length in days"=simtim);
  print("Number of beds"=n,"Poisson arrival rate per
  day"=lambda,"Mean (non reduced) length of stay in
  days"=beta2/mu, "Std dev of (non reduced) length of
  stay in
  days"=sqrt(beta3-beta2^2)/mu);
  print("Early departure parameter in days"=alpha);
  nocc:=0;
  cum[0]:=0;
  # COMPUTE THE NUMBER OF BEDS OCCUPIED AT TIME ZERO FROM
  THE INITIAL SCHEDULED DEPARTURE DATE FOR EACH BED
  for j from 1 to n do;
      if t0[j]<infinity then nocc:=nocc+1 end if;
      cum[j]:=0
  end do;
  print("Bed state at time zero - Scheduled departure times
  are"=[seq(t0[j],j=1..n)]);
```

```
# INITIALISE OTHER VARIABLES, SORT INITIAL DEPARTURE
DATES, SAMPLE TIME OF FIRST ARRIVAL, AND ADVANCE
SIMULATION TIME TO FIRST EVENT
q:=0;
na:=0;nout1:=0;nout2:=0;nin:=0;nrefer:=0;short:=0;
t:=sort(t0);
clock:=0;
randomize(seed1);
seed1:=rand();
r1:=evalf(seed1/10^12);
a:=clock-ln(r1)/lambda;
clockprev:=clock;
clock:=min(a,t[1]);
# CHECK SIMULATION FINISHED. IF NOT UPDATE CUMULATIVE TIME
THAT nocc BEDS HAVE BEEN OCCUPIED;
while clock<=simtim do;
    delta:=clock-clockprev;
    cum[nocc]:=cum[nocc]+delta;
# BOUND STATE CHANGE:ARRIVAL
  if clock=a then;
   na:=na+1;
   randomize(seed1);
   seed1:=rand();
   r1:=evalf(seed1/10^12);
   a:=clock-ln(r1)/lambda;
   q:=1;
  else # BOUND STATE CHANGE:DEPARTURE;
   t[1]:=infinity;
   t:=sort(t);
   nocc:=nocc-1;
   nout1:=nout1+1;
  end if;
# CONDITIONAL STATE CHANGE:NORMAL ADMISSION
   if q=1 and nocc<n then;
   nin:=nin+1;
   nocc:=nocc+1;
   randomize(seed2);
   seed2:=rand();
   r2:=evalf(seed2/10^12);
   t[nocc]:=clock+(-ln(r2))^beta1/mu;
   t:=sort(t);
   q:=0;
  end if;
# CONDITIONAL STATE CHANGE: ADMISSION OF PATIENT FOLLOWING
EARLY DEPARTURE
  if q=1 and nocc=n and t[1]-clock<=alpha then;
   nin:=nin+1;
   nout2:=nout2+1;
```

```
        short:=short+t[1]-clock;
        randomize(seed2);
        seed2:=rand();
        r2:=evalf(seed2/10^12);
        t[1]:=clock+(-ln(r(2))^beta1/mu;
        t:=sort(t);
        q:=0;
     end if;
# CONDITIONAL STATE CHANGE:REFERRAL OF ARRIVING PATIENT
     if q=1 and nocc=n and t[1]-clock>alpha then;
        randomize(seed2);
        seed2:=rand();# THIS RANDOM NUMBER IS GENERATED BUT
NOT USED! THIS IS DONE TO PRESERVE 1-1 CORRESPONDENCE
BETWEEN RANDOM NUMBERS AND PATIENTS' LENGTHS OF STAY (IN
CASE ANTITHETIC VARIATES ARE USED)
        nrefer:=nrefer+1;
        q:=0;
     end if;
# STORE CURRENT SIMULATION TIME AND ADVANCE IT TO THE
EARLIER OF TIME OF NEXT DEPARTURE AND OF NEXT ARRIVAL
     clockprev:=clock;
     clock:=min(a,t[1]);
     end do;
# SINCE clock NOW EXCEEDS REQUIRED SIMULATION TIME RESET
IT TO simtim, CALCULATE TIME SINCE PREVIOUS EVENT AND
UPDATE cum[nocc]
clock:=simtim;
delta:=clock-clockprev;
cum[nocc]:=cum[nocc]+delta;
# COMPUTE AVERAGE UTILIZATION, AVERAGE REFERRAL RATE,
AVERAGE REDUCTION IN LENGTH OF STAY FOR REFERRED PATIENTS,
AND LIST f FOR PLOTTING BED OCCUPANCY DISTRIBUTION
s:=0;
for j from 1 to n do;
   s:=s+j*cum[j];
end do;
f:=[seq([j,cum[j]/simtim],j=0..n)];
util:=s/n/simtim;
r:=evalf(nrefer/na);
ed:=evalf(nout2/na);
if nout2>0 then short:=short/nout2 end if;
print("_____OUTPUT
STATISTICS_____");
print("Utilisation"=util,"Referral proportion"=r,"Early
departure proportion"=ed,"Avg reduced stay in days for
early departures"=short);
[util,r,f];
end proc;
```

```
hospital_ward: = proc(simtim, n, t0, α, λ, μ, β, seeda, seedb)
local q, d, t, clock, r1, a, clockprev, cum, na, nout1, nin, r2, nout2, nrefer,
    δ, nocc, seed1, seed2, j, β1, f, s, util, r, β2, β3, short, ed;
    seed1: = seeda;
    seed2: = seedb;
    β1: = 1/β;
    β2: = evalf(Γ(β1 + 1));
    β3: = evalf(Γ(2/β + 1));
    print('_____INPUT DATA_____\
    _____');
    print('Arrival and service seeds' = seeda, seedb);
    print('Run length in days' = simtim);
    print('Number of beds' = n,'Poisson arrival rate per day' = λ,
        'Mean (non reduced) length of stay in days' = β2/μ,
        'Std dev of (non reduced) length of stay in days' = sqrt(β3 − β2^2)\μ);
    print ('Early departure parameter in days' = α);
    nocc: = 0;
    cum[0]: = 0;
    for j to n do if t0[j] < ∞ then nocc: = nocc + 1 end if; cum[j]: = 0 end do;
    print ('Bed state at time zero - Scheduled departure times are' =
        [seq(t0[j], j = 1..n)]);
    q: = 0;
    na: = 0;
    nout1: = 0;
    nout2: = 0;
    nin: = 0;
    nrefer: = 0;
    short: = 0;
    t: = sort(t0);
    clock: = 0;
    randomize(seed1);
    seed1: = rand( );
    r1: = evalf(1/1000000000000*seed1);
    a: = clock − ln(r1)/λ;
    clockprev: = clock;
    clock: = min(a, t[1]);
    while clock ≤ simtim do
        δ: = clock − clockprev;
        cum[nocc]: = cum[nocc] + δ;
        if clock = a then
            na  : = na + 1;
            randomize(seed1);
            seed1: = rand( );
            r1: = evalf(1/1000000000000*seed1);
            a: = clock − ln(r1)/λ;
            q: = 1;
```

```
            else t[1]: = ∞;  t: = sort(t);  nocc: = nocc-1;  nout1: = nout1 + 1
            end if;
            if q = 1 and nocc < n then
                    nin: = nin + 1;
                    nocc: = nocc + 1;
                    randomize(seed2);
                    seed2: = rand( );
                    r2: = evalf(1/1000000000000*seed2);
                    t[nocc]: = clock + (–ln(r2))^β1/μ;
                    t: = sort(t);
                    q: = 0
            end if;
            if q = 1 and nocc = n and t[1]-clock ≤ α then
                    nin: = nin + 1;
                    nout2: = nout2 + 1;
                    short: = short + t[1]-clock;
                    randomize(seed2);
                    seed2: = rand( );
                    r2: = evalf(1/1000000000000*seed2);
                    t[1]: = clock + (–ln(r2))^β1/μ;
                    t: = sort(t);
                    q: = 0;
            end if;
            if q = 1 and nocc = n and α < t[1]-clock then
                    randomize(seed2);  seed2: = rand( );  nrefer: = nrefer + 1;  q: = 0
            end if;
            clockprev: = clock;
            clock: = min(a,t[1])
    end do;
    clock: = simtim;
    δ: = clock – clockprev;
    cum[nocc]: = cum[nocc] + δ;
    s: = 0;
    for j to n do s: = s + j*cum[j] end do;
    f: = [seq([j,  cum[j]/simtim],  j = 0 .. n)];
    util: = s/(n*simtim);
    r: = evalf(nrefer/na);
    ed: = evalf(nout2/na);
    if 0 < nout2 then short: = short/nout2 end if;
    print('_____OUTPUT STATISTICS_____\
    _____');
    print('Utilisation' = util,'Referral proportion' = r,
            'Early departure proportion' = ed,
            'Avg reduced stay in days for early departures' = short);
    [util, r, f]
end proc
```

7.2.2 Verification against the M/M/n/n system

A partial verification of the simulation program is to test the output distribution of bed occupancy against $M/M/n/n$. This is done by setting $\beta = 1$ and $\alpha = 0$ (no early departures, so all arrivals to a full ward are referred). The stationary probabilities for the theoretical model are computed below in $p[0], \ldots, p[n]$. Without carrying out a formal test of hypothesis, it is apparent that the model results are in good agreement with theory.

```
> seeda:=87341:seedb:=64287:
  n:=10:
  t0:=[3.3,1.4,6.3,2.7,8.5,4.3,8.5,infinity,infinity,
  infinity]:
  simtim:=100000:alpha:=0:lambda:=1.2:mu:=0.15:beta:=1:
  v:=hospital_ward(simtim,n,t0,alpha,lambda,mu,beta,
  seeda,seedb):
  u:=op(3,v):d:=seq([[j,0],op(j+1,u)],j=0..n):# NB: d[j]
  is a list of two elements - these are the coordinates of
  two points, [j,0] and [j,proportion]. The line joining
  these two points is vertical and its height is the
  proportion of time that j beds are occupied;
      #
      # CALCULATE M/M/n/n stationary probabilities:
      #
  p[0]:=1:
  s:=p[0]:
  for j from 1 to n do:
      p[j]:=p[j-1]*lambda/mu/j:
      s:=s+p[j]:
  end do:
  for j from 0 to n do;
      p[j]:=p[j]/s:
      e:=seq([[j+0.1,0],[j+0.1,p[j]]],j=0..n):
  end do:
  PLOT(CURVES(d,COLOR(RGB, 1, 0,0)),CURVES(e,COLOR(RGB, 0,
  0, 1)),AXESLABELS("beds","proportion of simulated
  time"),AXESSTYLE(NORMAL),TITLE("Bed occupancy
  distribution, theory (red), \n simulation (blue)"));# PLOT
  A SET OF VERTICAL LINES DEFINED BY THE LIST 'd' AND A
  SLIGHTLY DISPLACED SET DEFINED BY THE LIST 'e'
```
`_____INPUT DATA_____\`

'Arrival and service seeds' = 87341, 64287
'Run length in days' = 100000
'Number of beds' = 10, 'Poisson arrival rate per day' = 1.2,
'Mean (non reduced) length of stay in days' = 6.666666667,

'Std dev of (non reduced) length of stay in days' = 6.666666667
'Early departure parameter in days' = 0
'Bed state at time zero - Scheduled departure times are' =
[3.3, 1.4, 6.3, 2.7, 8.5, 4.3, 8.5, ∞, ∞, ∞]

_____OUTPUT STATISTICS_____\
'

'Utilisation' = 0.7059855552, 'Referral proportion' = 0.1250620923,
'Early departure proportion' = 0., 'Avg reduced stay in days for early
departures' = 0

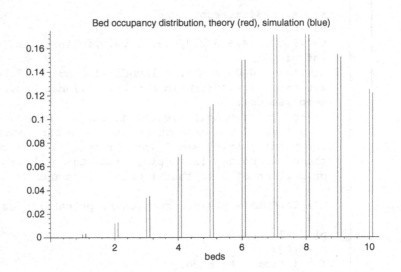

Bed occupancy distribution, theory (red), simulation (blue)

7.2.3 A 20 bed ward with nonexponential service durations and no early departure

```
> seeda:=87341:seedb:=64287:
  n:=20:
  t0:=[3.3,1.4,5.7,3.3,0.1,1.2,2.1,5.6,3.2,2,6,7.1,3.3,
  4.5,5.3,2.5,4.4,2.5,infinity,infinity,infinity]:
  simtim:=10000:
  alpha:=0:lambda:=2.5:mu:=0.15:beta:=2.5:
  v:=hospital_ward(simtim,n,t0,alpha,lambda,mu,beta,
  seeda,seedb):
  u:=op(3,v):d:=seq([[j,0], op(j+1,u)],j=0..n):
  PLOT(CURVES(d),AXESLABELS("beds","proportion of simulated
  time"),AXESSTYLE(NORMAL),TITLE("Bed occupancy
  distribution"));
```

'_____INPUT DATA_____\
'
'Arrival and service seeds' = 87341, 64287
'Run length in days' = 10000
'Number of beds' = 20, "Poisson arrival rate per day" = 2.5,
'Mean (non reduced) length of stay in days' = 5.915092117,
'Std dev of (non reduced) length of stay in days' = 2.531110333
'Early departure parameter in days' = 0
'Bed state at time zero - Scheduled departure times are' = [3.3, 1.4, 5.7, 3.3, 0.1,
1.2, 2.1, 5.6, 3.2, 2, 6, 7.1, 3.3, 4.5, 5.3, 2.5, 4.4, 2.5, ∞, ∞]
'_____OUTPUT STATISTICS_____\
'

'Utilisation' = 0.7082827055, 'Referral proportion' = 0.04285542795,
'Early departure proportion' = 0., 'Avg reduced stay in days for early
departures' = 0

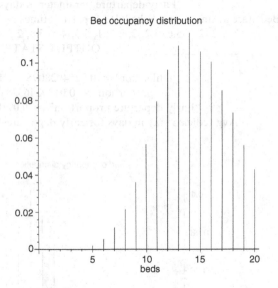

Bed occupancy distribution

7.2.4 As appendix 7.2.3 but with departures up to 1 day early

```
> seeda:=87341:seedb:=64287:
  n: = 20:
  t0: = [3.3,1.4,5.7,3.3,0.1,1.2,2.1,5.6,3.2,2,6,7.1,3.3,
  4.5,5.3,2.5,4.4,2.5,infinity,infinity,infinity]:
```

```
simtim:=10000:alpha:=1:lambda:=2.5:mu:=0.15: beta:=2.5:
v: =hospital_ward(simtim,n,t0,alpha,lambda,mu,
beta,seeda,seedb):
u:=op(3,v):d:=seq([[j,0],op(j+1,u)],j=0..n):
PLOT(CURVES(d),AXESLABELS("beds","proportion of simulated
time"),AXESSTYLE(NORMAL),TITLE("Bed occupancy
distribution"));
```

'_____INPUT DATA_____\
                              '
                  _____
'Arrival and service seeds' = 87341, 64287
'Run length in days' = 10000
'Number of beds' = 20, 'Poisson arrival rate per day' = 2.5,
'Mean (non reduced) length of stay in days' = 5.915092117,
'Std dev of (non reduced) length of stay in days' = 2.531110333
'Early departure parameter in days' = 1
'Bed state at time zero - Scheduled departure times are' = [3.3, 1.4, 5.7, 3.3, 0.1,
      1.2, 2.1, 5.6, 3.2, 2, 6, 7.1, 3.3, 4.5, 5.3, 2.5, 4.4, 2.5, ∞, ∞]
'_____OUTPUT STATISTICS_____\
                              '
                  _____
'Utilisation' = 0.7254928095, 'Referral
proportion' = 0.01292465288,
"Early departure proportion" = 0.09183305990,
"Avg reduced stay in days for early departures" = 0.4384030484

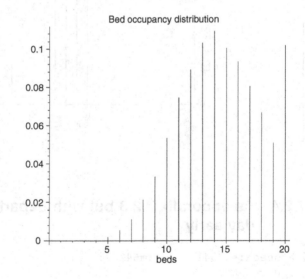

7.2.5 Plot performance as a function of the early departure parameter α

The procedure is now called repeatedly for values of α between 0 and 2 days, plotting the utilization and referral rate in each case. The print statements in 'hospital_ward' in Appendix 7.2.1 should be suppressed by inserting '#' before them.

```
> alphamax:=2:
  m:=floor(alphamax/0.1):
  simtim:=1000:lambda:= 4:mu:=0.15:beta:=2.5:
  n:=25:
  beta1:=1/beta:
  beta2:=evalf(GAMMA(beta1+1)):
  beta3:=evalf(GAMMA(2/beta+1)):
  t0:=[3.3,1.4,5.7,3.3,0.1,1.2,2.1,5.6,3.2,2,6,7.1,3.3,
  4.5,5.3, 2.5,4.4,2.5,4.8,3.5,infinity,infinity,infinity,
  infinity,infinity]:
  print("_____INPUT
  DATA_____");
  seeda:=87341:seedb:=64287:
  print("Arrival and service seeds"=seeda,seedb);
  print("Run length in days"=simtim);
  print("Number of beds"=n,"Poisson arrival rate per
  day"=lambda,"Mean (non reduced) length of stay in
  days"=beta2/mu,"Std dev of (non reduced) length of stay
  in days"=sqrt(beta3-beta2^2)/mu);
  for k from 0 to m do:
     alpha:=0.1*k:
     seeda:=87341:seedb:= 64287:
  v:=hospital_ward(simtim,n,t0,alpha,lambda,mu,beta,
  seeda,seedb):
     p1[k]:=[alpha,op(1,v)]:
     p2[k]:=[alpha,2*op(2,v)]:
  end do:
  v1:=seq(p1[k],k=0..m):
  v2:=seq(p2[k],k=0..m):
  PLOT(CURVES([v1],[v2]),VIEW(0..alphamax,0..1),AXESLABELS
  ("alpha", "Utilization (upper) and proportion referred*2
  (lower)"),AXESSTYLE(NORMAL),TITLE("Utilization and
  Referral rates as a function of alpha"));
  '_____INPUT DATA_____/
                       _____
                           ,
```

'Arrival and service seeds' = 87341, 64287
'Run length in days' = 1000
'Number of beds' = 25, 'Poisson arrival rate per day' = 4,
'Mean (non reduced) length of stay in days' = 5.915092117,
'Std dev of (non reduced) length of stay in days' = 2.531110333

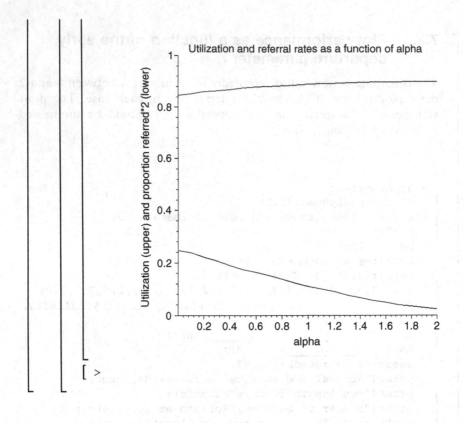

Appendix 8: Markov chain Monte Carlo

We wish to sample from a density $f(x) \propto \exp(-x^2/2)$ with a proposal density $q(y|x) = 1/(2a)$ on support $(x - a, x + a)$. Therefore, the acceptance probability is

$$\alpha(x, y) = \min\left[1, \frac{f(y)q(x|y)}{f(x)q(y|x)}\right] = \min\left[1, \exp\left(\frac{x^2 - y^2}{2}\right)\right].$$

The procedure 'mcmc' samples *iter* variates, given x_0.

```
> mcmc:=proc(x0,a,iter) local t,y,alpha,r,xold,xnew,b,count;
  xold:=x0;
  count:=0;
  b[0]:=[0,xold];
  for t from 1 to iter do;
      y:=xold-a+2*a*evalf(rand()/10^12);
      alpha:=min(1,exp(-0.5*(y^2-xold^2)));
      r:=evalf(rand()/10^12);
      if r<=alpha then xnew:=y else xnew:=xold end if;
      b[t]:=[t,xnew];
      xold:=xnew;
  end do;
  [seq(b[t],t=0..iter)];
  end proc:

> with(stats[statplots]):
```

Sample 500 variates starting with $x_0 = -4$ and $a = 0.5$. Note the long burn-in time and poor exploration of the state space (slow mixing of states) due to the small step length.

```
> seed:=randomize(59274);
  x:=mcmc(-4,0.5,500):
  PLOT(CURVES(x),TITLE("x-value against iteration
  number:a=0.5"),AXESSTYLE(NORMAL));
```

$$seed := 59274$$

x-value against iteration number: a = 0.5

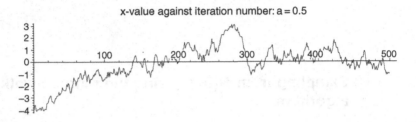

Now repeat with $a = 3$. The situation is much improved in both respects. Note that prospective variates are frequently rejected (horizontal sections of plot), resulting in no change in the variate from one iteration to the next.

```
> seed:=randomize(59274);
  x:=mcmc(-4,3,500):
  PLOT(CURVES(x),TITLE("x-value against iteration
  number:a=3"),AXESSTYLE(NORMAL));
```

$$seed := 59274$$

x-value against iteration number: a = 3

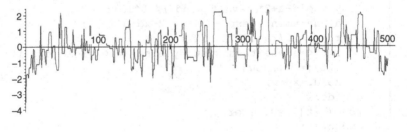

Sample 100 000 variates, starting with initial value of 0, to produce a histogram showing their distribution, together with sample mean and variance.

```
> seed:=randomize(59274);
  x:=mcmc(-4,3,100000):
  res:=[seq(op(2,x[t]),t=1..100000)]:
  histogram(res,title="distribution of sampled x-
  values",labels=["x","density"],numbars=100);
  Mean=describe[mean](res);StdDev=describe[standarddeviation
  ](res);
```

$$seed := 59274$$

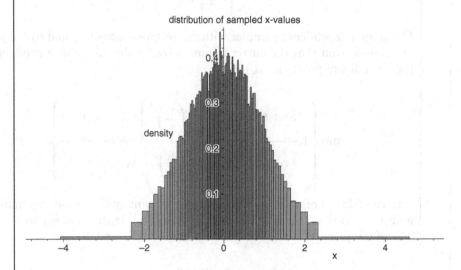

$$Mean = 0.0001202304356$$
$$StdDev = 1.006266364$$

```
> 
```

8.2 Reliability inference

We are interested in determining survival probabilities for components. The age at failure of the ith component is $X_i, i = 1, 2, \ldots$. These are i.i.d.

Weibull random variables with parameters α and β, where the joint prior distribution is

$$
g(\alpha, \beta) =
\begin{cases}
\dfrac{4}{1000}(\alpha - 1)\Gamma\left(\dfrac{1}{\alpha} + 1\right), & 1 < \alpha < 1.5, \\[3mm]
\dfrac{4}{1000}(2 - \alpha)\Gamma\left(\dfrac{1}{\alpha} + 1\right), & 1.5 < \alpha < 2,
\end{cases}
$$

and $2000/\Gamma(1/\alpha + 1) < \beta < 3000/\Gamma(1/\alpha + 1)$. Then the joint posterior density given the data $\{x_i, i = 1, \ldots, n\}$ is

$$
\pi(\alpha, \beta) \propto \alpha^n \beta^{-n\alpha} \exp\left[-\sum_{i=1}^{n}\left(\frac{x_i}{\beta}\right)^{\alpha}\right](x_1 \cdots x_n)^{\alpha - 1}g(\alpha, \beta).
$$

Using an independence sampler with the proposal density equal to the joint prior, and given that the current point is (α, β), the acceptance probability for a candidate point (α_c, β_c) is

$$
\min\left\{1, \frac{\alpha_c^n \beta_c^{-n\alpha_c} \exp\left[-\sum_{i=1}^{n}\left(\dfrac{x_i}{\beta_c}\right)^{\alpha_c}\right](x_1 \cdots x_n)^{\alpha_c - 1}}{\alpha^n \beta^{-n\alpha} \exp\left[-\sum_{i=1}^{n}\left(\dfrac{x_i}{\beta}\right)^{\alpha}\right](x_1 \cdots x_n)^{\alpha - 1}}\right\}
$$

The candidate point is generated from the joint prior by noting that the marginal prior of α_c is a symmetric triangular density on support $(1,2)$. Therefore, given two uniform random numbers R_1 and R_2, we set

$$
\alpha_c = 1 + 0.5(R_1 + R_2).
$$

Given α_c, the conditional prior density of β_c is

$$
U\left(\frac{2000}{\Gamma(1/\alpha_c + 1)}, \frac{3000}{\Gamma(1/\alpha_c + 1)}\right)
$$

so we set

$$
\beta_c = \frac{1000(2 + R_3)}{\Gamma(1/\alpha_c + 1)}.
$$

The procedure 'fail' performs k iterations of this independence sampler. It returns a matrix $\mathbf{C}$ having k rows and $q + 2$ columns. The first q elements of the ith row give sampled survivor probabilities at specified ages $y[1], \ldots, y[q]$. The last two elements of the ith row give the sampled alpha and beta at this ith iteration (point). The plots show that the burn-in time is negligible and that the mixing of states is quite reasonable. Bayes estimates are given for survival probabilities at ages 1000, 2000, and 3000 hours. Of course, if interval estimates are required it will be necessary to replicate these runs. It is instructive to see the effect of reducing the sample size (discard some of the original 43 ages at failure). The posterior density should then reflect the prior density more so, with the result that the acceptance probability becomes higher and successive α and β values are less correlated than previously. This is good from the viewpoint of reducing the variance of estimators.

```
> restart;
```

Here x is a list of 43 ages (in hours) at failure.

```
> x:= [293, 1902, 1272, 2987, 469, 3185, 1711, 8277, 356,
    822,2303, 317, 1066, 1181, 923, 7756, 2656, 879, 1232,
    697, 3368, 486, 6767, 484, 438, 1860, 113, 6062, 590,
    1633, 2425, 367, 712, 953, 1989, 768, 600, 3041, 1814,
    141, 10511, 7796, 1462];
```

$x := [293, 1902, 1272, 2987, 469, 3185, 1711, 8277, 356, 822, 2303, 317, 1066, 1181, 923, 7756, 2656, 879, 1232, 697, 3368, 486, 6767, 484, 438, 1860, 113, 6062, 590, 1633, 2425, 367, 712, 953, 1989, 768, 600, 3041, 1814, 141, 10511, 7796, 1462]$

The procedure 'fail' returns the simulation output from the MCMC.

```
> fail:=proc(k,a,b,n,x,q,y) local
    a1,b1,C,xp,sp,u,L1,i,r1,r2,r3,r4,r5,r6,mp,ap,rate,bp,L2,j;
    # k=# of iterations;
    # (a,b)=initial(alpha,beta);
    # x= failure data;
    # n=number of data items in x;
    # q=number of ages at which survivor probabilities are
    required;
    # y[i]=age at which a survivor probability is required for
    i=1..q
    a1:=a;b1:=b;
    C:=Matrix(k,q+2);
    # Compute log likelihood (L1) for current point (a1,b1);
    xp:=1;
    sp:=0;
    for u from 1 to n do;
        xp:=xp*x[u];
        sp:=sp+x[u]^a1;
    end do;
    xp:=ln(xp);
    L1:=evalf(n*ln(a1/b1^a1)+(a1-1)*xp-sp/b1^a1);
```

```
# Perform k iterations;
for i from 1 to k do;
r1:=evalf(rand()/10^12);r2:=evalf(rand()/10^12);r3:=evalf
(rand()/10^12);r4:=evalf(rand()/10^12);
    # Sample candidate point (ap,bp) and compute likelihood
(L2) for (ap,bp);
    ap:=1+0.5*(r1+r2);
    bp:=1000*(2+r3)/GAMMA(1/ap+1);
    sp:=0;
    for u from 1 to n do;
        sp:=sp+x[u]^ap;
    end do;
    L2:=evalf(n*ln(ap/bp^ap)+(ap-1)*xp-sp/bp^ap);
    # Decide whether to accept or reject candidate point;
    if ln(r4)<L2-L1 then a1:=ap; b1:=bp; L1:=L2; end if;
    # Enter survivor probs and alpha and beta values into
ith row of C;
    for j from 1 to q do;
        C[i,j]:=evalf(exp(-(y[j]/b1)^a1));
    end do;
    C[i,q+1]:=evalf(a1);C[i,q+2]:=evalf(b1);
    end do;
    C;
    end proc:
```

Set up parameter values. A sensible choice for initial α is the mode of the prior marginal. The initial β is set to the mode of its prior given α.

```
> n:=43;a:=1.5;b:=2500/GAMMA(1/a+1);k:=5000;q:=3;y:=Vector
  (q);y[1]:=1000;y[2]:=2000;y[3]:=3000;y;
```

$$n := 43$$
$$a := 1.5$$
$$b := 2769.330418$$
$$k := 5000$$
$$q := 3$$

$$y := \begin{bmatrix} 0 \\ 0 \\ 0 \end{bmatrix}$$

$$y_1 := 1000$$
$$y_2 := 2000$$
$$y_3 := 3000$$

$$\begin{bmatrix} 1000 \\ 2000 \\ 3000 \end{bmatrix}$$

Perform 5000 iterations.

```
> randomize(561293);d:=fail(k,a,b,n,x,q,y);
```

$$561293$$

$$d := \begin{bmatrix} 5000 \times 5 \text{ Matrix} \\ \text{Data Type: anything} \\ \text{Storage: rectangular} \\ \text{Order: Fortran\_order} \end{bmatrix}$$

```
> f:=Matrix(k,q+2);
```

$$f := \begin{bmatrix} 5000 \times 5 \text{ Matrix} \\ \text{Data Type: anything} \\ \text{Storage: rectangular} \\ \text{Order: Fortran\_order} \end{bmatrix}$$

```
> for i from 1 to k do:
  for j from 1 to q+2 do:
  f[i,j]:=[i,d[i,j]]:
  end do;
  end do:
```

Plot diagnostics for survivor probabilities and compute point estimates.

```
> f3:=[seq(f[i,3],i=1..k)]:PLOT(CURVES(f3),TITLE("survival
  probability at 3000 hours against iteration
  number:"),AXESSTYLE(NORMAL));
  with(stats[statplots]):dat3:=[seq(op(2,f3[i]),i=1..k)]:
  Estimate_Prob_survive_3000_hours=describe[mean](dat3);
```

survival probability at 3000 hours against iteration number:

Estimate_Prob_survive_3000_hours = 0.2867886198

```
> f2:=[seq(f[i,2],i=1..k)]:PLOT(CURVES(f2),TITLE("survival
  probability at 2000 hours against iteration
  number:"),AXESSTYLE(NORMAL));
  dat2:=[seq(op(2,f2[i]),i=1..k)]:Estimate_Prob_
  survive_2000_hours=describe[mean](dat2);
```

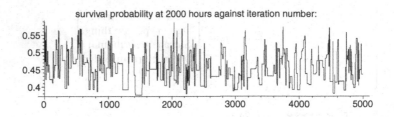

Estimate_Prob_survive_2000_hours = 0.4529730938

```
> f1:=[seq(f[i,1],i=1..k)]:PLOT(CURVES(f1),TITLE("survival
  probability at 1000 hours against iteration
  number:"),AXESSTYLE(NORMAL));
  dat1:=[seq(op(2,f1[i]),i=1..k)]:Estimate_Prob_survive_1000
  _hours=describe[mean](dat1);
```

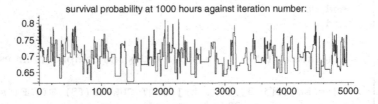

Estimate_Prob_survive_1000_hours = 0.6952910744

Plot diagnostics for α and β together with empirical posterior marginal densities.

```
> f5:=[seq(f[i,5],i=1..k)]:PLOT(CURVES(f5),TITLE("beta
  values against iteration number"),AXESSTYLE(NORMAL));
  dat5:=[seq(op(2,f5[i]),i=1..k)]:histogram(dat5);Bayes_
  estimate_beta=describe[mean](dat5);
```

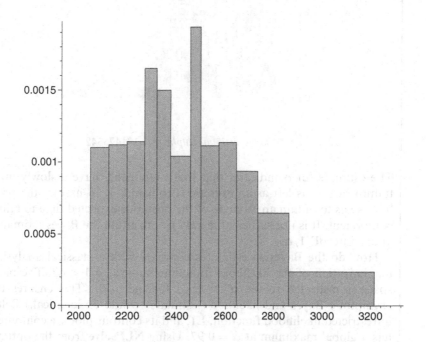

Bayes_estimate_beta = 2470.294876

```
> f4:=[seq(f[i,4],i=1..k)]:PLOT(CURVES(f4),TITLE("alpha
  values against iteration number"),AXESSTYLE(NORMAL));
  dat4:=[seq(op(2,f4[i]),i=1..k)]:histogram(dat4);
  Bayes_estimate_alpha=describe[mean](dat4);
```

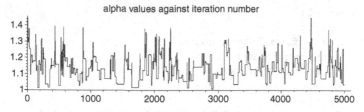

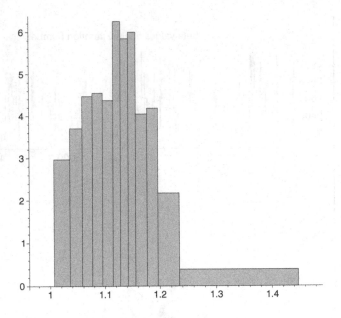

Bayes_estimate_alpha = 1.131475342

The estimate for α indicates that the components have a slowly increasing failure rate. It is left as an exercise (Problem 8.3) to modify the procedure 'fail' so as to obtain an estimate of the posterior expected time to failure of a component. It is not sufficient merely to substitute the Bayes estimates for α and β into $\beta\Gamma(1/\alpha+1)$.

How do the Bayesian estimates compare with a classical analysis, using maximization of the likelihood function, subject to $1 < \alpha$? (The parameter space is restricted to the space used for the prior. This ensures that the asymptotic Bayes and likelihood estimates would be identical.) Below, the unrestricted likelihood function, L1, and its contour plot are computed. This has a global maximum at $\hat{\alpha} = 0.97$. Using **NLPsolve** from the optimization package, the constrained maximum is at $\hat{\alpha} = 1$, $\hat{\beta} = 2201$. This represents a component with a constant failure rate (exponential life) and an expected time to failure of 2201 hours.

```
> with(plots):

> xp:=1:
  sp:=0:
  for u from 1 to n do:
      xp:=xp*x[u]:
      sp:=sp+x[u]^a1:
  end do:
  xp:=ln(xp):
  L1:=evalf(n*ln(a1/b1^a1)+(a1-1)*xp-sp/b1^a1):
```

```
> contourplot(exp(L1),a1=0.5..2,b1=1300..4000,contours=30,
  grid=[100,100]);
```

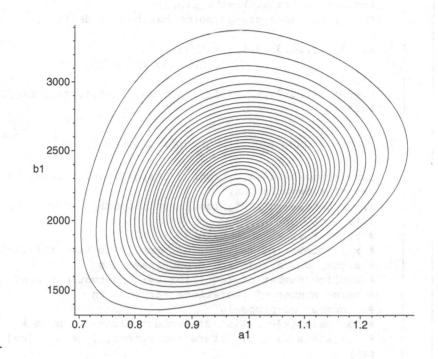

```
> with(Optimization);
```
[*ImportMPS, Interactive, LPSolve, LSSolve, Maximize,
Minimize, NLPSolve, QPSolve*]

```
> NLPSolve(-L1, a1=1..2,b1=1000..3000);
         [373.966224564945776, [b1 = 2201.48837205595191, a1 =1.]]
```

8.3 Pump failure

The data are from D.P. Gaver and I.G. O'Muircheartaigh (1987), Robust empirical Bayes analyses of event rates, *Technometrics* **29**, 1–15.

The number of failures for pump i in time $t[i]$ $(i = 1, \ldots, 10)$, $x[i]$, are assumed to be i.i.d. Poisson($\lambda_i t_i$) where λ_i are i.i.d. gamma $[\alpha, \beta]$ and β is a realization from gamma $[\delta,\text{gam}]$. The hyperparameters α, δ, and gam have been estimated as described in the text.

The procedure 'pump' returns, for each pump i, a list of simulated λ_i, $[i, \lambda_i]$, $(\alpha + x_i)/(\beta + t_i)$ values. The latter are individual Bayes estimates of λ_i.

```
> restart:with(stats):with(plots):
Warning, the name changecoords has been redefined.
```

```
> x:=[5,1,5,14,3,19,1,1,4,22];
                    x := [5, 1, 5, 14, 3, 19, 1, 1, 4, 22]
```

```
> t:=[94.32,15.72,62.88,125.76,5.24,31.44,1.048,1.048,2.096,
    10.48]:
```

The Gibbs sampling procedure 'pump' is as follows.

```
> pump:=proc(beta0,x,t,alpha,delta,gam,replic,burn) local
  bayeslam,bayeslambda,
  i,k,n,beta,a1,g1,j,jj,lam,lamval,lambda,lambdaval,s,lam5,
  tot;
  # beta0=initial beta value
  # x[i]=number of failures of pump in in time t[i],i=1..10
  # alpha, delta, gam are hyperparameters
  # replic= number of (equilibrium) observations used
  # burn= number of iterations for burn in
  # n=number of pumps=10
  # lambdaval[k]= a list of lambda values for pump k
  # lam[k]= a list of [iteration number, lambda value] for
  pump k
  n:=10;
  beta:=beta0;
  tot:=burn+replic;
  a1:=n*alpha+gam;
  for jj from 1 to tot do;
      j:=jj-burn;
      s:=0;
      for k from 1 to n do;
          g1:=random[gamma[x[k]+alpha,1/(t[k]+beta)]](1);
          s:=s+g1;
          if j>=1 then lambda[k,j]:=[j,g1];lambdaval[k,j]:=g1;
  bayeslam[k,j]:=(alpha+x[k])/(beta+t[k]);end if
      end do;
      s:=s+delta;
      beta:=random[gamma[a1,1/s]](1);
  end do;
  for k from 1 to n do;
  lam[k]:=[seq(lambda[k,j],j=1..replic)];
  lamval[k]:=[seq(lambdaval[k,j],j=1..replic)];
  bayeslambda[k]:=[seq(bayeslam[k,j],j=1..replic)];
  end do;
```

```
[[seq(lam[k],k=1..n)],[seq(lamval[k],k=1..n)],[seq(bayesla
mbda[k],k=1..n)]];
end proc:
```

```
> randomize(56342871);alpha:=0.54;delta:=1.11;gam:=2.20;beta
0:=0.25;replic:=2000;burn:=200;
```

$$56342871$$
$$\alpha := 0.54$$
$$\delta := 1.11$$
$$gam := 2.20$$
$$\beta0 := 0.25$$
$$replic := 2000$$
$$burn := 200$$

Print the Bayes failure rates and plot their posterior and prior densities. Note that the prior becomes unbounded as the failure rate approaches zero, so the failure rate axis does not extend to zero.

```
> v:=pump(beta0,x,t,alpha,delta,gam,replic,burn):
u:=evalf(lambda^(alpha-1)/(lambda+delta)^(alpha+gam)):
k:=int(u,lambda=0..infinity):
va:=op(1,v):vb:=op(2,v):vc:=op(3,v):
for kk from 1 to 10 do:
v1[kk]:=op(kk,va):
v2[kk]:=op(kk,vb):
v3[kk]:=op(kk,vc):
m[kk]:=max(seq(v2[kk][j],j=1..replic)):
ms[kk]:=min(seq(v2[kk][j],j=1..replic)):
failrate[kk]:=describe[mean](v3[kk]):
print("Bayes estimate failure rate
pump",kk,"is",failrate[kk]);
end do:
xa[1]:="lambda[1]":xa[2]:="lambda[2]":xa[3]:="lambda[3]":x
a[4]:="lambda[4]":xa[5]:="lambda[5]":
xa[6]:="lambda[6]":xa[7]:="lambda[7]":xa[8]:="lambda[8]":x
a[9]:="lambda[9]":xa[10]:="lambda[10]":
A:=array(1..5,1..2):
for i from 1 to 5 do:
A[i,1]:=display({

statplots[histogram](v2[2*i-1]),plot(u/k,lambda=ms[2*i-1]*
0.9..m[2*i-1]*1.1,labels=[xa[2*i-1],"prior and posterior

densities"],labeldirections=[HORIZONTAL,VERTICAL])},
tickmarks=[4,2]):
A[i,2]:=display({

statplots[histogram](v2[2*i]),plot(u/k,lambda=ms[2*i]*0.9.
.m[2*i]*1.1,labels=[xa[2*i],"prior and posterior
```

```
densities"],labeldirections=[HORIZONTAL,VERTICAL])},tickma
rks=[2,2]):
end do:
display(A);
```

'Bayes estimate failure rate pump', 1, 'is', 0.05809446755
'Bayes estimate failure rate pump', 2, 'is', 0.09192159100
'Bayes estimate failure rate pump', 3, 'is', 0.08666946370
'Bayes estimate failure rate pump', 4, 'is', 0.1146666174
'Bayes estimate failure rate pump', 5, 'is', 0.5657757990
'Bayes estimate failure rate pump', 6, 'is', 0.6016311355
'Bayes estimate failure rate pump', 7, 'is', 0.7643500150
'Bayes estimate failure rate pump', 8, 'is', 0.7643500150
'Bayes estimate failure rate pump', 9, 'is', 1.470380706
'Bayes estimate failure rate pump', 10, 'is', 1.958497332

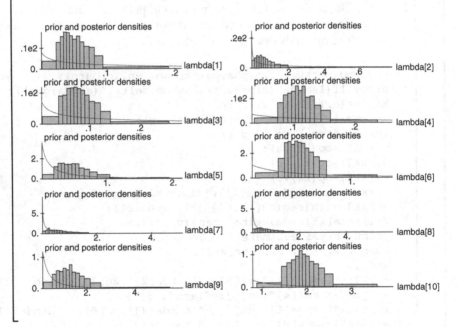

8.4 Capture–recapture

We are interested in estimating population size in a capture–recapture experiment consisting of two episodes. Let N = unknown population size (assumed the same in both episodes). The prior distribution of N is assumed to be Poisson(λ).

Let $n_{1,0}$, $n_{0,1}$, $n_{1,1}$ be the number of animals captured in the first only, second only, and both episodes respectively. Then the number of distinct animals captured is **nprime** $= n_{1,0} + n_{0,1} + n_{1,1}$ and the total number of animals captured in both episodes is **ncap** $= n_{1,0} + n_{0,1} + 2n_{1,1}$. Let **p** = probability that an animal has of being capured on an episode. The prior distribution of **p** is assumed to be U(0,1)

Let **notcap** $= N$-**nprime**. This is the number of animals in the population that were not captured in either of the episodes. We will use Gibbs sampling to estimate the posterior distributions of **notcap** and **p** and the Bayes estimate of **notcap**.

```
> restart;

> with(stats):with(stats[statplots]):

> gibbscapture:=proc(lambda,p0,notcap0,ncap,nprime,iter,
  burn)local i,ii,p,notcap,pa,notcapa,tot;
  # burn=number of iterations for burn in
  # iter=number of (equilibrium) iterations used
  # p0=initial p-value
  # notcap0=initial notcap-value
  p:=p0;
  notcap:=notcap0;
  pa[0]:=[0,p];
  notcapa[0]:=[0,notcap];
  tot:=iter+burn;
  for ii from 1 to tot do;
  i:=ii-burn;
  p:=stats[random,beta[ncap+1,2*notcap-ncap+2*nprime+1]](1);
  notcap:=stats[random,poisson[lambda*(1-p)^2]](1);
  if i>=1 then
      pa[i]:=[i,p];notcapa[i]:=[i,notcap];
  end if;
  end do;
  [[seq(pa[i],i=0..iter)],[seq(notcapa[i],i=0..iter)]];
  end proc:

> randomize(73179):
  res:=gibbscapture(250,0.3,120,160,130,500,200):
  pa:=res[1]:notcapa:=res[2]:

  PLOT(CURVES(pa),TITLE("p-value against iteration
  number"),AXESLABELS("Iteration
  Number","p"),AXESSTYLE(NORMAL));
  PLOT(CURVES(notcapa),TITLE("population-nprime against
  iteration number"),AXESLABELS("Iteration
  number","notcaptured"),AXESSTYLE(NORMAL));
```

```
l1:=[seq(op(2,pa[i]),i=1..500)]:# a list of p-values
l2:=[seq(op(2,notcapa[i]),i=1..500)]:# a list of notcap
values
l3:=[seq((1-l1[i])^2,i=1..500)]:# a list of (1-p)^2
values

histogram(l1,title="distribution of sampled p
values",labels=["p","density"]);
Mean=evalf(describe[mean](l1));
StdDev=evalf(describe[standarddeviation](l1));

histogram(l2,title="distribution of sampled
notcapvalues",labels=["notcap","density"]);
Mean=evalf(describe[mean](l2));
StdDev=evalf(describe[standarddeviation](l2));

Bayes_estimate_of_number_not_captured=250*evalf(describe
[mean](l3));
```

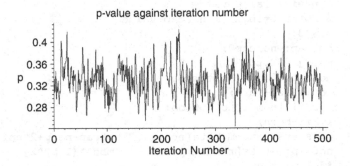

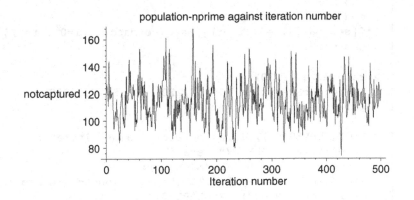

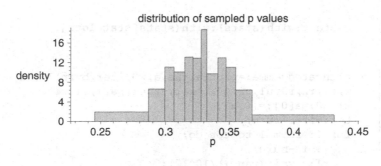

distribution of sampled p values

Mean = 0.3277568452
StdDev = 0.02987484843

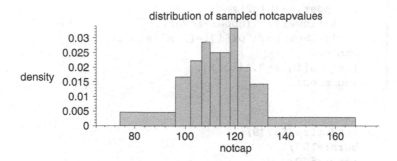

distribution of sampled notcapvalues

Mean = 114.5420000
StdDev = 14.69966789
Bayes_estimate_of_number_not_captured = 113.2008414

>

8.5 Slice sampler for truncated gamma density

Here, slice sampling is explored as applied to the generation of variates from a truncated gamma distribution with density proportional to $x^{\alpha-1} e^{-x}$ on support $[a, \infty)$ where $a > 0$.

```
[ > restart:with(stats):with(stats[statplots]):

[ > truncatedgamma:=proc(alpha,a,x0,iter,burn)local
    x,r1,r2,r3,u1,u2,u3,lower,upper,pa,i,ii,tot;
    x:=x0;pa[0]:=[0,x];
    tot:=iter+burn;
    for ii from 1 to tot do;
        i:=ii-burn;
        r1:=evalf(rand()/10^12);
        r2:=evalf(rand()/10^12);
        r3:=evalf(rand()/10^12);
        u1:=evalf(r1*x^(alpha-1));
        u2:=evalf(r2*exp(-x));
        lower:=max(a,u1^(1/(alpha-1)));
        upper:=-ln(u2);
        x:=lower+r3*(upper-lower);
        if i>=1 then pa[i]:=[i,x];end if;
    end do;
    [seq(pa[i],i=0..iter)];
    end proc:

[ > randomize(53179):
    burn:=100;
    iter:=500;
    alpha:=3;
    a:=9;
    x0:=10;
    pa:=truncatedgamma(alpha,a,x0,iter,burn):
    PLOT(CURVES(pa),TITLE("truncated gamma variate against
    iteration number"),AXESLABELS("Iteration
    number","variate"),AXESSTYLE(NORMAL));
    l1:=[seq(op(2,pa[i]),i=1..iter)]:
    histogram(l1,title="distribution of sampled x
    values",labels=["x","density"]);
    Mean:=evalf(describe[mean](l1));StdDev:=evalf
    (describe[standarddeviation[1]](l1));
```

$$burn := 100$$
$$iter := 500$$
$$\alpha := 3$$
$$a := 9$$
$$x0 := 10$$

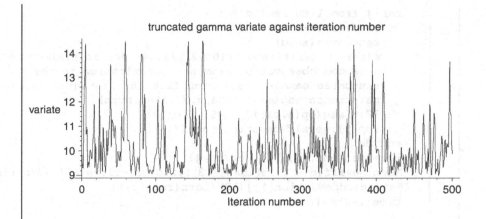

truncated gamma variate against iteration number

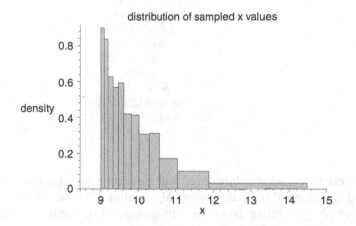

distribution of sampled x values

$$Mean := 10.16067140$$
$$StdDev := 1.147220533$$

In order to obtain an estimated standard error for the mean of the truncated gamma, the sample standard deviation is calculated of 100 independent replications (different seeds and starting states), each comprising 200 observations after a burn-in of 100 observations. In the code below the seed is changed for each new replication. It is not sufficient merely to use the last random number from one replication as the seed for the next replication. This would induce positive correlation between the two sample means.

```
> t1:=time();
  replic:=100;
  seed:=6318540;
  iter:=200;alpha:=3;a:=9;burn:=100;
```

```
for j from 1 to replic do:
    seed:=seed+113751:
    randomize(seed):
    x0:=a-ln(evalf(rand()/10^12))/1.04; #an initial state is
    a random observation from an approximation to the
    truncated gamma — estimated from the histogram above
    pa:=truncatedgamma(alpha,a,x0,iter,burn):
    l2:=[seq(op(2,pa[i]),i=1..iter)]:
    rep[j]:=evalf(describe[mean](l2)):
end do:
l3:=[seq(rep[j],j=1..replic)]:
Mean:=evalf(describe[mean](l3));Std_error1:=evalf (describe
[standarddeviation[1]](l3)/sqrt(replic));
time1:=time()-t1;
```

$$t1 := 3.266$$
$$replic := 100$$
$$seed := 6318540$$
$$iter := 200$$
$$\alpha := 3$$
$$a := 9$$
$$burn := 100$$
$$Mean := 10.205022061$$
$$Std\_error1 := 0.0178901125$$
$$time1 := 7.077$$

This time and standard error will be compared with that obtained for the generation of 100 independent observations. These are generated below by simply generating from the full gamma distribution, rejecting those not exceeding $a \, (= 9)$.

```
> t1:=time():
  j:=0;
  randomize(6318540);
  alpha:=3;
  a:=9;
  iter:=100;
  while j<iter do:
  y:=random[gamma[alpha]](1):
  if y>a then
      j:=j+1:
      v[j]:=y:
      pb[j]:=[j,y]
  end if:
  end do:
  l5:=[seq(pb[j],j=1..iter)]:
  b:=[seq(v[j],j=1..iter)]:
  Mean:=evalf(describe[mean](b));
```

```
StdDev:=evalf(describe[standarddeviation[1]](b)):
Std_error2:=evalf(StdDev/sqrt(iter));
time2:=time()-t1;
PLOT(CURVES(15),TITLE(" 100 independent truncated gamma
variates"),AXESLABELS("Iteration
number","variate"),AXESSTYLE(NORMAL));
```

$$j := 0$$
$$6318540$$
$$\alpha := 3$$
$$a := 9$$
$$iter := 100$$
$$Mean := 10.27322534$$
$$Std\_error2 := 0.1155931003$$
$$time2 := 5.750$$

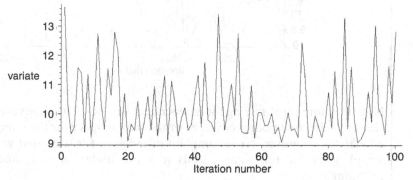

100 independent truncated gamma variates

```
> efficiency:=(Std_error2/Std_error1)^2*time2/time1;
```

$$efficiency := 33.92003894$$

For comparison of mixing 100 successive truncated variates from Gibbs sampling are displayed below.

```
> randomize(531577):
  burn:=100;
  iter:=100;
  alpha:=3;
  a:=9;
  x0:=10;
  pa:=truncatedgamma(alpha,a,x0,iter,burn):
  PLOT(CURVES(pa),TITLE("truncated gamma variate against
  iteration number"),AXESLABELS("Iteration
  number","variate"),AXESSTYLE(NORMAL));
  l1:=[seq(op(2,pa[i]),i=1..iter)]:
  histogram(l1,title="distribution of sampled x
```

```
values",labels=["x","density"]):
Mean:=evalf(describe[mean](l1)):StdDev:=evalf(describe[sta
ndarddeviation[1]](l1)):
```

$$burn := 100$$
$$iter := 100$$
$$\alpha := 3$$
$$a := 9$$
$$x0 := 10$$

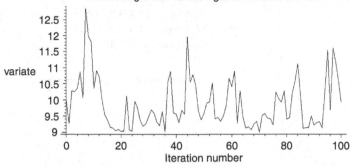

truncated gamma variate against iteration number

Note the more rapid mixing for independent variates. The Gibbs method has slower mixing and poorer coverage of the sample space for the same number of variates. However, naive sampling of independent truncated variates by sampling from a full gamma leads to a probability of acceptance of the following.

```
> evalf(int(x^(alpha-1)*exp(-x),x=a..infinity)/GAMMA(alpha));
                    0.006232195107
```

Because so many are rejected, the slice sampler comes out best, as the efficiency calculation above shows. Of course, a better method still is to choose an (exponential) envelope on support $[a, \infty)$ for the truncated gamma and use envelope rejection.

8.6 Pump failure revisited, an application of slice sampling

```
> restart:with(stats):with(plots):
  Warning, the name changecoords has been redefined
```

```
> x:=[5,1,5,14,3,19,1,1,4,22];
                    x := [5, 1, 5, 14, 3, 19, 1, 1, 4, 22]

> t:=[94.32,15.72,62.88,125.76,5.24,31.44,1.048,1.048,2.096,
  10.48]:

> slice:=proc(beta0,x,t,alpha,delta,gam,replic,burn) local
  beta,j,jj,bayeslam,bayeslambda,bmin,i,r,b,bmax,n,lam,
  lambda,k,g,lambdaval,lamval,tot;
  n:=10;
  beta:=beta0;
  tot:=burn+replic;
  for jj from 1 to tot do;
     j:=jj-burn;
     bmax:=infinity;
     for i from 1 to 10 do;
        r:=evalf(rand()/10^12);
        b:=(t[i]+beta)/r^(1/(x[i]+alpha))-t[i];
        if b<bmax then bmax:=b end if;
     end do;
     r:=evalf(rand()/10^12);
     b:=beta-ln(r)/delta;
     if b<bmax then bmax:=b end if;
     r:=evalf(rand()/10^12);
     bmin:=beta*r^(1/(n*alpha+gam-1));
     r:=evalf(rand()/10^12);
     beta:=bmin+(bmax-bmin)*r;
     if j>=1 then
        for k from 1 to 10 do;
           g:=random[gamma[x[k]+alpha,1/(t[k]+beta)]](1);
           lambda[k,j]:=[j,g];
           lambdaval[k,j]:=g;
           bayeslam[k,j]:=(alpha+x[k])/(beta+t[k])
        end do;
     end if;
  end do;
  for k from 1 to n do;
     lam[k]:=[seq(lambda[k,j],j=1..replic)];
     lamval[k]:=[seq(lambdaval[k,j],j=1..replic)];
     bayeslambda[k]:=[seq(bayeslam[k,j],j=1..replic)];
  end do;
  [[seq(lam[k],k=1..n)],[seq(lamval[k],k=1..n)],[seq
  (bayeslambda[k],k=1..n)]];
  end proc:

> randomize(56342871);alpha:=0.54;delta:=1.11;gam:=2.20;beta
  0:=0.25;replic:=2000;burn:=200;
```

$$56342871$$
$$\alpha := 0.54$$
$$\delta := 1.11$$
$$gam := 2.20$$
$$\beta 0 := 0.25$$
$$replic := 2000$$
$$burn := 200$$

```
> v:=slice(beta0,x,t,alpha,delta,gam,replic,burn):
  u:=evalf(lambda^(alpha-1)/(lambda+delta)^(alpha+gam)):
  k:=int(u,lambda=0..infinity):
  va:=op(1,v):vb:=op(2,v):vc:=op(3,v):
  for kk from 1 to 10 do:
  v1[kk]:=op(kk,va):
  v2[kk]:=op(kk,vb):
  v3[kk]:=op(kk,vc):
  m[kk]:=max(seq(v2[kk][j],j=1..replic)):
  ms[kk]:=min(seq(v2[kk][j],j=1..replic)):
  failrate[kk]:=describe[mean](v3[kk]):
  print("Bayes estimate failure rate
  pump",kk,"is",failrate[kk]);
  end do:
  xa[1]:="lambda[1]":xa[2]:="lambda[2]":xa[3]:="lambda[3]":x
  a[4]:="lambda[4]":xa[5]:="lambda[5]":
  xa[6]:="lambda[6]":xa[7]:="lambda[7]":xa[8]:="lambda[8]":x
  a[9]:="lambda[9]":xa[10]:="lambda[10]":
  A:=array(1..5,1..2):
  for i from 1 to 5 do:
  A[i,1]:=display({

  statplots[histogram](v2[2*i-1]),plot(u/k,lambda=ms[2*i-1]*
  0.9..m[2*i-1]*1.1,labels=[xa[2*i-1],"prior and posterior
  densities"],labeldirections=[HORIZONTAL,VERTICAL])},
  tickmarks=[4,2]):
  A[i,2]:=display({

  statplots[histogram](v2[2*i]),plot(u/k,lambda=ms[2*i]*0.9.
  .m[2*i]*1.1,labels=[xa[2*i],"prior and posterior
  densities"],labeldirections=[HORIZONTAL,VERTICAL])},
  tickmarks=[4,2]):
  end do:
  display(A);
```

'Bayes estimate failure rate pump', 1, 'is', 0.05815043965
'Bayes estimate failure rate pump', 2, 'is', 0.09241283960
'Bayes estimate failure rate pump', 3, 'is', 0.08679369040
'Bayes estimate failure rate pump', 4, 'is', 0.1147498208
'Bayes estimate failure rate pump', 5, 'is', 0.5734949755

'Bayes estimate failure rate pump', 6, 'is', 0.6033146695
'Bayes estimate failure rate pump', 7, 'is', 0.7929892845
'Bayes estimate failure rate pump', 8, 'is', 0.7929892845
'Bayes estimate failure rate pump', 9, 'is', 1.508423622
'Bayes estimate failure rate pump', 10, 'is', 1.973528342

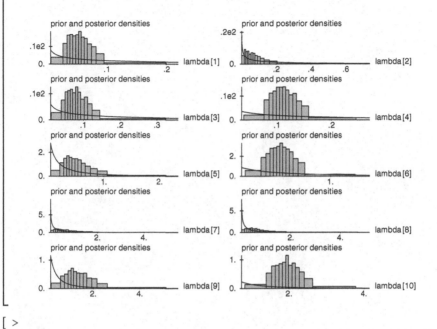

[>
[

References

Ahrens, J.H. and Dieter, U. (1980) Sampling from binomial and Poisson distributions: a method with bounded computation times. *Computing*, 25: 193–208.

Allen, L.J.S. (2003) *An Introduction to Stochastic Processes with Applications to Biology*. London: Pearson.

Anderson, S.L. (1990) Random number generators on vector supercomputers and other advanced architectures. *SIAM Review*, 32: 221–251.

Atkinson, A.C. (1979) The computer generation of Poisson random variables. *Applied Statistics*, 28: 29–35.

Atkinson, A.C. (1982) The simulation of generalised inverse Gaussian and hyperbolic random variables. *SIAM Journal of Scientific and Statistical Computing*, 3: 502–517.

Banks, J., Carson, J.S., Nelson, B.L. and Nicol, D.M. (2005) *Discrete Event System Simulation*, 4th edn. Upper Saddle River, New Jersey: Prentice Hall.

Barndorff-Nielson, O. (1977) Exponentially decreasing distributions for the logarithm of particle size. *Proceedings of the Royal Society*, A353: 401–419.

Beasley, J.D. and Springer, S.G. (1977) The percentage points of the normal distribution. *Applied Statistics*, 26: 118–121.

Besag, J. (1974) Spatial interaction and the statistical analysis of lattice systems (with discussion). *Journal of the Royal Statistical Society, Series B*, 36: 192–326.

Black, F. and Scholes, M. (1973) The pricing of options and corporate liabilities. *Journal of Political Economy*, 81: 637–654.

Borosh, I. and Niederreiter, H. (1983) Optimal multipliers for pseudo-random number generation by the linear congruential method. *BIT*, 23: 65–74.

Butcher, J.C. (1961) Random sampling from the normal distribution. *Computer Journal*, 3: 251–253.

Castledine, B.J. (1981) A Bayesian analysis of multiple-recapture sampling for a closed population. *Biometrika*, 67: 197–210.

Cheng, R.C.H. (1977) The generation of gamma variables with non-integral shape parameter. *Applied Statistics*, 26: 71–76.

Cheng, R.C.H. (1978) Generating beta variates with nonintegral shape parameters. *Communications of the Association of Computing Machinery*, 21: 317–322.

Cochran, W.G. (1977) *Sampling Techniques*. Chichester: John Wiley & Sons, Ltd.

Cox, D.R. and Miller, H.D. (1965) *The Theory of Stochastic Processes*. London: Methuen.

Dagpunar, J.S. (1978) Sampling of variates from a truncated gamma distribution. *Journal of Statistical Computation and Simulation*, 8: 59–64.

Dagpunar, J.S. (1988a) *Principles of Random Variate Generation*. Oxford: Oxford University Press.

Dagpunar, J.S. (1988b). Computer generation of random variates from the tail of t and normal distributions. *Communications in Statistics, Simulation and Computing*, 17: 653–661.

Dagpunar, J.S. (1989a) A compact and portable Poisson random variate generator. *Journal of Applied Statistics*, 16: 391–393.

Dagpunar, J.S. (1989b) An easily implemented generalised inverse Gaussian generator. *Communications in Statistics, Simulation and Computing*, 18: 703–710.

Devroye, L. (1986). *Non-uniform Random Variate Generation*. Berlin: Springer-Verlag.

Devroye, L. (1996). Random variate generation in one line of code. In J.M. Charnes, D.J. Morrice, D.T. Brunner and J.J Swain (eds), *Proceedings of the 1996 Winter Simulation Conference*, pp. 265–272. ACM Press

Eberlein, E. (2001) Application of generalized hyperbolic Lévy motions to finance. In O.E. Barndorff-Nielson, T. Mikosch and S. Resnick (eds), *Lévy Processes: Theory and Applications*, pp. 319–337. Birkhauser.

Entacher, K. (2000) A collection of classical pseudorandom number generators with linear structures – advanced version. In http://crypto.mat.sbg.ac.at/results/karl/server/.

Fishman, G.S. (1978) *Principles of Discrete Event Simulation*. Chichester: John Wiley & Sons, Ltd.

Fishman, G.S. (1979) Sampling from the binomial distribution on a computer. *Journal of the American Statistical Association*, 74: 418–423.

Fishman, G.S. and Moore, L.R. (1986) An exhaustive analysis of multiplicative congruential random number generators with modulus 2^31-1. *SIAM Journal on Scientific and Statistical Computing*, 7: 24–45.

Fouque, J.-P. and Tullie, T.A. (2002) Variance reduction for Monte Carlo simulation in a stochastic volatility environment. *Quantitative Finance*, 2: 24–30.

Gamerman, D. (1997) *Markov Chain Monte Carlo: Stochastic Simulation for Bayesian Inference*. London: Chapman and Hall.

Gaver, D.P. and O'Muircheartaigh, I.G. (1987) Robust empirical Bayes analyses of event rates. *Technometrics*, 29: 1–15.

Gelfand, A.E. and Smith, A.F.M. (1990) Sampling-based approaches to calculating marginal densities. *Journal of the American Statistical Association*, 85: 398–409.

Gentle, J.E. (2003) *Random Number Generation and Monte Carlo Methods*, 2nd edn. Berlin: Springer.

George, E.I. and Robert C.P. (1992) Capture–recapture estimation via Gibbs sampling. *Biometrika*, 79: 677–683.

Gilks, W.R. and Wild, P. (1992) Adaptive rejection sampling for Gibbs sampling. *Applied Statistics*, 41: 337–348.

Gilks, W.R., Best, N.G. and Tan, K.K.C. (1995) Adaptive rejection Metropolis sampling within Gibbs sampling. *Applied Statistics*, 44: 455–472.

Gilks, W.R., Richardson, S., Spiegelhalter, D.J. (1996) *Markov Chain Monte Carlo in Practice*. London: Chapman and Hall.

Glasserman, P. (2004) *Monte Carlo methods in Financial Engineering*. Berlin: Springer.

Glasserman, P., Heidelberger, P., Shahabuddin, P. (1999) Asymptotically optimal importance sampling and stratification for pricing path-dependent options. *Mathematical Finance*, 2: 117–152.

Hammersley, J.M. and Handscombe, D.C. (1964) *Monte Carlo Methods*. London: Methuen.

Hastings, W.K. (1970) Monte Carlo sampling methods using Markov chains and their applications. *Biometrika*, 57: 97–109.

Hobson, D.G. (1998) Stochastic volatility. In D.J. Hand and S.D. Jacka (eds), *Statistics and Finance*, Applications of Statistics Series, pp. 283–306. London: Arnold.

Hull, J.C. (2006) *Options, Futures and Other Derivatives*, 6th edn. Upper Saddle River, New Jersey: Prentice Hall.

Hull, T.E. and Dobell, A.R. (1962) Random number generators. *SIAM Review*, 4: 230–254.

Jöhnk, M.D. (1964) Erzeugung von Betaverteilten und Gammaverteilten Zufallszahlen. *Metrika*, 8: 5–15.

Johnson, N.L., Kotz, S. and Balakrishnan, N. (1995) *Continuous Univariate Distributions*, Volume 2, 2nd edn. Chichester: John Wiley & Sons, Ltd.

Karian, Z.A. and Goyal, R. (1994) Random Number Generation and Testing, *Maple Technical Newsletter,* Volume 1, Number 1, 32–37, Spring 1994. Birkhauser.

Kemp, A.W. (1981) Efficient generation of logarithmically distributed pseudo-random variables. *Applied Statistics,* 30: 249–253.

Kinderman, A.J. and Monahan, J.F. (1977) Computer generation of random variables using the ratio of uniform deviates. *ACM Transactions on Mathematical Software,* 3: 257–260.

Knuth, D.E. (1998) *The art of computer programming,Volume 2, Seminumerical Algorithms,* 3rd edn. Reading, Massachusetts: Addison-Wesley.

Lavenberg, S.S., Moeller, T.L. and Sauer, C.H. (1979) Concomitant control variables applied to the regenerative simulation of queueing systems. *Operations Research,* 27: 134–160.

Law, A.M. and Kelton, W.D. (2000) *Simulation Modeling and Analysis,* 3rd edn. London: McGraw-Hill.

L'Ecuyer, P. (1999) Tables of linear congruential generators of different sizes and good lattice structure. *Mathematics of Computation,* 68: 249–260.

Lewis, J.G. and Payne, W.H. (1973) Generalised feedback shift register pseudo-random number algorithm. *Journal of the Association of Computing Machinery,* 20: 456–458.

Lewis, P.A.W. and Shedler, G.S. (1976) Simulation of non-homogeneous Poisson Processes with log linear rate function. *Biometrika,* 63: 501–505.

Lewis, P.A.W. and Shedler, G.S. (1979a) Simulation of non-homogeneous Poisson Processes by thinning. *Naval Research Logistics Quarterly,* 26: 403–413.

Lewis, P.A.W. and Shedler, G.S. (1979b) Simulation of non-homogeneous processes with degree 2 exponential polynomial rate function. *Operations Research,* 27: 1026–1040.

Marsaglia, G. (1977) The squeeze method for generating gamma variates. *Computers and Mathematics with Applications,* 3: 321–325.

Marsaglia, G. (1972) The structure of linear congruential sequences. In S.K. Zaremba (ed.), *Applications of Number Theory to Numerical Analysis,* pp. 248–285. New York: Academic Press.

Marsaglia, G. and Zaman, A. (1993) Monkey tests for random number generators. *Computers and Mathematics with Applications,* 26: 1–10.

Marsaglia, G., Anathanarayanan, K. and Paul, N. (1972) *Random Number Generator Package – 'Super Duper'. Uniform, normal, and random number generator.* School of Computer Science, McGill University.

Merton, R.C. (1973) The theory of rational option pricing. *Bell Journal of Economics and Management Science,* 4: 141–183.

Metropolis, N., Rosenbluth, A.W., Rosenbluth, M.N., Teller, A.H. and Teller, E. (1953) Equation of state calculations by fast computing machine. *Journal of Chemical Physics,* 21: 1087–1091.

Morgan, B.J.T. (2000) *Applied stochastic modelling.* London: Arnold.

Moro, B. (1995) The full Monte. *Risk,* 8 (February): 57–58.

Neave, H.R. (1973) On using the Box–Müller transformation with multiplicative congruential pseudo-random number generators. *Applied Statistics,* 22: 92–97.

Percy, D.F. (2002) Bayesian enhanced strategic decision making for reliability. *European Journal of Operational Research,* 139: 133–145.

Pidd, M. (1998) *Computer Simulation in Management Science,* 4th edn. Chichester: John Wiley & Sons, Ltd.

Rand Corporation (1955) *A Million Random Digits with 100 000 Normal Deviates.* Glencoe: Free Press.

Ripley, B.D. (1983a) The lattice structure of pseudo-random generators. *Proceedings of the Royal Society,* A389: 197–204.

Ripley, B.D. (1983b) Computer generation of random variables: a tutorial. *International Statistical Review,* 51: 301–319.

Robert, C.P. (2001) *The Bayesian Choice: From decision theoretic foundations to computational implementation*, 2nd edn. New York and London: Springer-Verlag.

Robert, C.P. and Casella, G. (2004) *Monte Carlo Statistical Methods*, 2nd edn. New York: Springer-Verlag.

Ross, S.M. (2002) *Simulation*, 3rd edn. New York: Academic Press.

Rubinstein, R.Y. (1981) *Simulation and the Monte Carlo Method*. Chichester: John Wiley & Sons, Ltd.

Schrage, L. (1979) A more portable FORTRAN random number generator. *ACM Transactions on Mathematical Software*, 5: 132–138.

Shedler, G.S. (1993) *Regenerative Stochastic Simulation*. New York: Academic Press.

Tauseworthe, R.C. (1965) Random numbers generated by linear recurrence modulo two. *Mathematics of Computing*, 19: 201–209.

Tierney, L. (1994) Markov chains for exploring posterior distributions (with discussion). *Annals of Statistics*, 22: 1701–1728.

Tijms, H.C. (2003) *A first Course in Stochastic Models*. Chichester: John Wiley & Sons, Ltd.

Tin, M. (1965) Comparison of some ratio estimators. *Journal of the American Statistical Association*, 60: 294–307.

Tocher, K.D. (1963) *The Art of Simulation*. London: Hodder and Stoughton.

Toothill, J.P.R., Robinson, W.D. and Adams, A.G. (1971) The runs up-and-down performance of Tauseworthe pseudo-random number generators. *Journal of Association of Computing Machinery*, 18: 381–399.

Toothill, J.P.R., Robinson, W.D. and Eagle, D.J. (1973) An asymptotically random Tauseworthe sequence. *Journal of Association of Computing Machinery*, 20: 469–481.

Ulrich, G. (1984) Computer generation of distributions on the m-sphere. *Applied Statistics*, 33: 158–163.

Wichmann, B.A. and Hill, I.D. (1982) Algorithm AS 183. An efficient and portable pseudo-random number generator. *Applied Statistics*, 31: 188–190.

Wichmann, B.A. and Hill, I.D. (1984) Correction to algorithm AS 183. *Applied Statistics*, 33: 123.

Wilmott, P. (1998) *Derivatives. The Theory and Practice of Financial Engineering*. Chichester: John Wiley & Sons, Ltd.

Index

Acceptance probability
  envelope rejection 41
  independence sampler 165
  MCMC 159
  ratio method 46
  thinned Poisson process 141
  unity in Gibbs sampling 166
Activity 146
Adaptive rejection sampling 48–52, 57, 167
  Metropolis 176
Antithetic variates 79–82
  attenuated variance reduction with rejection
    methods 82
  option pricing 114
  queue simulation 148–149
  symmetric distributions 82
Arbitrage 112
Arithmetic average/mean 118–119, 133
Asian options 118–123
  geometric mean 120–121, 132, 133
  importance and stratified sampling for
    119–123
  viewed as a basket option 132
Asset price movements 109–111
Attribute 146
Autocorrelation 7, 148
Average, arithmetic/geometric 118, 132, 133
Average price or strike, see Asian options

Basket option 123–126, 133
Bayes estimate 165, 170, 171
Bayesian statistics 157–158
Beta distribution 67–68, 76, 77
  sum of i.i.d. variates 86–89
  symmetric 56, 70
Binomial distribution 74
Birth–death process simulation 143
Black–Scholes differential equation 112, 115

Bound state change 146, 149
Box–Müller 59–60, 75
  generalization of 77
Brownian motion 108–109, 153, 182
BUGS 157
Burn-in 7, 160, 162
  discrete event simulation 145, 151

Call, see European call/put
Candidate point in Metropolis–Hastings
  algorithm 159
Capture–recapture 171–172
Cauchy distribution
  inversion method 38, 52
  normal variate generation 53
  ratio method 55
  t-distribution 70
  variate as ratio of normal variates 55
  with parameter normally distributed 179
Chi-squared distribution 69
  non-central 69, 70–71
Cholesky decomposition 124
Completion of a density 178–179, 182, 183
Conditional Monte Carlo 101–103, 130
Conditional state change/event 146, 149–150
Conjugate prior 185
Control variates 98–101, 149
  linear regression 100
  stratification, compared with 101
Cost, of variate generation, fixed and variable
    48, see also Set-up time of variate
    generation
Currency, option on purchasing foreign 116
Customers, rejected 149–151, 155

Delta, see Hedging parameter (delta)
Delta hedging, see Hedging
Derivative, financial 1, 107, 111, 129

Dimension, problems of increasing   26, 28, 107
Discount factor   113
Discrete event simulation   135–156
Drift of Wiener process   108

Efficiency of Monte Carlo experiment   81, 95
Efficiency of variate generation algorithm   43–44, 48
Empirical Bayes   169
Empty and idle state   150
Entity   146
Envelope rejection   40–44
    acceptance probability   41
    antithetic variates, effectiveness of   82
    beta variate   67–68
    discrete variates   53–54
    gamma tail variate generation   177
    gamma variate   65–66
    generalized inverse Gaussian variate   71–73
    Metropolis–Hastings   180–181
    normal variate   41–44, 53, 61–62
    Poisson distribution, logistic envelope   73
    variates from stratum of normal distribution   122, 132
Epidemic   1, 153
Erlang distribution, special   64
Euler approximation for simulating a stochastic differential equation   111
European call/put   111, 113–115
Event   142, 146
Event based approach   146
Exceedance probability   86–89
Exercise date   111
Exercise price, see Strike price
Exotic options   114
Expiration date, see Exercise date
Exponential distribution, see Negative exponential distribution
Exponential polynomial intensity (in Poisson process)   140
Extend   135

Failure rate   163
    estimating Poisson   167–171
Fibonacci generator   34
Forward contract   111, 131
Full conditional distribution   166

Gamma distribution   64–66
    adaptive rejection, illustration of   49–52
    beta variate generation   67

conjugate prior   185
envelope rejection   65–66
Erlang, special distribution   64
life distribution with prior distributed parameters   183–185
negative binomial variates   75
Poisson failure rate follows a   167
shape parameter less than one   65, 76
tail variate generation   177, 183
Weibull scale parameter follows   174
Generalized hyperbolic distribution   71
Generalized inverse Gaussian distribution   71–73
Geometric average/mean   119–122, 132, 133
Geometric Brownian motion   109–111, 153
Geometric distribution
    inversion method   39
    negative binomial variates   74
Gibbs sampling   165–167
    slice sampling, completions   176–179
Growth rate of asset value   109
    expected   111, 112, 113, 116, 117

Hedging   112, 115–117, 129, 131–132
Hedging parameter (delta)   112, 130
Hyperparameters   167, 169, 172, 174
Hyperprior density   172

Importance sampling   4, 82–86
    Asian option   119–121
    basket option   124–125
    upper bound on variance   84, 104–105, 120
Income, asset earning   115–117
Independence sampler   161–162
Integration, multiple   1, 4–5, 82, 158
    Monte Carlo compared with other numerical methods   107, 172
Integration, one dimensional   2–4
Inverse Gaussian variate generation   182
Inversion method   37–39
    Cauchy variate   38, 52
    discrete variates   39, 54, 73, 74, 75
    exponential   38
    geometric   39
    logistic   38
    Markov Chains   142–143
    triangular   52
    unstored   73, 74
    Weibull variate   38, 52, 92
Itô's lemma   109, 127

Lattice, linear congruential generator
26–28
Likelihood function   158
Linear Congruential generators   18–27
bit shifting   21–22, 23
dimension, problems of increasing   26–28
double precision   20
full period   18
increment   18
lattice   26–28
mixed   18–22
modulus   18
multiplicative   22–25
multiplier   18
overflow   20–21, 22
prime modulus, maximum period   22–25,
33
rand() Maple generator   23
recommended   28
seed   18–19, 32
Super-Duper   21
List   146
Little's formula   147
Log concave distribution   48, 57, 176, 185
Log-logistic proposal distribution   65, 67–68
Logistic distribution
as envelope for Poisson generator   73
inversion method   38
Lognormal distribution   57, 62–63
jumps in a geometric Brownian motion
153–154
price of asset   75–76, 110
sum of i.i.d. variates   104–105

Maple
binomial generator   9
colon in   11
correlation in   81
execution group   8
global variable   19
histogram in   14
list   8
local variable   9
mean in   10
normal generator   13, 14
op   10
PLOT   10
Poisson generator   9
procedure   8
seed in   9
semicolon in   8

sequence   8
sorting list in   14
statplots subpackage   14
stats package   8
Market price of risk   129
Markov Chain
Metropolis–Hastings algorithm   159
simulation, discrete state, homogeneous
142–143
stationary distribution   6, 7
Markov Chain Monte Carlo   157–186
Mean reverting random walk   127, 133
Mersenne prime   24
Metropolis algorithm (sampler)   161
Metropolis–Hastings algorithm   159–160
Microsaint   135
Minimal repair   172–176
Mixing in MCMC   161, 163
Monte Carlo method   1, 5
Multinomial distribution   171
Multivariate p.d.f.   82, 158, 166

Neave effect   60
Negative binomial distribution   74–75
Negative exponential distribution
inversion method   38
ratio method   46
Normal distribution
bivariate   63
Box–Müller (polar) method   59–60
envelope rejection   41–44, 61–62
Maple generator   14
multivariate, sampling   124, 161
prior for a Cauchy parameter   179
ratio method   47
sampling from a conditional multivariate
122
sampling from a stratum of   121–122, 132
sampling from a tail of   132

Option
Asian   118–123
basket   123–126
definition   111
path-dependent   118
Order statistics and simulation Poisson
processes   138–139, 141
Ornstein–Uhlenbeck process   127–129

Path dependent option, *see* Option
Payoff   111, 113, 118, 119

Poisson distribution
 failures follow a 167
 negative binomial variates 75
 non-central chi-squared distribution 71
 variate generation 73, 138
Poisson process simulation 137–141
 over a plane 141–142, 152
 simple 137–139
 time dependent 140–141, 142
Posterior distribution 158
Posterior predictive survivor function
  164–165, 170, 183
Precision 2
Prime modulus generator 22, 33
Primitive root 22, 24, 33
Prior distribution 157
Prior expectation and variance 174–175
Prior predictive survivor function 164
Probability mixture 57, 75
Process based simulation approach 146
Project
 cost, stochastic 102
 duration, stochastic 86
 planning 5
Proposal density
 envelope rejection 40, 42
 Gibbs sampling 165
 MCMC 159, 165
Prospective variate 40, see also Candidate
  point in Metropolis–Hastings algorithm
Pseudo-code 7
Pump failure example 167–171
 slice sampling 177–178
Put, see European call/put
Put-call parity 115

Queue 4, 58, 103
 control variates 149
 G/G/1 system 145–149, 154
 hospital ward 149–151, 155
 M/G/1 system 145
 M/M/1 system 14, 148–149, 154
 M/M/n/n system 151
 separate random number streams for arrivals
  and service durations 82, 149

Random numbers 1, 3, 17–35
 collision test 31
 combining streams 31, 34
 coupon collector's test 31
 frequency test 29–30, 35

gap test 31
 overlapping, nonoverlapping k-tuples 26,
  30, 34
 period 18
 poker test 31
 pseudo 17
 reproducible 17
 runs test 31
 seed 18, 32
 serial test 30
 several sampling activities 82, 149
 shuffled 28–29, 34–35
 tests, empirical for 30–31
 tests, theoretical for 25–28
Random Walk sampler 161
Randomize 9
Ratio estimator 145
Ratio of uniforms method 44–48
 acceptance probability 46
 beta, symmetric 56
 Cauchy 55
 exponential 46
 normal 47, 55
 relocated mode 55–56
Regeneration points 145, 154
 G/G/1 queue 145, 147
 M/G/1 queue 145
 M/M/1 queue 154
Regenerative analysis 144–146, 155–156
Regression, effectiveness of
  stratification/control variable 92, 98, 101
Regression, linear, connection with control
  variates 100
Reliability inference 163–165, 167–171,
  172–176, 183
Renewal reward process/theorem 144
Return on asset 109, 110, 123
Risk-free interest (growth) rate 112, 114,
  117
Risk neutral drift 129
Risk-neutral world 113
Riskless asset 112

Set-up time of variate generation 167,
  see also Cost, of variate generation,
  fixed and variable
Simscript II.5 135
Simul8 135
Simulation, definition 5
Simulation, different approaches for discrete
  event 146

Slice sampling   176–178, 182
Spot price   123, 126
Squeeze technique   62
Step length in MCMC   161
Stochastic differential equation, Itô   109
Stochastic volatility   126–130
Stratification variable   89, 93, 97, 100, 121, 126
Stratified sampling   89–98
   Asian option   119, 121–123
   basket option   124
   dimensionality, problems of   95
   effectiveness compared with control variate
      101–112
   post stratification   96–98, 122, 124
   proportional   90
Strike price   111, 114
Sum of
   i.i.d. beta random variables   86–89
   i.i.d. lognormal random variables   104–105
   Weibull random variables   93, 96
Symmetric distributions
   antithetic sampling   82
   beta   56, 70

Target density   44
t-distribution   69–71
   Cauchy distribution   70
   doubly noncentral   70–71
   non-central   70
   symmetric beta distribution   70

Thinning   140–141, 152
Three-phase approach   146, 149
Tilted distribution   89, 104
Time scale transformation (Poisson process)
      140, 152
Traffic intensity   148
Transition Kernel (in Markov Chain)   159
Triangular distribution
   inversion method   52
   symmetric, variate generation   163

Utilization   151

Vanilla options   114
Variance reduction ratio   80, 123, 126
Visual interactive simulations   135
Volatility   111, 114, 115, 126–130
von Mises distribution and completion
      sampling   183

Wear of equipment   86
Weibull distribution
   first and second moments   164, 174
   G/G/1 queue   148
   inversion method   38, 92
   length of stay in hospital ward   149
   reliability   163, 172–174
   sum of i.i.d. variates   92, 96
Wiener process, see Brownian motion
WINBUGS   157
Witness   135